Results and Problems in Cell Differentiation

Volume 59

More information about this series at http://www.springer.com/series/400

Swathi Arur
Editor

Signaling-Mediated Control of Cell Division

From Oogenesis to Oocyte-to-Embryo Development

Springer

Editor
Swathi Arur
Dept. of Genetics, Genes & Dev. Program
UT MD Anderson Cancer Center
Houston, Texas
USA

ISSN 0080-1844 ISSN 1861-0412 (electronic)
Results and Problems in Cell Differentiation
ISBN 978-3-319-44819-0 ISBN 978-3-319-44820-6 (eBook)
DOI 10.1007/978-3-319-44820-6

Library of Congress Control Number: 2016961937

Printed on acid-free paper

This Springer imprint is published by Springer Nature
The registered company is Springer International Publishing AG
The registered company address is: Gewerbestrasse 11, 6330 Cham, Switzerland

Preface

Germ cells sustain life through generations. Germ cells are set aside from somatic cells in the embryo and go through the specialized cell division process of meiosis to produce the gametes—the oocyte or the sperm. The process whereby germ cells develop into gametes is governed via several developmental, nutritional, and other environmental cues or signals. Unsurprisingly, due to a variety of causes, sometimes these signals are spatially and/or temporally anomalous during organismal development, and such anomalies can manifest as birth defects in the progeny. Thus, elucidating how a multitude of molecular mediators integrate the various developmental and environmental signals to coordinate germ cell formation, maintenance, and function remains an active area of investigation.

In both invertebrate and vertebrate animals, environmentally activated signaling pathways regulate production of gametes with nutritional availability. Much work has focused on understanding the mechanisms that regulate oocyte fitness and quality in humans and other vertebrate model organisms such as mice. However, in the mammalian female, germ cells enter and arrest in meiosis I during fetal development, and initiation of meiosis II—ovulation—occurs after birth and only when the progeny attains sexual maturity. In contrast, the invertebrate female germline stem cells produce oocytes continuously throughout development. Therefore, understanding the cellular and molecular mechanisms that govern germline stem cell populations and meiosis I is relatively more tractable using invertebrate model organisms such as *Caenorhabditis elegans* and *Drosophila melanogaster*. Together, studies using both vertebrate and invertebrates are beginning to provide insights into the mechanisms that ensure oocyte quality and minimize errors during meiosis and meiotic maturation. In this volume, experts working on diverse model organisms come together to discuss the state-of-the-art knowledge base on the role of signaling pathways in regulating germline stem cell biology and meiotic progression of oocytes until fertilization and birth and identify exciting new avenues and questions for future research.

We begin with a description of epigenetic mechanisms that regulate *Drosophila* germline stem cell gene expression. *Rangan and coauthors* elaborate the role of several histone variants and nucleosome remodeling complexes in maintaining germline stem cells. They further provide insights into the interplay between epigenetic and signaling mechanisms that drive differentiation into the meiotic program.

The decision of a stem cell to differentiate or self-renew is of fundamental importance to organismal development and tissue maintenance. *Singh and Hansen* discuss the signaling circuitry, including feedback loops between differentiated cells and stem cells, which enables a germline stem cell to maintain a fine balance between self-renewal (enough to maintain a pool of stem cells) and differentiation (to maintain oocyte numbers). They also discuss the similarities and unique aspects of stem cell renewal and differentiation in worms, flies, and male mice.

Oocyte quality is influenced significantly by nutritional and other environmental cues or signals.

Laws and *Drummond-Barbosa* elaborate on the strong influence of diet and physiology on reproductive fitness. The authors critically review in vivo studies in model organisms (worms, flies, and mice) and discuss how diet-dependent signals can control the proliferation of germline stem cells and their maintenance and subsequent meiotic progression and production of oocytes.

I then provide a mechanistic view on how nutrition-induced signals regulate oogenic meiosis I progression *via* the ERK/MAPK pathway in *C. elegans* and signaling and second messenger pathways that regulate oocyte meiotic maturation in *C. elegans*, *Drosophila* and mammals. Through detailed discussions of several evolutionarily conserved signaling pathways, I synthesize common themes in the regulation of key meiotic events during oocyte growth and maturation.

Meiosis-specific processes such as chromosomal pairing and recombination generate diversity in a species. Failure in the coordinated regulation of these meiotic processes also results in aneuploidy in the progeny. *Smolikove and colleagues* examine the interplay of signaling with meiosis I that is critical for normal oocyte development. They also examine the role of increasing maternal age on oocyte quality.

Oocyte growth, development, and maturation occur under a molecular environment of transcriptional quiescence. Transcription is not reinitiated until embryogenesis. Thus, posttranscriptional and translational mechanisms are critical to normal oocyte development. *Subramaniam and colleagues* discuss the translational and posttranslational mechanisms that control the decision of stem cells to enter meiosis, oocyte development, and maturation.

Mammalian oocytes undergo meiotic arrest in two phases: meiotic I arrest and meiotic II arrest. The second meiotic arrest is coordinated with embryo development by the activity of the cell division protein Cdc6. *Kubiak and colleagues* discuss mechanisms utilized by Cdc6 to exquisitely coordinate DNA replication after meiotic II arrest with that in the embryo. *Yeste and colleagues* weave in the importance of sperm- and oocyte-derived factors that regulate meiosis II completion—after resumption from the meiotic II arrest—and prevent polyspermy, thereby ensuring a normal zygote. They discuss the role of lipid signaling in these processes and contemplate the impact of artificial oocyte activating factors in clinical settings dedicated to helping humans with oocyte activation deficiency. They also discuss options for better and more endogenous alternatives to artificial oocyte activating factors.

Hoang and Miller focus on the specific mechanisms that enable sperm guidance in the female reproductive tract. They examine the role of pheromones and nutritional signals that affect the sperm guidance machinery and impact sperm motility and fertility and discuss implications to human reproduction and oncologic treatments.

Fassnacht and Ciosk analyze our current understanding on the critical reprogramming mechanisms that underlie the progression of an oocyte to a single totipotent embryo. They highlight the role of developmental reprogramming with implications to varied aspects of human development and regenerative medicine.

The authors synthesize information derived from various different model systems to provide critical insights into the progression of female germline stem cells through meiotic maturation and development of a healthy embryo, highlighting the impact on human health. Additionally, we discuss the critical gaps in knowledge and questions that remain to be addressed, understanding of which will influence human reproduction, development, and regenerative medicine.

Houston, TX
2016

Swathi Arur

Contents

1 **Role of Chromatin Modifications in *Drosophila* Germline Stem Cell Differentiation** 1
Pooja Flora, Alicia McCarthy, Maitreyi Upadhyay, and Prashanth Rangan

2 **Regulation of the Balance Between Proliferation and Differentiation in Germ Line Stem Cells** 31
Ramya Singh and Dave Hansen

3 **Control of Germline Stem Cell Lineages by Diet and Physiology** . . . 67
Kaitlin M. Laws and Daniela Drummond-Barbosa

4 **Signaling-Mediated Regulation of Meiotic Prophase I and Transition During Oogenesis** 101
Swathi Arur

5 **Prophase I: Preparing Chromosomes for Segregation in the Developing Oocyte** 125
Rachel Reichman, Benjamin Alleva, and Sarit Smolikove

6 **Translational Control of Germ Cell Decisions** 175
Kumari Pushpa, Ganga Anil Kumar, and Kuppuswamy Subramaniam

7 **Role of Cdc6 During Oogenesis and Early Embryo Development in Mouse and *Xenopus laevis*** 201
Ewa Borsuk, Joanna Jachowicz, Malgorzata Kloc, Jean-Pierre Tassan, and Jacek Z. Kubiak

8 **Oocyte Activation and Fertilisation: Crucial Contributors from the Sperm and Oocyte** 213
Marc Yeste, Celine Jones, Siti Nornadhirah Amdani, and Kevin Coward

9 **Sperm Navigation Mechanisms in the Female Reproductive Tract** . . . 241
Hieu D. Hoang and Michael A. Miller

10 Cell Fate Maintenance and Reprogramming During the Oocyte-to-Embryo Transition 269
Christina Fassnacht and Rafal Ciosk

Index 287

Chapter 1
Role of Chromatin Modifications in *Drosophila* Germline Stem Cell Differentiation

Pooja Flora, Alicia McCarthy, Maitreyi Upadhyay, and Prashanth Rangan

Abstract During *Drosophila* oogenesis, germline stem cells (GSCs) self-renew and differentiate to give rise to a mature egg. Self-renewal and differentiation of GSCs are regulated by both intrinsic mechanisms such as regulation of gene expression in the germ line and extrinsic signaling pathways from the surrounding somatic niche. Epigenetic mechanisms, including histone-modifying proteins, nucleosome remodeling complexes, and histone variants, play a critical role in regulating intrinsic gene expression and extrinsic signaling cues from the somatic niche. In the GSCs, intrinsic epigenetic modifiers are required to maintain a stem cell fate by promoting expression of self-renewal factors and repressing the differentiation program. Subsequently, in the GSC daughters, epigenetic regulators activate the differentiation program to promote GSC differentiation. During differentiation, the GSC daughter undergoes meiosis to give rise to the developing egg, containing a compacted chromatin architecture called the karyosome. Epigenetic modifiers control the attachment of chromosomes to the nuclear lamina to aid in meiotic recombination and the release from the lamina for karyosome formation. The germ line is in close contact with the soma for the entirety of this developmental process. This proximity facilitates signaling from the somatic niche to the developing germ line. Epigenetic modifiers play a critical role in the somatic niche, modulating signaling pathways in order to coordinate the transition of GSC to an egg. Together, intrinsic and extrinsic epigenetic mechanisms modulate this exquisitely balanced program.

Epigenetics is the study of how environmental factors influence gene expression. Historically, the term epigenetics has been used to describe events that could not be explained by classic genetics. C.H. Waddington's metaphor illustrates that during differentiation, a cell encounters a variable landscape that determines its fate (Goldberg et al. 2007). This definition of epigenetics has evolved to encompass

P. Flora • A. McCarthy • M. Upadhyay • P. Rangan (✉)
Department of Biological Sciences/RNA Institute, University at Albany SUNY, Albany, NY
University at Albany SUNY, 1400 Washington Avenue, Albany, NY 12222, USA
e-mail: prangan@albany.edu

S. Arur (ed.), *Signaling-Mediated Control of Cell Division*, Results and Problems in Cell Differentiation 59, DOI 10.1007/978-3-319-44820-6_1

both mitotically and meiotically heritable changes in gene expression, which are not consequences of changes in DNA sequence. The epigenome is the combination of chemical modifications in a cell that modulate gene expression. These chemical modifications include DNA methylation, histone posttranslational modifications, and histone variants, which can play pivotal roles in biological processes such as cellular reprogramming and DNA replication (Boros 2012).

The phenotypic effects of epigenetic modifications were first discovered in *Drosophila melanogaster*, paving the way for further studies using *Drosophila* as a model system (Muller 1930). One well-studied aspect of epigenetics is histone modifications (Yasuhara and Wakimoto 2008; Riddle et al. 2011). Histone modifications are deposited during early development, ensuring accurate cell fate determination of the next generation (Horard and Loppin 2015). As male *Drosophila* mostly lack histones, the female *Drosophila* germ line is the prime model for epigenetic studies (Rathke et al. 2014). *Drosophila* oogenesis consists of temporal stages that can be easily visualized and tracked (Fig. 1.1a) (Bastock and St Johnston 2008). During *Drosophila* egg production, there are several interactions between the soma and germ line that are essential for proper development, which is similar to mammalian

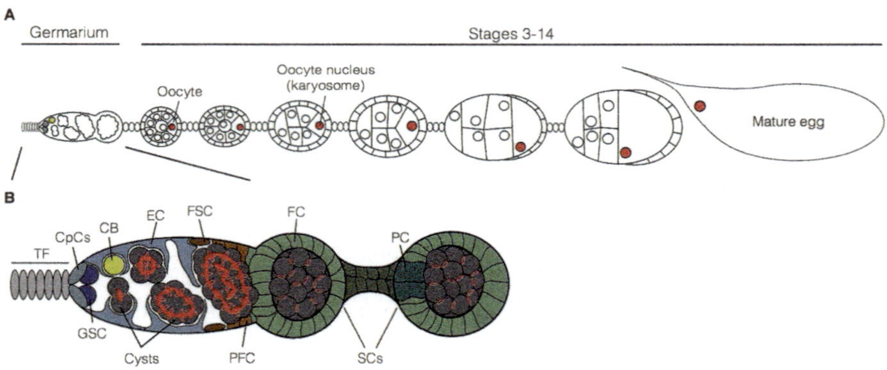

Fig. 1.1 (**a**) Schematic of *Drosophila* oogenesis. *Drosophila* females have two ovaries that consist of 16–20 ovarioles (assembly lines for mature eggs). Oogenesis can be split into 14 developmentally distinct stages, initiating with the division of GSCs that reside at the anterior region of the ovary in a structure, called the germarium. The GSC daughter then undergoes four incomplete rounds of mitosis, eventually giving rise to a 16-cell cyst. One of the 16 cells will become the egg, while the other 15 cells of the cyst will become polyploid nurse cells, which support the growth of the egg. The syncytium of nurse cell and oocyte is surrounded by a single layer of somatic follicle cells, constituting an egg chamber. The oocyte nucleus, also called the karyosome, assumes a characteristic structure in stage 3 of oogenesis. As development progresses, the nurse cells deposit mRNAs and proteins into the growing oocyte. This oocyte eventually becomes a mature egg (stage 14). (**b**) Inset germarium showing the somatic cell types that surround the germ line. TF and CpCs present at the anterior end of the germarium are intimately connected to the GSCs. Upon differentiation, the differentiating CB and 2- to 16-cell cysts make extensive contact with ECs. The 16-cell cyst then comes into contact with two FSCs. FSCs are progenitors of PFCs, FCs, SCs, and PCs. Of these, PFCs and FCs surround the developing egg chamber, SCs connect these separate egg chambers, and PCs present at both ends of the egg chambers specify polarity of the follicle epithelium

oogenesis (Gilboa and Lehmann 2006; Albertini and Bromfield 2010). Therefore, epigenetic studies in *Drosophila* can be extrapolated to higher organisms.

Drosophila oogenesis begins in the anterior-most part of the ovary, in a structure called the germarium (Fig. 1.1b) (Kirilly and Xie 2007; Spradling et al. 2011). The germarium consists of germ line stem cells (GSCs) and a specialized microenvironment called the somatic niche, which supports the GSCs. The somatic niche includes the terminal filament (TF), cap cells (CpCs), and escort cells (ECs) (Spradling et al. 2001, 2008; Lin 2002; Kirilly and Xie 2007). GSCs are attached to the niche via the CpCs and receive self-renewing signals from both the TF and CpCs. GSCs divide asymmetrically to both self-renew and give rise to a differentiated progeny, called the cystoblast (CB) (Chen 2003). After division, the CB comes into close proximity to ECs, which provide differentiation signals (Decotto and Spradling 2005). The CB expresses differentiation factor bag of marbles (Bam), which is necessary and sufficient for differentiation (McKearin and Spradling 1990; McKearin and Ohlstein 1995). Upon Bam expression, the CB undergoes four incomplete mitotic divisions to create a 16-cell cyst. One of the 16 cells becomes the oocyte, while the others become nurse cells that nourish the developing oocyte (Xie and Spradling 2000; Gilboa and Lehmann 2004) (Fig. 1.2).

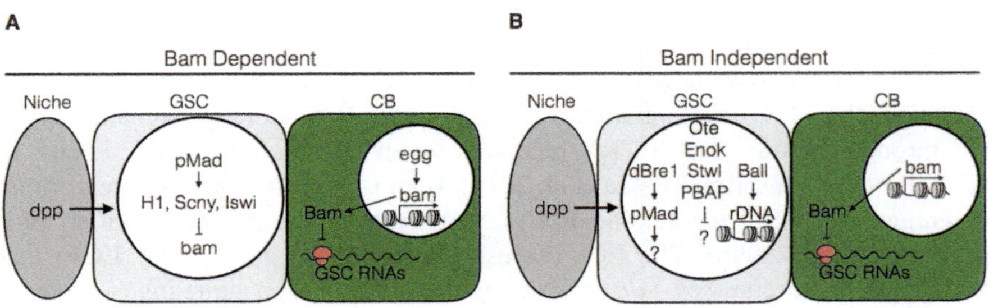

Fig. 1.2 Schematic of mode of function of epigenetic regulators in the GSC and CB during *Drosophila* oogenesis. The somatic niche provides Dpp signaling to the GSC which promotes pMad expression required for GSC self-renewal. pMad expression has shown to downregulate bam transcription in order to suppress a differentiation program. Upon GSC division, the CB turns on transcription of Bam which translationally suppresses GSC maintaining RNAs. However, several epigenetic factors can help mediate GSC self-renewal and differentiation in a Bam-dependent (A) or a Bam-independent (B) pathway. (A) Epigenetic factors H1, Scny, and Iswi act downstream of the incoming Dpp signaling to repress transcription of Bam, ensuring stem cell fate maintenance. On the other hand, dSETDB1/Egg expression in the CB is required to promote transcription of Bam to initiate a differentiation program. (B) dBRE1 expression in the GSC is required for proper pMad levels, although its downstream effector is not Bam. Ote, Enok, Stwl, and PBAP also repress a differentiation program by targeting expression of differentiating factors other than Bam. Ball has been implicated in promoting ribosomal biogenesis, which is crucial for GSC self-renewal and maintenance

1.1 GSC Self-renewal Is Associated with Epigenetic Silencing of Differentiation Program

GSC self-renewal and differentiation are controlled by several intrinsic and extrinsic regulatory factors. Extrinsic factors include extracellular molecules that trigger signal transduction pathways such as Decapentaplegic (Dpp) signaling, which promotes GSC self-renewal. Dpp signaling is initiated by the cytokine-like ligands, Unpaired (Upd) and Unpaired-2 (Upd-2), produced by the TF. Upd binds to Janus Kinase (JAK) surface receptors present on the CpCs (Lopez-Onieva et al. 2008; Wang et al. 2008), leading to the phosphorylation and activation of signal transducer and activator of transcription 92e (Stat92e). This activated Stat92e triggers the production of signaling molecules by the CpCs called Decapentaplegic (Dpp) or Glass bottom boat (Gbb), which are similar to Bone morphogenetic proteins (BMPs) in mammals (Xie and Spradling 1998; Kawase et al. 2004; Wang et al. 2008). Dpp binds to the Type I receptors Thick veins (Tkv) and Saxophone (Sax) on GSCs, leading to the phosphorylation of mothers against Dpp (pMad), a *Drosophila* Smad3 protein. pMad then complexes with Medea (Med), a *Drosophila* Smad4 protein, to transcriptionally silence the differentiation factor Bam (Brummel et al. 1994; Nguyen et al. 1998; Song 2004). The somatic niche also expresses Division abnormally delayed (Dally), a glypican protein which is responsible for distributing and stabilizing Dpp signaling (Guo and Wang 2009; Hayashi et al. 2009). However, Dpp signaling in the GSCs also results in a negative-feedback circuit that induces the expression of Daughters against Dpp (Dad), which dampens Dpp activity (Tsuneizumi et al. 1997). The interplay between Dpp signaling and the GSC's response to this incoming signaling controls self-renewal and differentiation.

Differentiation is initiated after GSC division, when the CB is displaced from the somatic niche. GSC differentiation requires downregulation of Dpp signaling in the CB through the turnover of TKV receptors by an E3 ubiquitin ligase, Smurf, to allow for *bam* transcription (Xia et al. 2010). Bam is a translational suppressor that downregulates GSC-maintaining RNAs, thereby initiating a differentiation program (Li et al. 2009b). The ECs express Epidermal Growth Factor (EGF) that transcriptionally represses *dally*, restricting the range of Dpp signaling to promote an efficient CB differentiation program (Schulz et al. 2002). Although, signaling pathways play an indispensable role in providing cues for maintaining GSCs and differentiation, epigenetic changes in the germ line and the surrounding somatic niche are also critical.

Gene expression is controlled by both epigenetic state and nuclear organization. The epigenome is a catalogue of chemical modifications that affect gene expression without directly altering the encoded genetic material (Goldberg et al. 2007). Several epigenetic modifications can occur directly on DNA or its associated proteins that constitute the chromatin (Cedar and Bergman 2009). The basic unit of chromatin is the nucleosome, consisting of five histones H2A, H2B, H3, H4, and a linker histone H1. The nucleosome octamer core contains one tetramer of H3 and

H4 and two heterodimers of H2A and H2B (Luger et al. 1997). Approximately, 150 base pairs of DNA wrap around each octomer, providing an epigenetic landscape that can be manipulated via different histone modifications (Jenuwein and Allis 2001). Each histone protein has a tail that can be modified to alter the local chromatin structure and regulate gene expression (Luger et al. 1997). Several modifications can be deposited onto histone tails such as methylation, acetylation, ribosylation, ubiquitination, and phosphorylation (Kouzarides 2007). These modifications together result in a specific readout of either gene activation or repression. For example, the trimethylation of lysine 9 and 27 residue of H3 (H3K9me3, H3K27me3) can result in gene repression, while trimethylation of lysine 4 and acetylation of lysine 9 of H3 (H3K4me3, H3K9ac) can result in gene activation (Bannister and Kouzarides 2011). H1 and its variants are essential for compaction of nucleosomes for chromatin remodeling (Bednar et al. 1998; Misteli et al. 2000; Bustin et al. 2005). Loss of H1 not only alters chromatin compaction but can also affect several histone modifications such as H3 lysine 4 (H3K4), lysine 9 (H3K9), and lysine 27 (H3K27), demonstrating its vital role in maintaining the epigenetic landscape for modulating gene expression (Zhang et al. 2012). Therefore, the combination of nuclear organization and chromatin modification is crucial for proper gene regulation. Direct DNA methylation can also contribute to the regulation of gene expression, but meaningful DNA methylation is not present in *Drosophila* (Raddatz et al. 2013), so it will not be addressed in this chapter.

1.1.1 Histones and Histone-Modifying Enzymes in the Germ Line that Regulate GSC Self-renewal and Differentiation

GSCs have a unique epigenetic state that is constantly altered to ensure proper differentiation (Buszczak and Spradling 2006; Li and Zhao 2008). H1 linker histone is intrinsically required for GSC maintenance, as depletion of H1 in the germ line results in loss of GSCs over time. Depletion of H1 in the GSC, however, does not alter response to Dpp signaling, as Mad is normally phosphorylated. However, in spite of normal pMad levels, Bam is prematurely expressed. Depletion of H1 also increases H4 lysine 16 acetylation (H4K16ac) in the GSCs, a mark associated with active chromatin. Additionally, overexpression of a specific H4K16 acetyl transferase, *male absent on the first* (*mof*), leads to GSC loss and premature expression of Bam (Sun et al. 2015). This suggests that H1 is responsible for silencing differentiation genes, while Mof activates differentiation genes to ensure remodeling of the chromatin to an active state for efficient GSC differentiation (Akhtar and Becker 2000). These studies indicate that histones modulate accessibility of differentiation genes via nucleosome compaction in the GSCs.

In addition to nucleosome compaction, GSC fate is also maintained at the chromatin level by histone modifications. H2B monoubiquitination (Ub-H2B) plays an important role during chromatin regulation and is required for H3K4 methylation, which is an indicator of gene activation (Dover 2002; Sun and Allis 2002). In *Drosophila*, the gene *scrawny* (*scny*) encodes an H2B ubiquitin protease that deubiquitinates H2B to suppress methylation at lysine residues, resulting in gene repression (Gardner et al. 2005). GSCs express Scny at high levels, and loss of *scny* results in increased acetylation of H3 tails and early expression of Bam, causing a loss of GSCs (Buszczak et al. 2009). On the other hand, it has been shown that *dBre1* encodes an ubiquitin ligase required for the monoubiquitination of H2B and is expressed throughout the germarium (Bray et al. 2005). Loss of *dBre1* in the GSC results in GSCs with low levels of H3K4 trimethylation (H3K4me3). Although, GSCs lacking *dBre1* display low levels of pMad and Dad expression, *bam* transcription remains unchanged. This suggests that dBre1 modulates GSC maintenance by targeting differentiation factors other than Bam (Xuan et al. 2013). Collectively, these two studies have shown that an ubiquitin protease and an ubiquitin ligase maintain H3K4me3 levels in the GSCs to ensure the proper balance of gene expression.

A histone methyltransferase, *dSETDB1* or *eggless* (*egg*), has also been shown to regulate GSC fate during *Drosophila* oogenesis. dSETDB1 deposits the repressive H3K9me3 mark, which recruits heterochromatin protein 1 (Hp1) to form heterochromatin (Schotta 2002; Ebert 2004). H3K9me3 is present at low levels in the GSC, but increases in the CB once division is complete. This is concurrent with the finding that dSETDB1 in the GSCs is mostly cytoplasmic, but becomes increasingly nuclear after division (Rangan et al. 2011; Clough et al. 2014). Generation of *dSETDB1*-deficient GSCs leads to loss of self-renewal (Wang et al. 2011). And although the *dSETDB1*-deficient GSCs differentiate properly, they arrest mid-oogenesis. The targets of dSETDB1-mediated H3K9me3 marks were identified as Piwi-interacting small RNA (piRNA) clusters (Rangan et al. 2011). piRNA clusters produce small RNAs (piRNAs) that target transposable elements (TEs) for degradation (Gunawardane et al. 2007; Brennecke et al. 2007; Li et al. 2009a; Malone et al. 2009). Loss of *egg* in the germ line results in the loss of piRNAs and a failure to degrade TEs, leading to double-strand breaks (DSBs) in the genome (Rangan et al. 2011). A critical step prior to heterochromatin formation is the removal of methylations on H3K4 by *Lysine-specific demethylase 1* (*dLsd1*) (Rudolph et al. 2007; Di Stefano et al. 2007). Intriguingly, loss of *dLsd1* in the GSCs does not affect their maintenance (Eliazer et al. 2011). Germ line clones of *dLsd1* mutants also differentiate normally, with no observed block in the differentiation program, but result in sterile females (Eliazer et al. 2011). Together, these results indicate that methylation of histones mediated by intrinsic factors is imperative for GSC maintenance and fertility.

Although histone-modifying enzymes in the GSCs are responsible for promoting a repressive chromatin state, a recent study has shown that GSC self-renewal also depends on deposition of active marks (Xin et al. 2013). Acetylation of histones is associated with euchromatin formation, which makes the DNA more accessible to transcription factors (An 2007). A putative acetyl transferase, Enoki mushroom (Enok), has been shown to act intrinsically to maintain GSCs during *Drosophila* oogenesis. Depletion of *enok* in the GSCs results in GSC loss over time. This loss of GSC phenotype does not result from inappropriate pMad levels or from premature Bam expression. This suggests that in the GSC, Enok acts intrinsically downstream of Dpp signaling and independent of the Bam pathway to control genes that may promote self-renewal (Xin et al. 2013). Cumulatively, these observations give insight into how histone-modifying enzymes mediate the balance of gene expression in the early stages of oogenesis to ensure the onset of differentiation.

1.1.2 Nucleosome Remodeling Complexes in the Germ Line that Regulate GSC Self-renewal and Differentiation

Various protein partners that interact with histone-modifying enzymes also contribute to the epigenetic state of a cell. Stonewall (Stwl), a DNA-binding protein that acts as a dominant Suppressor of variegation [Su(var)], is highly expressed in GSCs. Su(var)s usually modulate gene expression by depositing heterochromatin and promoting chromosomal rearrangement to inactivate the genes (Schotta 2002). Loss of *stwl* results in loss of GSCs, while overexpression results in a significant increase in GSC-like cells. These observations indicate that Stwl is an intrinsic GSC maintaining factor. However, loss of GSCs in a *stwl* mutant does not result from early expression of Bam, indicating that Stwl targets other differentiating factors for silencing in the GSC (Maines et al. 2007). Nanos (Nos) and Pumilio (Pum) are conserved RNA translational suppressors that suppress translation of differentiating mRNAs in the GSCs (Lin and Spradling 1997; Forbes and Lehmann 1998; Wang and Lin 2004). Interestingly, it was found that Stwl transcriptionally inhibits the genes that are translationally regulated by Nos and Pum (Maines et al. 2007). This shows that proteins such as Stwl interact with histone-modifying enzymes to maintain a repressive chromatin state that suppresses the transcription of differentiating mRNAs in the GSCs.

Chromatin remodelers, along with chromatin organization and epigenetic marks, are critical for GSC maintenance and differentiation. Chromatin remodeling proteins regulate gene expression by modulating nucleosome position, which allows transcription factors to access the DNA (Armstrong and Emerson 1998). In *Drosophila*, there are five chromatin remodeling factors: Brahma (Brm), Imitation SWI (Iswi), Domino (Dom), Kismet (Kis), and dMi-2 (Tamkun et al. 1992; Kehle 1998; Daubresse et al. 1999; Ruhf et al. 2001; Corona 2002). To remodel chromatin, Brm interacts with either Osa, a Trithorax group (TrxG) protein, to form the Brahma associated protein (BAP) complex or with Polybromo to form the Polybromo associated protein (PBAP) complex (Mohrmann et al. 2004; Moshkin et al. 2006). Brm is expressed throughout the germarium, in both the somatic niche and the germ cells, and loss of *brm* in the germ line results in GSC maintenance defects. Loss of Bap180, a PBAP complex member, in the germ line resulted in loss of GSCs over time. However, mutations in the BAP associated protein Osa did not affect GSCs, suggesting that PBAP but not BAP is required for GSC maintenance. Although GSC loss is usually associated with misregulation of Dpp signaling and early expression of Bam, this phenomenon was not observed in the *bap180* mutants (He et al. 2014). These findings reveal that chromatin remodeling is critical for GSC maintenance, but their downstream targets in GSCs have yet to be identified.

Along with Brm, other chromatin remodelers have been shown to function in the GSCs: Iswi regulates transcription by compacting the chromatin, while Dom is responsible for exchanging histone variants to induce transcriptional repression (Deuring et al. 2000; Ruhf et al. 2001; Corona 2002; Badenhorst et al. 2002). Iswi protein levels are high in the nuclei of all germ cells, while Dom is present in high levels in the nuclei of GSCs and in low levels in the nuclei of somatic stem cells (SSCs). *iswi* and *dom* mutant oocytes arrest in mid-oogenesis, suggesting they are essential during *Drosophila* oogenesis. Intriguingly, it was reported that only loss of *iswi*, but not *dom*, resulted in the loss of GSCs. GSCs lacking *iswi* showed downregulation of Dad, but pMad levels were not altered concurrent with premature expression of Bam. This indicates that Iswi functions downstream of Dpp signaling to maintain GSCs by repressing Bam (Xi 2005). Although, Kismet's role during oogenesis has yet to be investigated, dMi-2 has been shown to play an important role in oocyte specification and karyosome formation, discussed in Sect. 1.2.3.

1.1.3 Nuclear Organization in the Germ Line that Regulates GSC Self-renewal and Differentiation

The cytoplasm and the nucleoplasm are separated by the nuclear envelope. Embedded beneath the nuclear envelope is a network of filamentous proteins, referred to as the nuclear lamina, which helps organize the chromatin during interphase (Gruenbaum et al. 2005). An important family of lamin-interacting proteins are the LEM domain (LEM-D) proteins, named after LAP2, Emerin, and MAN1 (Lin 2000; Mansharamani and Wilson 2005; Wagner and Krohne 2007). Protein domains in LEM-D proteins interact directly with transcriptional repressors as well as signaling cascade effectors, such as Smads, implying that they broadly contribute to gene regulation and chromatin organization (Holaska et al. 2003; Haraguchi et al. 2004; Lin et al. 2005; Mansharamani and Wilson 2005; Bakay 2006; Jiang et al. 2008). LEM-D proteins are constitutively expressed. However, they display tissue-specific pathological roles. For example, mutations in the LEM-D protein, Emerin, result in age enhanced, X-linked muscular dystrophy in humans (Manilal 1996, 1998). These observations suggest that diseases associated with LEM-D may result from dysregulation of adult stem cell maintenance. Otefin (Ote), the *Drosophila* homolog of the LEM-D protein Emerin, localizes to the nuclear envelope and physically interacts with nuclear lamin proteins B and YA

(Ashery-Padan et al. 1997; Goldberg et al. 1998). However, RNAi knockdown of Ote in *Drosophila* cell lines showed no defects in the organization of the nuclear envelope, suggesting a role beyond nuclear architecture (Wagner 2004). Jiang et al. (2008) showed that Ote is expressed in both the germ line and somatic cells of the *Drosophila* gonad. A mutation in the *ote* allele resulted in sterile females with rudimentary ovaries lacking GSCs with few differentiated germ cells and abnormal egg chambers. Germ line specific loss of *ote* by clone induction results in loss of GSC, suggesting that Ote is required intrinsically for GSC maintenance. And ectopic overexpression of Ote increased the number of GSCs in the ovary, suggesting a role in modulating differentiation. Concurrent Ote and Dpp antagonist Dad overexpression resulted in complete loss of GSCs, indicating that Ote acts as an agonist of the Dpp pathway. It was reported that loss of *ote* also resulted in Bam expression in the GSC, thereby promoting premature differentiation. Additionally, this group demonstrated that Ote physically interacts with Med and together associate with the silencer element in the *bam* gene to suppress its transcription, thereby conserving GSC fate (Jiang et al. 2008). In contrast to this work, studies conducted by Barton et al. reported that loss of *ote* during oogenesis results in differentiation defects with no upregulation of Bam expression. They also established that GSCs lacking *ote* failed to undergo proper differentiation despite ectopic expression of Bam, indicating that Ote does not target *bam* for GSC maintenance (Barton et al. 2013). Even though there is conflicting evidence on how Ote mediates GSC maintenance, it is evident that a LEM-D containing protein is indispensable for intrinsically maintaining GSC fate. Studies in vertebrates have shown that the LEM-D protein MAN1 binds to Smad2 and Smad3 to antagonize BMP signaling (Lin 2000; Ishimura 2006; Cohen et al. 2007). Altogether, these studies suggest that different LEM-D proteins may have opposing functions in maintaining homeostasis in various adult stem cell niches. Therefore, along with histone modifications, chromatin organization is required for GSC fate maintenance.

Studies have shown that the *Drosophila* gene *bällchen* (*ball*), also known as Nucleosomal Histone Kinase-1 (NHK-1), encodes a protein kinase orthologous to mammalian Vaccinia-related Kinase 1 (VRK1), which is required for both chromatin organization and appropriate histone modification patterns in the oocyte (Aihara et al. 2004). The Ball protein is expressed in the GSC and is required for long-term GSC self-renewal (Herzig et al. 2014). However, Ball neither affects the Dpp signaling pathway in the GSCs nor regulates Bam expression in the differentiating germ cells, suggesting that its function in GSC maintenance is independent of Bam regulation. It was found that Ball not only localized with the entire chromatin but was also highly enriched in the nucleolus. The nucleolus assembles around the chromatin region containing ribosomal DNA (rDNA) repeats. The GSCs have a large nucleolus as compared to those of the differentiating germ line cysts, which re-expand after the cysts begin to polyplodize (Neumüller et al. 2008). In the absence of Ball, half the GSCs exhibit fragmentation and disintegration of the nucleoli (Herzig et al. 2014). This indicates that Ball is not required for initiating but rather for maintaining nucleolar expansion of the GSCs. Overall, these studies indicate that Ball is an intrinsic factor required to maintain nucleolar organization for increased ribosome biogenesis in the GSC. Studies conducted by Zhang et al. have shown that high levels of ribosomal RNA transcription is dependent on *underdeveloped* (*udd*) in the GSCs. Udd interacts with RNA polymerase I transcription-promoting factors to mediate GSC maintenance (Seifarth et al. 1991; Zomerdijk et al. 1994). Udd localization in the GSCs is highly nuclear with increased rRNA synthesis when compared to the CB. This increased rRNA synthesis leads to proper translation of GSC-fate maintaining RNAs, such as Mad, which allows the cell to promote its self-renewal properties (Zhang et al. 2014). Altogether, these studies show the importance of nucleolar assembly in the GSCs in order to promote rRNA transcription that ensures gene expression for self-renewal.

1.2 Epigenetic Mechanisms for Oocyte Specification and Maintenance

After GSC self-renewal and differentiation, the stem cell daughter undergoes four rounds of incomplete division. This produces an egg chamber cyst with a single cell that is a fully determined oocyte and 15 endocycling nurse cells. As the cyst migrates posteriorly, there is encapsulation by follicle cells (FCs). These encapsulation events lead to the formation of 14 distinct developmental stages or egg chambers (Spradling 1993).

Despite the oocyte being directly connected to the nurse cells, the developing oocyte has a unique cell cycle from that of the nurse cells. Nurse cells fluctuate expression of cyclin E (*CycE*) and *dacapo* (*dap*) to achieve an altered cell cycle where they increase their DNA content without dividing (Robinson and Cooley 1996; Su and O'Farrell 1997). The current hypothesis on how oocytes are specified proposes that fusomes and ring canals accumulate in cysts, facilitating RNA transport to promote this specification. Before the oocyte is specified, in meiotic prophase its chromosomes condense and align to form synaptonemal complexes (SCs) (Koch et al. 1970; Carpenter 1975; Schmekel and Daneholt 1995). Until recombination in prophase I is complete, the oocyte can be identified by the presence of SC members, such as crossover suppressor on 3 of Gowen (C(3)G) (Page 2001).

A growing body of research suggests that chromatin architecture and remodeling play a critical role in specifying the oocyte and modulating cell cycle and meiosis. Chromosomes store genetic information for propagation to subsequent generations. During meiosis, chromosomes have functions that demand opposing epigenetic states. Meiotic chromosomes are maintained in a haploid state for extended periods of time, so they must undergo dynamic structural changes to balance both transcriptional activation and silencing. Homologous meiotic chromosomes also undergo DSBs to allow recombination to take place. This process of homologous recombination (HR) is stabilized by the SC (Page and Hawley 2004), and the assembly of the SC is mediated, in part, by epigenetic modifications (Ivanovska and Orr-Weaver 2014). Following recombination, a conserved, prolonged prophase I arrest occurs. In prophase I arrest, *Drosophila* chromosomes condense within the oocyte nucleus to form a structure called the karyosome, which is conserved in higher organisms. Although it has been postulated that karyosome formation serves to keep chromosomes close together in the growing oocyte nucleus, the reason for a chromosomal rearrangement in metazoan prophase I is, however, unclear (Gruzova and Parfenov 1993).

1.2.1 Histone-Modifying Enzymes and Histone Variants in Oocyte Specification and Maintenance

Histone-modifying complexes such as polycomb repressive complex 2 (PRC2) have been shown to be involved in oocyte specification and by extension karyosome formation. Enhancer of zeste [E(z)], the enzymatic subunit of PRC2, has previously been shown to deposit H3K27me3 marks to suppress gene expression (Czermin et al. 2001). Depletion of E(z) results in loss of oocyte markers, Orb and Bicaudal (BicD). By stage 3, when the meiotic chromosomes first condense to form the karyosome, the single C(3)G-positive cell in depleted ovaries lacked this karyosome structure. Chromatin immunoprecipitation (ChIP) assays using anti-E(z) and anti-H3K27me3 followed by qualitative real-time PCR (qPCR) assaying for *CycE* and *dap* loci showed that both loci are enriched for E(z) and H3K27me3 deposition. These data suggest that E(z) suppresses the nurse cell-like fate in the oocyte. This is mediated by the silencing of CycE and Dap expression. This in turn prevents improper endocycling before the oocyte becomes fully specified. And while PRC2 plays an important role in oocyte fate, loss of polycomb repressive complex 1 (PRC1) do not cause infertility or oocyte defects (Li et al. 2010; Gandille et al. 2010; Iovino et al. 2013) suggesting that PRC2 specifically has an epigenetic role in karyosome formation.

Gene regulation, modulated by histone modifiers, is critical for oocyte specification and karyosome formation. To characterize the oocyte epigenome, an antibody screen against 21 histone marks was carried out, and it was identified that the oocyte epigenome is unique, has both active and repressive marks, and is temporally dynamic (Navarro-Costa et al. 2016). One such dynamic mark that is deposited early during oogenesis and maintained is H3K4me3, which is found at transcription start sites of active genes (Santos-Rosa et al. 2002; Barski et al. 2007; Navarro-Costa et al. 2016). It was identified, through an RNAi screen, that *Drosophila* histone demethylase KDM5 (dKDM5) is responsible for regulating early H3K4me3 levels (Navarro-Costa et al. 2016). Tightly regulated H3K4me3 levels are crucial, as early H3K4me3 levels mediate karyosome remodeling and prophase I arrested-oocyte transcriptional reactivation (King 1970; Navarro-Costa et al. 2016). Thus, dKDM5 is required for timely transcriptional reactivation, proper karyosome remodeling, and fertility.

During meiotic recombination, there is an activation of a meiotic checkpoint due to the formation of DSBs (Su 2006). Ataxia telangiectasia-mutated (ATM) and ataxia telangiectasia-related (ATR/MEI-41) kinases are activated by these DSBs and phosphorylate several targets, including checkpoint-1 (Chk1), checkpoint-2 (Chk2/MNK), and histone H2A variant (H2Av) (Kurz and Lees-Miller 2004; Shiloh 2006; Joyce et al. 2011). It has been proposed that phosphorylation of H2Av recruits chromatin-remodeling complexes, such as the histone acetlytransferase Tip60, to DSB sites (Kusch et al. 2004). This suggests that DSBs recruit factors necessary for remodeling the local chromatin architecture for meiotic progression. When there is defective DSB repair, the meiotic checkpoint is activated for a prolonged period, which in turn results in inefficient accumulation of a TGFα-like protein called Gurken (Grk) causing alteration in the dorsal-ventral (DV) patterning in *Drosophila* oocytes (Ghabrial et al. 1998; Abdu et al. 2002).

Some members of the *spindle* group (which include *spindle* (*spn*)-*A*, *spn-B*, *spn-C*, *spn-D*, *spn-E*, *vasa*, *maelstrom*, *aubergine*, and *okra*) have been shown to have a role in meiotic checkpoint progression, oocyte specification, and chromatin organization. *spn-A*, *spn-B*, *spn-C*, *spn-D*, and *okra* were identified to be important in meiosis, as mutants for these genes had prolonged meiotic checkpoints, as demonstrated by H2Av accumulation, resulting in altered DV patterning phenotypes (Huynh and St Johnston 2004). Despite having similar phenotypes as the other *spindle*-group members, *spn-E*, *vasa*, *aubergine*, and *maelstrom* mutants did not show direct involvement during meiosis progression (Ghabrial et al. 1998; Abdu et al. 2002; Findley 2003; Tomancak et al. 2007). Thus, prolonged exposure to DSBs produces altered chromatin architecture that leads to defects in oocyte specification.

1.2.2 Nuclear Organization in Oocyte Specification and Maintenance

Chromosomes are anchored to the nuclear envelope during meiotic recombination (Su and O'Farrell 1997; Tange et al. 1998; Phillips and Dernburg 2006; Scherthan et al. 2007; Phillips et al. 2009). This association must be dynamic, given that after completion of meiotic recombination, chromosomes must detach from the nuclear envelope to form the karyosome (Cullen et al. 2005). SC formation takes place prior to recombination (McKim and Hayashi-Hagihara 1998). It was found that NHK-1, or Bällchen, a histone H2A kinase, is required for proper SC disassembly, histone H3 lysine 14 and histone H4 lysine 5 acetylation, and karyosome localization to the interior of the oocyte nucleus (Ivanovska 2005; Cullen et al. 2005). In *nhk-1* mutants, a recombination subunit of condensin, structural maintenance of chromosomes 4 (SMC4) does not localize to the karyosome (Chan et al. 2004; Ivanovska 2005). These findings suggest that NHK-1 plays a critical role in the formation and maintenance of the karyosome (Ivanovska 2005; Ivanovska and Orr-Weaver 2014).

Barrier to Autointegration Factor (BAF) binds to LEM-D proteins in the inner nuclear membrane and anchors DNA (Shumaker 2001; Furukawa 2003; Schmid and Nanda 2004; Mansharamani and Wilson 2005). Intriguingly, VRK1 in mammals has a high degree conservation of the kinase domain and structure to NHK-1 and has been shown to phosphorylate BAF (Nichols 2006; Gorjánácz et al. 2007). *Drosophila* karyosome and mouse oocytes exhibit similar chromatin architecture. *Vrk1* mutant oocytes exhibit nuclear organization defects, as compared to wild-type mouse oocytes. This suggests that VRK1 phosphorylation of BAF is critical for the release of DNA to form the karyosome (Schober et al. 2011). In *Drosophila*, mutation of the BAF phosphorylation site leads to continuous association of DNA with the inner nuclear membrane and loss of karyosome formation (Lancaster et al. 2007). Interestingly, *mnk, nhk-1* double mutants did not rescue the karyosome defect, suggesting that this defect is not a result of meiotic checkpoint activation. In *spn*-A, *spn*-B, and *spn*-D ovaries, the phosphorylation of histone H2A threonine 119, a mark deposited by NHK-1, was greatly reduced. This indicated that unrepaired DSBs suppress NHK-1 phosphorylation on histone H2A (Lancaster et al. 2010). Together, these findings support the model that NHK-1 kinase activity is suppressed during meiotic recombination while the oocyte chromosomes are anchored to the nuclear envelope. Once the oocyte has exited the meiotic checkpoint, NHK-1 phosphorylates BAF, leading to the release of chromosomes from the nuclear envelope (Ivanovska 2005; Lancaster et al. 2007, 2010; Ivanovska and Orr-Weaver 2014) (Fig. 1.3). Thus, NHK-1 is a critical factor mediating the dynamic architecture, which allows chromosomes to attach and detach from the nuclear envelope during meiosis.

1.2.3 Role of a Nucleosome Remodeling Complex in Oocyte Specification and Maintenance

The role of histone-modifying complexes in chromatin modification during oocyte specification and karyosome formation is still not fully understood (Kehle 1998; Brehm et al. 2000). dMi-2 is a member of multiple chromatin-remodeling complexes, including the NuRD complex (Kunert 2009). It been shown that loss of *dMi-2* causes chromosome decondensation and disrupts the association of cohesion with polytene chromosomes (Fasulo et al. 2012). While both dMi-2 and NHK-1 depleted oocytes had undercondensed chromosomes in mature oocytes, they most likely are involved in distinct pathways given that *dMi-2* oocytes show normal karyosome organization before nuclear envelope breakdown, unlike *nhk-1* mutants (Ivanovska 2005; Cullen et al. 2005; Nikalayevich and Ohkura 2015). Despite the fact that the mechanism of NuRD action in chromosome condensation has not yet been described, several suggestions have been offered. It has been proposed that perhaps nucleosome remodeling or even the histone deacetylase activity of NuRD is required for chromosome condensation; however, these proposals have yet to be tested (Nikalayevich and Ohkura 2015).

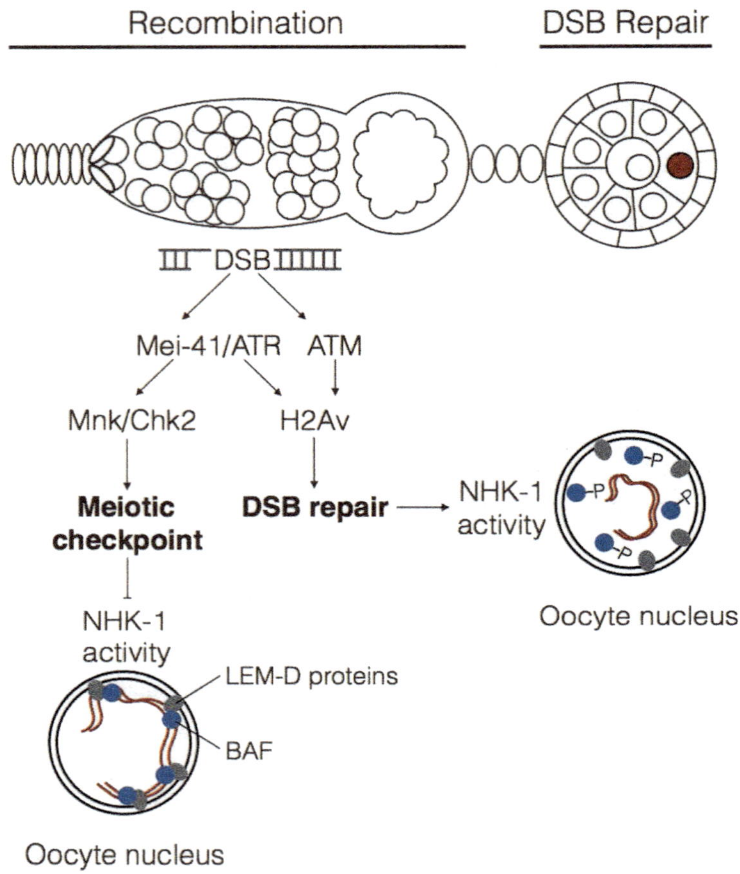

Fig. 1.3 Schematic of NHK-1 activity in the ovary. Upon double-strand break formation, meiotic recombination checkpoint is activated, NHK-1 kinase activity is suppressed, and BAF is unphosphorylated. This anchors chromosomes to the nuclear envelope via LEM-D proteins. Once recombination is complete, NHK-1 phosphorylates BAF and the chromosomes dissociate from the nuclear envelope, allowing karyosome formation. Adapted from Lancaster et al. (2007)

1.3 Chromatin Regulation in Somatic Niche Cells Regulates Oogenesis

Oogenesis requires proper GSC self-renewal and maintenance while also requiring close contact between the germ line and surrounding somatic niche. The soma–germ line contact is conserved from *C. elegans* to humans. In *C. elegans*, contact between the somatic sheath and germ cells is critical for oogenesis (Hall et al. 1999; Erickson 2008), and in mammals, somatic granulosa cells surround the oocytes to promote maturation (Buccione et al. 1990; Canipari 2000; Matzuk et al. 2002; Gilchrist et al. 2004). A similar interaction between *Drosophila* germ cells and the somatic niche is critical for GSC self-renewal and differentiation (Morgan and Mahowald 1996; Xie and Spradling 2000).

In *Drosophila*, somatic niche cells such as the terminal filament (TF), cap cells (CpCs), and escort cells (ECs) surround the germ line and play an important role in GSC self-renewal and differentiation (Fig. 1.1b) (Song 2002). The TF consists of approximately seven to eight cells connecting the germarium with the sheath cells surrounding the ovariole (Lin and Spradling 1993; Sahut-Barnola et al. 1995). Attached to the TF is a cluster of CpCs (Forbes et al. 1996). GSC self-renewal requires adhesion to the CpCs, mediated by adherens junction molecules such as DE-Cadherin and Armadillo/β-catenin. Not only do CpCs provide somatic structural support but also are a source of Dpp signaling that is critical for GSC self-renewal (Song 2002).

Upon GSC division, the CB loses contact with the CpCs, turns on Bam, and undergoes four rounds of incomplete divisions to form a 16-cell cyst, as described earlier (Xie and Spradling 2000). The ECs form long processes that make extensive contact with the CB and all the differentiating cysts. These EC processes are required for proper differentiation (Schulz et al. 2002; Decotto and Spradling 2005). The 16-cell cyst containing the specified oocyte comes into contact with follicular stem cells (FSCs) (Margolis and Spradling 1995; Nystul and Spradling 2007). FSCs generate multiple somatic cell types, such as epithelial follicle cells, pre-follicular cells (PFCs), stalk cells (SCs), and polar cells (PCs) (Tworoger et al. 1999; Nystul and Spradling 2010). Epithelial follicle cells surround each maturing egg chamber and are required for proper follicle budding, oocyte development, and patterning (Huynh and St Johnston 2004; Nystul and Spradling 2010). The separate egg chambers are connected by an array of stalk cells and have specialized cells known as the polar cells at the anterior and posterior ends (Wu et al. 2008). All of these somatic cell types coordinate transition from GSC to the mature egg.

1.3.1 Chromatin Modifiers Regulate Signaling Pathways in the Somatic Niche

Numerous chromatin modifiers have been identified in the somatic niche cells that play a critical role in oogenesis. These molecules either act in a specific cell type or in multiple cell types to regulate different steps of this process. Female sterile (1) Yb (fs(1)Yb), also referred to as Yb, encodes a protein that has domains homologous to DEAD/DEAH box RNA helicase and Tudor domain (Szakmary et al. 2009). Fs(1)Yb interacts with other proteins of the piRNA pathway to form Yb bodies in somatic cells of the gonad (King and Lin 1999). These Yb bodies have been identified as a major site for piRNA biogenesis (Qi et al. 2011). In the TF and CpCs, fs(1)Yb controls GSC maintenance by regulating two pathways: Piwi and Hedgehog (Hh) (King and Lin 1999; King et al. 2001).

Loss of Yb results in mislocalization of Piwi from Yb bodies in the somatic niche cells. Piwi, a nucleoplasmic factor that regulates chromatin state, is expressed in both the soma and germ line (Cox et al. 1998). It plays a pivotal role during oogenesis by promoting piRNA production (Thomson and Lin 2009). Loss of *piwi* in the germarium leads to pleotropic phenotypes, such as accumulation of GSC-like cells, loss of GSCs, and differentiation defects. *piwi* depletion in only the CpCs did not result in GSC tumors, but exhibited loss of GSCs suggesting that *piwi* is required in the CpCs for GSC maintenance (Cox et al. 1998, 2000; Ma et al. 2014). Corto, a chromodomain protein expressed in the TF and CpCs, binds to methylated lysine residues in histones. Corto has been shown to interact with both the Polycomb Group (PcG) genes that mediate gene silencing and the Trithorax Group (TrxG) genes that confer gene activation (Salvaing et al. 2003; Smulders-Srinivasan et al. 2010). It was shown that mutations in *corto* suppressed the GSC maintenance defects of *Yb* and *piwi* mutants. This suggests that an as yet unidentified regulatory mechanism acting downstream of Yb and Piwi in the CpCs is required for GSC maintenance (Smulders-Srinivasan et al. 2010).

piwi depletion in the ECs results in tumors containing GSC-like cells. This suggests that loss of *piwi* results in elevated Dpp signaling in the ECs (Jin et al. 2013). Further studies showed that reduction of *dpp* in these *piwi* depleted ECs did not drastically rescue the differentiation defects, indicating that Dpp upregulation is not the major cause of these defects. *piwi* is required in EC for survival and formation of its processes (Ma et al. 2014; Upadhyay et al. 2016). It has been shown that Piwi is also required for expression of Wingless-type mouse mammary tumor virus (MMTV) integration site family member 4 (dWnt4) in the ECs. dWnt4 signaling in the ECs is required for both proliferation and processes formation (Wang et al. 2015). Thus, in the ECs, Piwi is required for regulating Dpp and dWnt4 signaling to promote GSC differentiation.

Loss of *Yb* also phenocopies loss of *Hh* in the TF and CpCs, where FSCs under-proliferate resulting in egg chamber defects (Forbes et al. 1996; King and Lin 1999). Loss of *piwi* does not affect Hh expression in the TF and CpCs, suggesting that piwi and Hh work in parallel, downstream of Yb (King et al. 2001). Studies have shown that Hh affects FSC proliferation and cell fate specification by coupling with the Hippo pathway (Huang and Kalderon 2014) or by regulating two transcription factors: Cubitus interruptus (Ci) and Castor (Cas) (Chang et al. 2013; Zhang and Kalderon 2001). Thus, Yb independent of *piwi* regulates Hh signaling to promote proper differentiation.

1.3.2 *Histone-Modifying Enzymes Regulate Signaling Pathways in the Somatic Niche*

Enok, a putative histone acetyltransferase, is required in both the GSCs and in the somatic niche for proper GSC maintenance. *Enok* depleted CpCs showed loss of GSC concomitant with reduced levels of pMAD, dally, and Notch reporters; however, Dpp levels were not altered. This suggests Enok acts downstream of Dpp signaling to control self-renewal. As Enok genetically interacts with Notch and Notch signaling has been shown to regulate CpC numbers (Ward et al. 2006; Song et al. 2007), it is thought that Enok contributes to GSC maintenance by maintaining CpC numbers. Thus, Enok not only controls the size of the niche by potentially activating the expression of Notch but also regulates the reach of Dpp signaling by modulating levels of Dally (Xin et al. 2013).

dBre1 and *dSet1* are required for regulating levels of H3K4me3, which is required for activating transcription (Ardehali et al. 2011; Hallson et al. 2012). Loss of *dBre1* or *dSet1* in the CpCs and ECs resulted in low levels of H3K4me3 in both of these cell types. *dBre1* depleted CpCs resulted in GSC loss over a period of time, indicating its role in GSC maintenance. Additionally, *dBre1* genetically interacts with *dSet1* to control GSC self-renewal. In TF and CpCs depleted of these enzymes, GSCs were found to have significantly reduced pMad expression, suggesting that Dpp signaling was compromised. However, *dally* mRNA levels were reduced, as opposed to *dpp* mRNA levels, suggesting that impaired dally expression is the cause for GSC loss. Furthermore, it was shown that *dBre1/dSet1* led to reduction in Adherens junctions (AJ) protein components, DE-Cadherin and Armadillo. This suggests that *dBre1/dSet1* regulates the accumulation of adhesion molecules at the CpC–GSC junction, thereby contributing to GSC maintenance. Thus, transcriptional activation mediated by *dBre1/dSet1* in the TF and CpCs is required both to modulate the reach of Dpp signaling by modulating levels of Dally and to regulate adhesion of CpCs to GSCs by modulating AJ components (Xuan et al. 2013).

dLsd1, a lysine demethylase, removes mono and dimethyl groups from H3K4, thus causing transcriptional silencing (Shi et al. 2004). dLsd1 is expressed in both the germ line and in somatic cells, but it is only required in the ECs for GSC differentiation. Loss of *dLsd1* in ECs leads to accumulation of pMad positive GSC-like cells. ChiP-seq assay revealed that dLsd1 binds to *engrailed* promoter in the ECs. Engrailed is normally only expressed in the TF and CpCs and encodes for a homeobox transcription factor (Morata and Lawrence 1975; Kornberg 1981; Desplan et al. 1985). Engrailed is known to promote Hh expression, which in turn drives Dpp expression (Lee et al. 1992; Tabata et al. 1992; Morata 2001; Alexandre and Vincent 2003). Loss of *dLsd1* resulted in ectopic expression of Engrailed in the ECs, and loss of *engrailed* in *dLsd1* ECs suppressed *dLsd1* phenotype. Additionally, *Hh* depletion in *dLsd1* mutants also suppressed *dLsd1* phenotype. This suggests that loss of GSC differentiation is due to the ectopic expression of Engrailed in the ECs, which promotes ectopic expression of Dpp (Fig. 1.4) (Eliazer et al. 2011,

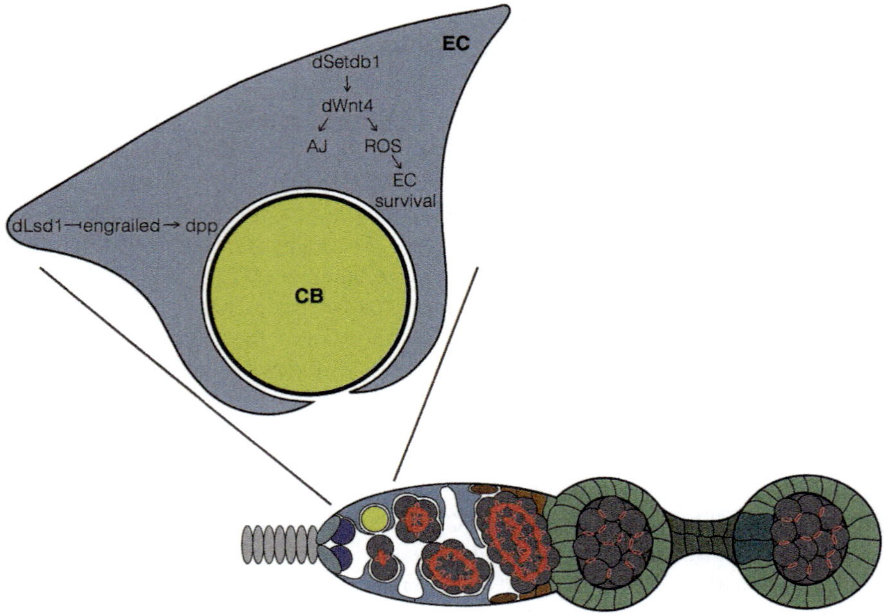

Fig. 1.4 Illustration of an escort cell (EC) enveloping the cystoblast (CB). Examples of two chromatin modifiers acting in the ECs are shown. dLsd1, a lysine demethylase represses engrailed and thereby dpp expression allowing proper differentiation. dSETDB1, a trimethylase regulates dWnt4 which promotes differentiation by regulating adherens junction proteins required for enveloping the CB and ROS required for EC survival in parallel

2014). Thus, silencing of self-renewal signaling in the ECs by dLsd1 is critical for promoting differentiation.

dSETDB1 or eggless (egg) transfers methyl groups on lysine 9 (K9) residues of Histone 3 and is also required for transcriptional silencing. Loss of *dSETDB1* in the ECs leads to loss of heterochromatin, upregulation of TEs, and Bam-dependent GSC differentiation defects (Rangan et al. 2011; Wang et al. 2011). Intriguingly, it was found that dWnt4 acts downstream of dSETDB1 and TEs to affect GSC differentiation. dWnt4 controls GSC differentiation by regulating AJ proteins and promoting the formation of cytoplasmic processes that envelope the differentiating daughter (Upadhyay et al. 2016). Furthermore, depletion of *disheveled* and *armadillo*, the downstream targets of Wnt signaling, leads to increased reactive oxygen species (ROS) levels in ECs, resulting in accumulation of CBs and EC loss. Overexpression of enzymes that help eliminate hydrogen peroxide, such as glutathione-S-transferase (Gst), catalase, or superoxide dismutase1, could rescue the differentiation defect and EC numbers. This suggests that Wnt signaling in the ECs maintains the reduced redox state and is responsible for germ cell differentiation and EC maintenance (Fig. 1.4) (Wang et al. 2015). Thus, silencing of TEs by dSETDB1 is critical to promote dWnt4 signaling, which promotes EC proliferation and process formation.

1.3.3 Chromatin Remodeling Factors Regulate Signaling Pathways in the Somatic Niche

Chromatin remodeling factors that reposition nucleosomes are also required in the somatic niche for GSC maintenance and proper differentiation. As previously discussed, Polybromo interacts with Bap180 to form the PBAP complex in the germ line and is required for GSC maintenance. It was also shown that Polybromo associates with Bap170 in the TF and CpCs to control GSC differentiation. Loss of *Polybromo* and *Bap170* in the TF and CpCs resulted in lower pMAD levels leading to GSC loss over time (He et al. 2014). However, it is not known if it directly regulates Dpp signaling or indirectly by regulating dally levels.

PcG that can remodel chromatin is known to form two complexes, PRC1 and PRC2. Pc, Psc, Ph, Sce, Scm, and Su(z)2 together form the PRC1 complex (Shao et al. 1999; Saurin et al. 2001; Fritsch et al. 2003). The PRC2 complex is formed of Esc and E(z) (Ng et al. 2000; Tie et al. 2001). Of all these, only Ph, Sce, Scm, Psc, and Su(z)2 are critical for oocyte development. Ph is expressed both in the germ line and soma. Germ line clones for *ph* mutants revealed that Ph is not required in the germ line for oogenesis. Somatic clones of *ph* mutants showed that pre-follicular cells failed to form protrusions to encapsulate the 16-cell cyst and instead showed round morphology. These mutants also exhibited follicle stem cell, stalk and polar cell differentiation defects, and polar and stalk cell proliferation defects. Mutant *Sce* and *Scm* somatic mitotic clones showed multicyst follicles with several oocytes, disorganized encapsulation of the 16-cell cyst, and abnormal number of interfollicular stalk cells. However, no defect in polar cells was observed (Narbonne 2004). Psc functions with Su(z)2 to regulate FSC self-renewal. Although loss of *Psc* and *Su(z)2* shows germ line defects, these mutants are capable of differentiating into germ line cysts and follicular epithelia. This implies that these proteins are required for maintenance of GSCs and FSCs. Further studies showed that loss of these genes leads to sustained canonical and noncanonical Wnt signaling, resulting in loss of FSC self-renewal and causing basal epithelial extrusion, respectively (Li et al. 2010). All these studies suggest that PcG components also have a function independent of the canonical PRC1 complex in the somatic cells to regulate oogenesis.

1.4 Conclusion

The transition from GSC to a mature egg requires the fine balance of several opposing processes, which is inherently difficult. Transcription of differentiation factors must be silenced within the GSCs, only to be immediately upregulated in the

GSC daughter to counter the surrounding self-renewal signaling. While this chapter has outlined several repressive epigenetic mechanisms required for silencing in the GSCs, very little is known about what activates the GSC self-renewal program and the differentiation program in the daughter. However, it is known that a repressive epigenetic program is required in the GSC daughter to initiate the piRNA pathway that protects the genome against TEs. Upon differentiation, the specified oocyte enters meiosis. This transition is mediated by several epigenetic mechanisms to balance the chromosome's need to be anchored to the nuclear membrane to promote recombination and the chromosome's need to be released to proceed into prophase arrest for making an egg. The chromatin state of the oocyte nucleus is not known, nor is it known how this state changes during development to give rise to the karyosome. Lastly, the somatic niche plays a critical role in mediating these transitions. Key epigenetic regulators in the various somatic cells of the niche tune the signaling pathways to promote both self-renewal and differentiation. However, how these various signaling pathways are integrated to produce a specific epigenetic state in the egg is yet to be determined.

Acknowledgments We would like to acknowledge members of the Rangan Lab. A.M provided the illustrations. P.R is funded by NIH/NIGMS RO1 1119484-1-68857.

References

Abdu U, Brodsky M, Schüpbach T (2002) Activation of a meiotic checkpoint during Drosophila oogenesis regulates the translation of Gurken through Chk2/Mnk. Curr Biol 12:1645–1651

Aihara H, Nakagawa T, Yasui K, Ohta T, Hirose S, Dhomae N, Takio K, Kaneko M, Takeshima Y, Muramatsu M, Ito T (2004) Nucleosomal histone kinase-1 phosphorylates H2A Thr 119 during mitosis in the early Drosophila embryo. Genes Dev 18:877–888

Akhtar A, Becker PB (2000) Activation of transcription through histone H4 acetylation by MOF, an acetyltransferase essential for dosage compensation in Drosophila. Mol Cell 5:367–375

Albertini D, Bromfield J (2010) Soma-germline interactions in the ovary: an evolutionary perspective. Oogenesis: 103–113

Alexandre C, Vincent J-P (2003) Requirements for transcriptional repression and activation by Engrailed in Drosophila embryos. Development 130(4):729–739

An W (2007) Histone acetylation and methylation. In: Kundu TK, Bittman R, Dasgupta D, Engelhardt H, Flohe L, Herrmann H, Holzenburg A, Nasheuer H-P, Rottem S, Wyss M, Zwickl P (eds) Chromatin and Disease, Subcellular Biochemistry. Springer Netherlands, Dordrecht, pp 355–374

Ardehali MB, Mei A, Zobeck KL, Caron M, Lis JT, Kusch T (2011) Drosophila Set1 is the major histone H3 lysine 4 trimethyltransferase with role in transcription. EMBO J 30(14):2817–2828, http://doi.org/10.1038/emboj.2011.19

Armstrong JA, Emerson BM (1998) Transcription of chromatin: these are complex times. Curr Opin Genet Dev 8:165–172

Ashery-Padan R, Weiss AM, Feinstein N, Gruenbaum Y (1997) Distinct regions specify the targeting of otefin to the nucleoplasmic side of the nuclear envelope. J Biol Chem 272:2493–2499

Badenhorst P, Voas M, Rebay I, Wu C (2002) Biological functions of the ISWI chromatin remodeling complex NURF. Genes Dev 16:3186–3198

Bakay M (2006) Nuclear envelope dystrophies show a transcriptional fingerprint suggesting disruption of Rb-MyoD pathways in muscle regeneration. Brain 129:996–1013

Bannister AJ, Kouzarides T (2011) Regulation of chromatin by histone modifications. Cell Res 21:381–395

Barski A, Cuddapah S, Cui K, Ty R, Schones DE, Wang Z, Wei G, Chepelev I, Zhao K (2007) High-resolution profiling of histone methylations in the human genome. Cell 129:823–837. doi:10.1016/j.cell.2007.05.009

Barton LJ, Pinto BS, Wallrath LL, Geyer PK (2013) The Drosophila nuclear lamina protein otefin is required for germline stem cell survival. Dev Cell 25:645–654

Bastock R, St Johnston D (2008) Drosophila oogenesis. Curr Biol 18:R1082–R1087

Bednar J, Horowitz RA, Grigoryev SA, Carruthers LM, Hansen JC, Koster AJ, Woodcock CL (1998) Nucleosomes, linker DNA, and linker histone form a unique structural motif that directs the higher-order folding and compaction of chromatin. Proc Natl Acad Sci 95:14173–14178

Boros IM (2012) Histone modification in Drosophila. Brief Funct Genomics 11:319–331

Bray S, Musisi H, Bienz M (2005) Bre1 is required for notch signaling and histone modification. Dev Cell 8:279–286

Brehm A, Längst G, Kehle J, Clapier CR, Imhof A, Eberharter A, Müller J, Becker PB (2000) dMi-2 and ISWI chromatin remodelling factors have distinct nucleosome binding and mobilization properties. EMBO J 19:4332–4341

Brennecke J, Aravin AA, Stark A, Dus M, Kellis M, Sachidanandam R, Hannon GJ (2007) Discrete small RNA-generating loci as master regulators of transposon activity in Drosophila. Cell 128:1089–1103

Brummel TJ, Twombly V, Marqués G, Wrana JL, Newfeld SJ, Attisano L, Massagué J, O'Connor MB, Gelbart WM (1994) Characterization and relationship of Dpp receptors encoded by the saxophone and thick veins genes in Drosophila. Cell 78:251–261

Buccione R, Schroeder AC, Eppig JJ (1990) Interactions between somatic cells and germ cells throughout mammalian oogenesis. Biol Reprod 43(4):543–547, http://doi.org/10.1095/biolreprod43.4.543

Bustin M, Catez F, Lim J-H (2005) The dynamics of histone H1 function in chromatin. Mol Cell 17:617–620

Buszczak M, Spradling AC (2006) Searching chromatin for stem cell identity. Cell 125:233–236

Buszczak M, Paterno S, Spradling AC (2009) Drosophila stem cells share a common requirement for the histone H2B ubiquitin protease scrawny. Science 323:248–251

Canipari R (2000) Oocyte—granulosa cell interactions. Hum Reprod Update 6(3):279–289, http://doi.org/10.1093/humupd/6.3.279

Carpenter ATC (1975) Electron microscopy of meiosis in Drosophila melanogaster females: II: The recombination nodule—a recombination-associated structure at pachytene? Proc Natl Acad Sci 72:3186–3189

Cedar H, Bergman Y (2009) Linking DNA methylation and histone modification: patterns and paradigms. Nat Rev Genet 10:295–304

Chan RC, Severson AF, Meyer BJ (2004) Condensin restructures chromosomes in preparation for meiotic divisions. J Cell Biol 167:613–625

Chang YC, Jang ACC, Lin CH, Montell DJ (2013) Castor is required for Hedgehog-dependent cell-fate specification and follicle stem cell maintenance in Drosophila oogenesis. Proc Natl Acad Sci 110:E1734–E1742

Chen D (2003) A discrete transcriptional silencer in the bam gene determines asymmetric division of the Drosophila germline stem cell. Development 130:1159–1170

Clough E, Tedeschi T, Hazelrigg T (2014) Epigenetic regulation of oogenesis and germ stem cell maintainance by the *Drosophila* histone methyltransferae Eggless/dSetDB1. Dev Biol 388:181–191

Cohen TV, Kosti O, Stewart CL (2007) The nuclear envelope protein MAN1 regulates TGFbeta signaling and vasculogenesis in the embryonic yolk sac. Development 134:1385–1395

Corona DFV (2002) Modulation of ISWI function by site-specific histone acetylation. EMBO Rep 3:242–247

Cox DN, Chao A, Baker J, Chang L, Qiao D, Lin H (1998) A novel class of evolutionarily conserved genes defined by piwi are essential for stem cell self-renewal. Genes Dev 12:3715–3727

Cox DN, Chao A, Lin H (2000) piwi encodes a nucleoplasmic factor whose activity modulates the number and division rate of germline stem cells. Development 127(3):503–514

Cullen CF, Brittle AL, Ito T, Ohkura H (2005) The conserved kinase NHK-1 is essential for mitotic progression and unifying acentrosomal meiotic spindles in Drosophila melanogaster. J Cell Biol 171:593–602

Czermin B, Schotta G, Hulsmann BB, Brehm A, Becker PB, Reuter G, Imhof A (2001) Physical and functional association of SU(VAR)3-9 and HDAC1 in Drosophila. EMBO Rep 2:915–919

Daubresse G, Deuring R, Moore L, Papoulas O, Zakrajsek I, Waldrip WR, Scott MP, Kennison JA, Tamkun JW (1999) The Drosophila kismet gene is related to chromatin-remodeling factors and is required for both segmentation and segment identity. Development 126:1175–1187

Decotto E, Spradling AC (2005) The Drosophila ovarian and testis stem cell niches: similar somatic stem cells and signals. Dev Cell 9:501–510

Desplan C, Theis J, O'Farrell PH (1985) The Drosophila developmental gene, engrailed, encodes a sequence-specific DNA binding activity. Nature 318(6047):630–635

Deuring R, Fanti L, Armstrong JA, Sarte M, Papoulas O, Prestel M, Daubresse G, Verardo M, Moseley SL, Berloco M, Tsukiyama T, Wu C, Pimpinelli S, Tamkun JW (2000) The ISWI chromatin-remodeling protein is required for gene expression and the maintenance of higher order chromatin structure in vivo. Mol Cell 5:355–365

Di Stefano L, Ji J-Y, Moon N-S, Herr A, Dyson N (2007) Mutation of Drosophila Lsd1 disrupts H3-K4 methylation, resulting in tissue-specific defects during development. Curr Biol 17:808–812

Dover J (2002) Methylation of histone H3 by COMPASS requires ubiquitination of histone H2B by Rad6. J Biol Chem 277:28368–28371

Ebert A (2004) Su(var) genes regulate the balance between euchromatin and heterochromatin in Drosophila. Genes Dev 18:2973–2983

Eliazer S, Shalaby NA, Buszczak M (2011) Loss of lysine-specific demethylase 1 nonautonomously causes stem cell tumors in the Drosophila ovary. Proc Natl Acad Sci 108(17):7064–7069, http://doi.org/10.1073/pnas.1015874108

Eliazer S, Palacios V, Wang Z, Kollipara RK, Kittler R, Buszczak M (2014). In: Perrimon N (ed) Lsd1 restricts the number of germline stem cells by regulating multiple targets in escort cells. PLoS Genet 10:e1004200

Erickson G (2008) An analysis of follicle development and ovum maturation. Semin Reprod Med 4:233–254

Fasulo B, Deuring R, Murawska M, Gause M, Dorighi KM, Schaaf CA, Dorsett D, Brehm A, Tamkun JW (2012) The *Drosophila* Mi-2 chromatin-remodeling factor regulates higher-order chromatin structure and cohesin dynamics *in vivo*. PLoS Genet 8:e1002878

Findley SD (2003) Maelstrom, a Drosophila spindle-class gene, encodes a protein that colocalizes with Vasa and RDE1/AGO1 homolog, Aubergine, in nuage. Development 130:859–871

Forbes A, Lehmann R (1998) Nanos and Pumilio have critical roles in the development and function of Drosophila germline stem cells. Development 125:679–690

Forbes AJ, Lin H, Ingham PW, Spradling AC (1996) hedgehog is required for the proliferation and specification of ovarian somatic cells prior to egg chamber formation in Drosophila. Development 122:1125–1135

Fritsch C, Beuchle D, Müller J (2003) Molecular and genetic analysis of the Polycomb group gene Sex combs extra/Ring in Drosophila. Mech Dev 120(8):949–954

Furukawa K (2003) Barrier-to-autointegration factor plays crucial roles in cell cycle progression and nuclear organization in Drosophila. J Cell Sci 116:3811–3823

Gandille P, Narbonne-Reveau K, Boissonneau E, Randsholt N, Busson D, Pret A-M (2010). In: Cotterill S (ed) Mutations in the polycomb group gene polyhomeotic lead to epithelial instability in both the ovary and wing imaginal disc in Drosophila. PLoS ONE 5:e13946

Gardner RG, Nelson ZW, Gottschling DE (2005) Ubp10/Dot4p regulates the persistence of ubiquitinated histone H2B: distinct roles in telomeric silencing and general chromatin. Mol Cell Biol 25:6123–6139

Ghabrial A, Ray RP, Schupbach T (1998) okra and spindle-B encode components of the RAD52 DNA repair pathway and affect meiosis and patterning in Drosophila oogenesis. Genes Dev 12:2711–2723

Gilboa L, Lehmann R (2004) How different is Venus from Mars? The genetics of germ-line stem cells in Drosophila females and males. Development 131:4895–4905

Gilboa L, Lehmann R (2006) Soma–germline interactions coordinate homeostasis and growth in the Drosophila gonad. Nature 443:97–100

Gilchrist RB, Ritter LJ, Armstrong DT (2004) Oocyte–somatic cell interactions during follicle development in mammals. Anim Reprod Sci 82–83:431–446, http://doi.org/10.1016/j.anireprosci.2004.05.017

Goldberg M, Lu H, Stuurman N, Ashery-Padan R, Weiss AM, Yu J, Bhattacharyya D, Fisher PA, Gruenbaum Y, Wolfner MF (1998) Interactions among Drosophila nuclear envelope proteins lamin, otefin, and YA. Mol Cell Biol 18:4315–4323

Goldberg AD, Allis CD, Bernstein E (2007) Epigenetics: a landscape takes shape. Cell 128:635–638

Gorjánácz M, Klerkx EPF, Galy V, Santarella R, López-Iglesias C, Askjaer P, Mattaj IW (2007) Caenorhabditis elegans BAF-1 and its kinase VRK-1 participate directly in post-mitotic nuclear envelope assembly. EMBO J 26:132–143

Gruenbaum Y, Margalit A, Goldman RD, Shumaker DK, Wilson KL (2005) The nuclear lamina comes of age. Nat Rev Mol Cell Biol 6:21–31

Gruzova MN, Parfenov VN (1993) Karyosphere in oogenesis and intranuclear morphogenesis. An analysis of follicle development and ovum maturation. Int Rev Cytol 144:1–52

Gunawardane LS, Saito K, Nishida KM, Miyoshi K, Kawamura Y, Nagami T, Siomi H, Siomi MC (2007) A slicer-mediated mechanism for repeat-associated siRNA 5' end formation in Drosophila. Science 315:1587–1590

Guo Z, Wang Z (2009) The glypican Dally is required in the niche for the maintenance of germline stem cells and short-range BMP signaling in the Drosophila ovary. Development 136:3627–3635

Hall DH, Winfrey VP, Blaeuer G, Hoffman LH, Furuta T, Rose KL, Hobert O, Greenstein D (1999) Ultrastructural features of the adult hermaphrodite gonad of Caenorhabditis elegans: relations between the germ line and soma. Dev Biol 212:101–123

Hallson G, Hollebakken RE, Li T, Syrzycka M, Kim I, Cotsworth S et al (2012) dSet1 is the main H3K4 di- and tri-methyltransferase throughout Drosophila development. Genetics 190 (1):91–100, http://doi.org/10.1534/genetics.111.135863

Haraguchi T, Holaska JM, Yamane M, Koujin T, Hashiguchi N, Mori C, Wilson KL, Hiraoka Y (2004) Emerin binding to Btf, a death-promoting transcriptional repressor, is disrupted by a missense mutation that causes Emery-Dreifuss muscular dystrophy. Eur J Biochem 271:1035–1045

Hayashi Y, Kobayashi S, Nakato H (2009) Drosophilaglypicans regulate the germline stem cell niche. J Cell Biol 187:473–480

He J, Xuan T, Xin T, An H, Wang J, Zhao G, Li M (2014). In: Singh SR (ed) Evidence for chromatin-remodeling complex PBAP-controlled maintenance of the Drosophila ovarian germline stem cells. PLoS ONE 9:e103473

Herzig B, Yakulov TA, Klinge K, Gunesdogan U, Jackle H, Herzig A (2014) Ballchen is required for self-renewal of germline stem cells in Drosophila melanogaster. Biol Open 3:510–521

Holaska JM, Lee KK, Kowalski AK, Wilson KL (2003) Transcriptional repressor germ cell-less (GCL) and barrier to autointegration factor (BAF) compete for binding to emerin in vitro. J Biol Chem 278:6969–6975

Horard B, Loppin B (2015) Histone storage and deposition in the early Drosophila embryo. Chromosoma 124:163–175

Huang J, Kalderon D (2014) Coupling of Hedgehog and Hippo pathways promotes stem cell maintenance by stimulating proliferation. J Cell Biol 205:325–338

Huynh J-R, St Johnston D (2004) The origin of asymmetry: early polarisation of the Drosophila germline cyst and oocyte. Curr Biol 14:R438–R449

Iovino N, Ciabrelli F, Cavalli G (2013) PRC2 controls Drosophila oocyte cell fate by repressing cell cycle genes. Dev Cell 26:431–439

Ishimura A (2006) Man1, an inner nuclear membrane protein, regulates vascular remodeling by modulating transforming growth factor signaling. Development 133:3919–3928

Ivanovska I (2005) A histone code in meiosis: the histone kinase, NHK-1, is required for proper chromosomal architecture in Drosophila oocytes. Genes Dev 19:2571–2582

Ivanovska I, Orr-Weaver TL (2014) Histone modifications and the chromatin scaffold for meiotic chromosome architecture. Cell Cycle 5:2064–2071

Jenuwein T, Allis CD (2001) Translating the histone code. Science 293:1074–1080

Jiang X, Xia L, Chen D, Yang Y, Huang H, Yang L, Zhao Q, Shen L, Wang J, Chen D (2008) Otefin, a nuclear membrane protein, determines the fate of germline stem cells in Drosophila via interaction with Smad complexes. Dev Cell 14:810

Jin Z, Flynt AS, Lai EC (2013) Drosophila piwi mutants exhibit germline stem cell tumors that are sustained by elevated Dpp signaling. Curr Biol 23:1442–1448

Joyce EF, Pedersen M, Tiong S, White-Brown SK, Paul A, Campbell SD, McKim KS (2011) DrosophilaATM and ATR have distinct activities in the regulation of meiotic DNA damage and repair. J Cell Biol 195:359–367

Kawase E, Wong MD, Ding BC, Xie T (2004) Gbb/Bmp signaling is essential for maintaining germline stem cells and for repressing bam transcription in the Drosophila testis. Development 131:1365–1375

Kehle J (1998) dMi-2, a hunchback-interacting protein that functions in polycomb repression. Science 282:1897–1900

King FJ, Lin H (1999) Somatic signaling mediated by fs(1)Yb is essential for germline stem cell maintenance during Drosophila oogenesis. Development 126:1833–1844

King FJ, Szakmary A, Cox DN, Lin H (2001) Yb modulates the divisions of both germline and somatic stem cells through piwi- and hh-mediated mechanisms in the Drosophila ovary. Mol Cell 7:497–508

King RC (1970) Ovarian development in *Drosophila melanogaster*. Academic Press, New York

Kirilly D, Xie T (2007) The Drosophila ovary: an active stem cell community. Cell Res 17:271

Koch EA, Smith PA, King RC (1970) Variations in the radiosensitivity of oocyte chromosomes during meiosis in Drosophila melanogaster. Chromosoma 30:98–108

Kornberg T (1981) Engrailed: a gene controlling compartment and segment formation in Drosophila. Proc Natl Acad Sci 78(2):1095–1099, http://doi.org/10.1073/pnas.78.2.1095

Kouzarides T (2007) Chromatin modifications and their function. Cell 128:693–705

Kunert N, Wagner E, Murawska M, Klinker H, Kremmer E, Brehm A (2009) dMec: a novel Mi-2 chromatin remodelling complex involved in transcriptional repression. EMBO J 28:533–544. doi:10.1038/emboj.2009.3

Kurz EU, Lees-Miller SP (2004) DNA damage-induced activation of ATM and ATM-dependent signaling pathways. DNA Repair 3:889–900

Kusch T, Florens L, Macdonald WH, Swanson SK, Glaser RL, Yates JR, Abmayr SM, Washburn MP, Workman JL (2004) Acetylation by Tip60 is required for selective histone variant exchange at DNA lesions. Science 306:2084–2087

Lancaster OM, Cullen CF, Ohkura H (2007) NHK-1 phosphorylates BAF to allow karyosome formation in the Drosophilaoocyte nucleus. J Cell Biol 179:817–824

Lancaster OM, Breuer M, Cullen CF, Ito T, Ohkura H (2010). In: Hawley RS (ed) The meiotic recombination checkpoint suppresses NHK-1 kinase to prevent reorganisation of the oocyte nucleus in Drosophila. PLoS Genet 6:e1001179

Lee JJ, von Kessler DP, Parks S, Beachy PA (1992) Secretion and localized transcription suggest a role in positional signaling for products of the segmentation gene hedgehog. Cell 71(1):33–50
Li X, Zhao X (2008) Epigenetic regulation of mammalian stem cells. Stem Cells Dev 17:1043–1052
Li C, Vagin VV, Lee S, Xu J, Ma S, Xi H, Seitz H, Horwich MD, Syrzycka M, Honda BM, Kittler ELW, Zapp ML, Klattenhoff C, Schulz N, Theurkauf WE, Weng Z, Zamore PD (2009a) Collapse of germline piRNAs in the absence of Argonaute3 reveals somatic piRNAs in flies. Cell 137:509–521
Li Y, Minor NT, Park JK, McKearin DM, Maines JZ (2009b) Bam and Bgcn antagonize Nanos-dependent germ-line stem cell maintenance. Proc Natl Acad Sci U S A 106:9304–9309
Li X, Han Y, Xi R (2010) Polycomb group genes Psc and Su(z)2 restrict follicle stem cell self-renewal and extrusion by controlling canonical and noncanonical Wnt signaling. Genes Dev 24:933–946
Lin F (2000) MAN1, an inner nuclear membrane protein that shares the LEM domain with lamina-associated polypeptide 2 and emerin. J Biol Chem 275:4840–4847
Lin H (2002) The stem-cell niche theory: lessons from flies. Nat Rev Genet 3:931–940
Lin H, Spradling AC (1993) Germline stem cell division and egg chamber development in transplanted Drosophila germaria. Dev Biol 159:140–152
Lin H, Spradling AC (1997) A novel group of pumilio mutations affects the asymmetric division of germline stem cells in the Drosophila ovary. Development 124:2463–2476
Lin F, Morrison JM, Wu W, Worman HJ (2005) MAN1, an integral protein of the inner nuclear membrane, binds Smad2 and Smad3 and antagonizes transforming growth factor-beta signaling. Hum Mol Genet 14:437–445
Lopez-Onieva L, Fernandez-Minan A, Gonzalez-Reyes A (2008) Jak/Stat signalling in niche support cells regulates dpp transcription to control germline stem cell maintenance in the Drosophila ovary. Development 135:533–540
Luger K, Mäder AW, Richmond RK, Sargent DF, Richmond TJ (1997) Crystal structure of the nucleosome core particle at 2.8 A resolution. Nature 389:251–260
Ma X, Wang S, Do T, Song X, Inaba M, Nishimoto Y, Liu L-P, Gao Y, Mao Y, Li H, McDowell W, Park J, Malanowski K, Peak A, Perera A, Li H, Gaudenz K, Haug J, Yamashita Y, Lin H, Ni J-Q, Xie T (2014). In: Bökel C (ed) Piwi is required in multiple cell types to control germline stem cell lineage development in the Drosophila ovary. PLoS ONE 9: e90267
Maines JZ, Park JK, Williams M, McKearin DM (2007) Stonewalling Drosophila stem cell differentiation by epigenetic controls. Development 134:1471–1479
Malone CD, Brennecke J, Dus M, Stark A, McCombie WR, Sachidanandam R, Hannon GJ (2009) Specialized piRNA pathways act in germline and somatic tissues of the Drosophila ovary. Cell 137:522–535
Manilal S (1996) The Emery-Dreifuss muscular dystrophy protein, emerin, is a nuclear membrane protein. Hum Mol Genet 5:801–808
Manilal S (1998) Mutations in Emery-Dreifuss muscular dystrophy and their effects on emerin protein expression. Hum Mol Genet 7:855–864
Mansharamani M, Wilson KL (2005) Direct binding of nuclear membrane protein MAN1 to emerin in vitro and two modes of binding to barrier-to-autointegration factor. J Biol Chem 280:13863–13870
Margolis J, Spradling A (1995) Identification and behavior of epithelial stem cells in the Drosophila ovary. Development 121:3797–3807
Matzuk MM, Burns KH, Viveiros MM, Eppig JJ (2002) Intercellular communication in the mammalian ovary: oocytes carry the conversation. Science 296(5576):2178–2180, http://doi.org/10.1126/science.1071965
McKearin D, Ohlstein B (1995) A role for the Drosophila bag-of-marbles protein in the differentiation of cystoblasts from germline stem cells. Development 121:2937–2947

McKearin DM, Spradling AC (1990) bag-of-marbles: a Drosophila gene required to initiate both male and female gametogenesis. Genes Dev 4:2242–2251

McKim KS, Hayashi-Hagihara A (1998) mei-W68 in Drosophila melanogaster encodes a Spo11 homolog: evidence that the mechanism for initiating meiotic recombination is conserved. Genes Dev 12:2932–2942

Misteli T, Gunjan A, Hock R, Bustin M, Brown DT (2000) Dynamic binding of histone H1 to chromatin in living cells. Nature 408:877–881

Mohrmann L, Langenberg K, Krijgsveld J, Kal AJ, Heck AJR, Verrijzer CP (2004) Differential targeting of two distinct SWI/SNF-Related Drosophila chromatin-remodeling complexes. Mol Cell Biol 24:3077–3088

Morata G (2001) How Drosophila appendages develop. Nat Rev Mol Cell Biol 2(2):89–97, http://doi.org/10.1038/35052047

Morata G, Lawrence PA (1975) Control of compartment development by the engrailed gene in Drosophila. Nature 255(5510):614–617, http://doi.org/10.1038/255614a0

Morgan MM, Mahowald AP (1996) Multiple signaling pathways establish both the individuation and the polarity of the oocyte follicle in Drosophila. Arch Insect Biochem Physiol 33:211–230

Moshkin YM, Mohrmann L, van Ijcken WFJ, Verrijzer CP (2006) Functional differentiation of SWI/SNF remodelers in transcription and cell cycle control. Mol Cell Biol 27:651–661

Muller HJ (1930) Types of visible variations induced by X-rays in Drosophila. J Genet 22:299–334

Narbonne K (2004) polyhomeotic is required for somatic cell proliferation and differentiation during ovarian follicle formation in Drosophila. Development 131:1389–1400

Navarro-Costa P, McCarthy A, Prudêncio P, Greer C, Guilgur LG, Becker JD, Secombe J, Rangan P, Martinho RG (2016) Early programming of the oocyte epigenome temporally controls late prophase I transcription and chromatin remodelling. Nat Commun. doi:10.1038/ncomms12331

Neumüller RA, Betschinger J, Fischer A, Bushati N, Poernbacher I, Mechtler K, Cohen SM, Knoblich JA (2008) Mei-P26 regulates microRNAs and cell growth in the Drosophila ovarian stem cell lineage. Nature 454:241–245

Ng J, Hart CM, Morgan K, Simon JA (2000) A Drosophila ESC-E(Z) protein complex is distinct from other polycomb group complexes and contains covalently modified ESC. Mol Cell Biol 20(9):3069–3078, http://doi.org/10.1128/MCB.20.9.3069-3078.2000

Nguyen M, Park S, Marqués G, Arora K (1998) Interpretation of a BMP activity gradient in Drosophila embryos depends on synergistic signaling by two type I receptors, SAX and TKV. Cell 95:495–506

Nichols RJ (2006) The vaccinia-related kinases phosphorylate the N' terminus of BAF, regulating its interaction with DNA and its retention in the nucleus. Mol Biol Cell 17:2451–2464

Nikalayevich E, Ohkura H (2015) The NuRD nucleosome remodelling complex and NHK-1 kinase are required for chromosome condensation in oocytes. J Cell Sci 128:566–575

Nystul T, Spradling A (2007) An epithelial niche in the Drosophila ovary undergoes long-range stem cell replacement. Cell Stem Cell 1:277–285

Nystul T, Spradling A (2010) Regulation of epithelial stem cell replacement and follicle formation in the Drosophila ovary. Genetics 184:503–515

Page SL (2001) c(3)G encodes a Drosophila synaptonemal complex protein. Genes Dev 15:3130–3143

Page SL, Hawley RS (2004) The genetics and molecular biology of the synaptonemal complex. Annu Rev Cell Dev Biol 20:525–558

Phillips CM, Dernburg AF (2006) A family of zinc-finger proteins is required for chromosome-specific pairing and synapsis during meiosis in C. elegans. Dev Cell 11:817–829

Phillips CM, Meng X, Zhang L, Chretien JH, Urnov FD, Dernburg AF (2009) Identification of chromosome sequence motifs that mediate meiotic pairing and synapsis in C. elegans. Nat Cell Biol 11:1163

Qi H, Watanabe T, Ku HY, Liu N, Zhong M, Lin H (2011) The Yb body, a major site for Piwi-associated RNA biogenesis and a gateway for Piwi expression and transport to the nucleus in somatic cells. J Biol Chem 286:3789–3797

Raddatz G, Guzzardo PM, Olova N, Fantappié MR, Rampp M, Schaefer M et al (2013) Dnmt2-dependent methylomes lack defined DNA methylation patterns. Proc Natl Acad Sci U S A 110 (21):8627–8631, http://doi.org/10.1073/pnas.1306723110

Rangan P, Malone CD, Navarro C, Newbold SP, Hayes PS, Sachidanandam R, Hannon GJ, Lehmann R (2011) piRNA production requires heterochromatin formation in Drosophila. Curr Biol 21:1373–1379

Rathke C, Baarends WM, Awe S, Renkawitz-Pohl R (2014) Chromatin dynamics during spermiogenesis. Biochim Biophys Acta (BBA) Gene Regul Mech 1839:155–168

Riddle NC, Minoda A, Kharchenko PV, Alekseyenko AA, Schwartz YB, Tolstorukov MY, Gorchakov AA, Jaffe JD, Kennedy C, Linder-Basso D, Peach SE, Shanower G, Zheng H, Kuroda MI, Pirrotta V, Park PJ, Elgin SCR, Karpen GH (2011) Plasticity in patterns of histone modifications and chromosomal proteins in Drosophila heterochromatin. Genome Res 21:147–163

Robinson DN, Cooley L (1996) Stable intercellular bridges in development: the cytoskeleton lining the tunnel. Trends Cell Biol 6:474–479

Rudolph T, Yonezawa M, Lein S, Heidrich K, Kubicek S, Schäfer C, Phalke S, Walther M, Schmidt A, Jenuwein T, Reuter G (2007) Heterochromatin formation in Drosophila is initiated through active removal of H3K4 methylation by the LSD1 homolog SU(VAR)3-3. Mol Cell 26:103–115

Ruhf ML, Braun A, Papoulas O, Tamkun JW, Randsholt N, Meister M (2001) The domino gene of Drosophila encodes novel members of the SWI2/SNF2 family of DNA-dependent ATPases, which contribute to the silencing of homeotic genes. Development 128:1429–1441

Sahut-Barnola I, Godt D, Laski FA, Couderc J-L (1995) Drosophila ovary morphogenesis: analysis of terminal filament formation and identification of a gene required for this process. Dev Biol 170:127–135

Salvaing J, Lopez A, Boivin A, Deutsch JS, Peronnet F (2003) The Drosophila Corto protein interacts with Polycomb-group proteins and the GAGA factor. Nucleic Acids Res 31:2873–2882

Santos-Rosa H, Schneider R, Bannister AJ, Sherrif J, Bernstein BE, Emre NC, Schreiber S, Mellor J, Kouzarides T (2002) Active genes are tri-methylated at K4 of histone H3. Nature 419:407–411. doi:10.1038/nature01080

Saurin AJ, Shao Z, Erdjument-Bromage H, Tempst P, Kingston RE (2001) A Drosophila Polycomb group complex includes Zeste and dTAFII proteins. Nature 412(6847):655–660, http://doi.org/10.1038/35088096

Scherthan H, Wang H, Adelfalk C, White EJ, Cowan C, Cande WZ, Kaback DB (2007) Chromosome mobility during meiotic prophase in Saccharomyces cerevisiae. Proc Natl Acad Sci 104:16934–16939

Schmekel K, Daneholt B (1995) The central region of the synaptonemal complex revealed in three dimensions. Trends in Cell Biol 5:239–242

Schmid M, Nanda I (eds) (2004) Chromosomes today. Springer Netherlands, Dordrecht

Schober CS, Aydiner F, Booth CJ, Seli E, Reinke V (2011) The kinase VRK1 is required for normal meiotic progression in mammalian oogenesis. Mech Dev 128:178–190

Schotta G (2002) Central role of Drosophila SU(VAR)3-9 in histone H3-K9 methylation and heterochromatic gene silencing. EMBO J 21:1121–1131

Schulz C, Wood CG, Jones DL, Tazuke SI, Fuller MT (2002) Signaling from germ cells mediated by the rhomboid homolog stet organizes encapsulation by somatic support cells. Development 129:4523–4534

Seifarth W, Petersen G, Kontermann R, Rival M, Huet J, Bautz EKF (1991) Identification of the genes coding for the second-largest subunits of RNA polymerases I and III of Drosophila melanogaster. Mol Gen Genet 228:424–432

Shao Z, Raible F, Mollaaghababa R, Guyon JR, Wu CT, Bender W, Kingston RE (1999) Stabilization of chromatin structure by PRC1, a polycomb complex. Cell 98(1):37–46, http://doi.org/10.1016/S0092-8674(00)80604-2

Shi Y, Lan F, Matson C, Mulligan P, Whetstine JR, Cole PA et al (2004) Histone demethylation mediated by the nuclear amine oxidase homolog LSD1. Cell 119(7):941–953, http://doi.org/10.1016/j.cell.2004.12.012

Shiloh Y (2006) The ATM-mediated DNA-damage response: taking shape. Trends Biochem Sci 31:402–410

Shumaker DK (2001) LAP2 binds to BAFDNA complexes: requirement for the LEM domain and modulation by variable regions. EMBO J 20:1754–1764

Smulders-Srinivasan TK, Szakmary A, Lin H (2010) A Drosophila chromatin factor interacts with the Piwi-interacting RNA mechanism in niche cells to regulate germline stem cell self-renewal. Genetics 186:573–583

Song X (2002) Germline stem cells anchored by adherens junctions in the Drosophila ovary niches. Science 296:1855–1857

Song X (2004) Bmp signals from niche cells directly repress transcription of a differentiation-promoting gene, bag of marbles, in germline stem cells in the Drosophila ovary. Development 131:1353–1364

Song X, Call GB, Kirilly D, Xie T (2007) Notch signaling controls germline stem cell niche formation in the Drosophila ovary. Development 134(6):1071–1080, http://doi.org/10.1242/dev.003392

Spradling A (1993) Germline cysts: communes that work. Cell 72:649–651

Spradling A, Drummond-Barbosa D, Kai T (2001) Stem cells find their niche. Nature 414:98–104

Spradling AC, Nystul T, Lighthouse D, Morris L, Fox D, Cox R, Tootle T, Frederick R, Skora A (2008) Stem cells and their niches: integrated units that maintain Drosophila tissues. Cold Spring Harb Symp Quant Biol 73:49–57

Spradling A, Fuller MT, Braun RE, Yoshida S (2011) Germline stem cells. Cold Spring Harb Perspect Biol 3:a002642

Su TT (2006) Cellular responses to DNA damage: one signal, multiple choices. Annu Rev Genet 40:187–208

Su TT, O'Farrell PH (1997) Chromosome association of minichromosome maintenance proteins in Drosophila mitotic cycles. J Cell Biol 139:13–21

Sun Z-W, Allis CD (2002) Ubiquitination of histone H2B regulates H3 methylation and gene silencing in yeast. Nature 418:104–108

Sun J, Wei H-M, Xu J, Chang J-F, Yang Z, Ren X, Lv W-W, Liu L-P, Pan L-X, Wang X, Qiao H-H, Zhu B, Ji J-Y, Yan D, Xie T, Sun F-L, Ni J-Q (2015) Histone H1-mediated epigenetic regulation controls germline stem cell self-renewal by modulating H4K16 acetylation. Nat Commun 6:8856

Szakmary A, Reedy M, Qi H, Lin H (2009) The Yb protein defines a novel organelle and regulates male germline stem cell self-renewal in Drosophila melanogaster. J Cell Biol 185(4):613–627, http://doi.org/10.1083/jcb.200903034

Tabata T, Eaton S, Kornberg TB (1992) The Drosophila hedgehog gene is expressed specifically in posterior compartment cells and is a target of engrailed regulation. Genes Dev 6 (12B):2635–2645

Tamkun JW, Deuring R, Scott MP, Kissinger M, Pattatucci AM, Kaufman TC, Kennison JA (1992) brahma: a regulator of Drosophila homeotic genes structurally related to the yeast transcriptional activator SNF2SWI2. Cell 68:561–572

Tange Y, Horio T, Shimanuki M, Ding D-Q, Hiraoka Y, Niwa O (1998) A novel fission yeast gene, tht1+, is required for the fusion of nuclear envelopes during karyogamy. J Cell Biol 140:247–258

Thomson T, Lin H (2009) The biogenesis and function of PIWI proteins and piRNAs: progress and prospect. Annu Rev Cell Dev Biol 25(1):355–376, http://doi.org/10.1146/annurev.cellbio.24.110707.175327

Tie F, Furuyama T, Prasad-Sinha J, Jane E, Harte PJ (2001) The Drosophila polycomb group proteins ESC and E(Z) are present in a complex containing the histone-binding protein p55 and the histone deacetylase RPD3. Development 128(2):275–286

Tomancak P, Berman BP, Beaton A, Weiszmann R, Kwan E, Hartenstein V, Celniker SE, Rubin GM (2007) Global analysis of patterns of gene expression during Drosophila embryogenesis. Genome Biol 8:R145

Tsuneizumi K, Nakayama T, Kamoshida Y, Kornberg TB, Christian JL, Tabata T (1997) Daughters against dpp modulates dpp organizing activity in Drosophila wing development. Nature 389:627–631

Tworoger M, Larkin MK, Bryant Z, Ruohola-Baker H (1999) Mosaic analysis in the Drosophila ovary reveals a common hedgehog-inducible precursor stage for stalk and polar cells. Genetics 151:739–748

Upadhyay M, Cortez YM, Wong-Deyrup S, Tavares L, Schowalter S, Flora P, Hill C, Nasrallah M, Chittur S, Rangan P (2016) Transposon dysregulation modulates dWnt4 signaling to control germline stem cell differentiation in *Drosophila*. PLoS Genet 12(3):e1005918

Wagner N (2004) Two novel LEM-domain proteins are splice products of the annotated gene CG9424. Eur J Cell Biol 82:605–616

Wagner N, Krohne G (2007) LEM-domain proteins: new insights into lamin-interacting proteins. In: KW Jeon (ed) International review of cytology. Elsevier, San Diego, CA, pp 1–46

Wang Z, Lin H (2004) Nanos maintains germline stem cell self-renewal by preventing differentiation. Science 303:2016–2019

Wang L, Li Z, Cai Y (2008) The JAK/STAT pathway positively regulates DPP signaling in the Drosophila germline stem cell niche. J Cell Biol 180:721–728

Wang X, Pan L, Wang S, Zhou J, McDowell W, Park J, Haug J, Staehling K, Tang H, Xie T (2011). In: Perrimon N (ed) Histone H3K9 trimethylase eggless controls germline stem cell maintenance and differentiation. PLoS Genet 7:e1002426

Wang S, Gao Y, Song X, Ma X, Zhu X, Mao Y, Yang Z, Ni J, Li H, Malanowski KE, Anoja P, Park J, Haug J, Xie T (2015) Wnt signaling-mediated redox regulation maintains the germ line stem cell differentiation niche. eLife 4:408

Ward EJ, Shcherbata HR, Reynolds SH, Fischer KA, Hatfield SD, Ruohola-Baker H (2006) Stem cells signal to the niche through the Notch pathway in the Drosophila ovary. Curr Biol 16 (23):2352–2358, http://doi.org/10.1016/j.cub.2006.10.022

Wu X, Tanwar PS, Raftery LA (2008) Drosophila follicle cells: morphogenesis in an eggshell. Semin Cell Dev Biol 19:271–282

Xi R (2005) Stem cell self-renewal controlled by chromatin remodeling factors. Science 310:1487–1489

Xia L, Jia S, Huang S, Wang H, Zhu Y, Mu Y, Kan L, Zheng W, Wu D, Li X, Sun Q, Meng A, Chen D (2010) The Fused/Smurf complex controls the fate of Drosophila germline stem cells by generating a gradient BMP response. Cell 143:978–990

Xie T, Spradling AC (1998) decapentaplegic is essential for the maintenance and division of germline stem cells in the Drosophila ovary. Cell 94:251–260

Xie T, Spradling AC (2000) A niche maintaining germ line stem cells in the Drosophila ovary. Science 290:328–330

Xin T, Xuan T, Tan J, Li M, Zhao G, Li M (2013) The Drosophila putative histone acetyltransferase Enok maintains female germline stem cells through regulating Bruno and the niche. Dev Biol 384:1–12

Xuan T, Xin T, He J, Tan J, Gao Y, Feng S, He L, Zhao G, Li M (2013) dBre1/dSet1-dependent pathway for histone H3K4 trimethylation has essential roles in controlling germline stem cell maintenance and germ cell differentiation in the Drosophila ovary. Dev Biol 379:167–181

Yasuhara JC, Wakimoto BT (2008) Molecular landscape of modified histones in Drosophila heterochromatic genes and euchromatin-heterochromatin transition zones. PLoS Genet 4:e16

Zhang Y, Kalderon D (2001) Hedgehog acts as a somatic stem cell factor in the Drosophila ovary. Nature 410:599–604

Zhang Y, Cooke M, Panjwani S, Cao K, Krauth B, Ho P-Y, Medrzycki M, Berhe DT, Pan C, McDevitt TC, Fan Y (2012). In: Copenhaver GP (ed) Histone H1 depletion impairs embryonic stem cell differentiation. PLoS Genet 8:e1002691

Zhang Q, Shalaby NA, Buszczak M (2014) Changes in rRNA transcription influence proliferation and cell fate within a stem cell lineage. Science 343:298–301

Zomerdijk J, Beckmann H, Comai L, Tjian R (1994) Assembly of transcriptionally active RNA polymerase I initiation factor SL1 from recombinant subunits. Science 266:2015–2018

Chapter 2
Regulation of the Balance Between Proliferation and Differentiation in Germ Line Stem Cells

Ramya Singh and Dave Hansen

Abstract In many animals, reproductive fitness is dependent upon the production of large numbers of gametes over an extended period of time. This level of gamete production is possible due to the continued presence of germ line stem cells. These cells can produce two types of daughter cells, self-renewing daughter cells that will maintain the stem cell population and differentiating daughter cells that will become gametes. A balance must be maintained between the proliferating self-renewing cells and those that differentiate for long-term gamete production to be maintained. Too little proliferation can result in depletion of the stem cell population, while too little differentiation can lead to a lack of gamete formation and possible tumor formation. In this chapter, we discuss our current understanding of how the balance between proliferation and differentiation is achieved in three well-studied germ line model systems: the Drosophila female, the mouse male, and the *C. elegans* hermaphrodite. While these three systems have significant differences in how this balance is regulated, including differences in stem cell population size, signaling pathways utilized, and the use of symmetric and/or asymmetric cell divisions, there are also similarities found between them. These similarities include the reliance on a predominant signaling pathway to promote proliferation, negative feedback loops to rapidly shutoff proliferation-promoting cues, close association of the germ line stem cells with a somatic niche, cytoplasmic connections between cells, projections emanating from the niche cell, and multiple mechanisms to limit the spatial influence of the niche. A comparison between different systems may help to identify elements that are essential for a proper balance between proliferation and differentiation to be achieved and elements that may be achieved through various mechanisms.

R. Singh • D. Hansen (✉)
Department of Biological Sciences, University of Calgary, Calgary, AB, Canada, T2N 1N4
e-mail: dhansen@ucalgary.ca

S. Arur (ed.), *Signaling-Mediated Control of Cell Division*, Results and Problems in Cell Differentiation 59, DOI 10.1007/978-3-319-44820-6_2

2.1 Introduction

In order for a species to attain reproductive success in an environment, it must be able to produce sufficient numbers of viable offspring that can survive to reproductive maturity and produce their own offspring. The germ line of an organism must be able to produce gametes in a sustained manner during the reproductive lifespan of the organism in order for optimal reproductive fitness to be achieved. In many species, the reproductive fitness of an animal relies on the continuous production of large numbers of gametes. A stem cell population of undifferentiated cells provides the source material for the extended production of these gametes. Germ line Stem Cells (GSCs) are a special subset of adult stem cells that can only give rise to more GSCs and meiotic cells/gametes (sperm or oocytes).

GSCs divide mitotically to replenish themselves and/or give rise to differentiating daughter cells. The amount of stem cell proliferation, or self-renewal, relative to the number of cells that enter a differentiation pathway leading to the formation of gametes must be balanced. If the GSC population undergoes too little self-renewal, the stem cell population may become depleted, adversely affecting gamete production (Kimble and White 1981; Austin and Kimble 1987; Meng et al. 2000; Kai and Spradling 2003). Similarly, excess self-renewal will diminish the number of cells that enter the differentiation pathway, as well as potentially lead to tumor formation (Berry et al. 1997; Xie and Spradling 1998; Meng et al. 2000; Hansen et al. 2004a). Therefore, the balance between the amount of self-renewal (proliferation) and the amount of differentiation is vital. A stem cell divides mitotically to produce two daughter cells. In some instances, both daughter cells will adopt the same fate (symmetric division), either as stem cells or differentiating cells, while in other instances one daughter cell will remain a stem cell while the other enters a pathway to differentiation (asymmetric division). However, at least at a population level, a balance must be maintained between the daughter cells that remain stem cells and the daughter cells that differentiate. Cells that are to form gametes must progress through meiosis, which produces four haploid products; depending on the species and sex, some or all of these products will become functional gametes. Therefore, at the level of a single cell in the germ line, this balance translates into a single decision made by each GSC, whether to maintain its undifferentiated stem cell character or to enter a differentiation pathway that results in the formation of gametes. It is this decision that underlies the vital proliferation/differentiation balance and ensures that an adequate numbers of gametes are made at any given time, while maintaining a population of stem cells necessary for future gamete production. It is important to note that in some systems differentiating cells may divide mitotically; however, for the purposes of this review the term "proliferation" will be reserved for mitotically dividing GSCs that are considered undifferentiated.

2.1.1 GSCs Proliferate in the Niche by Either Asymmetric or Symmetric Divisions

In most organisms, GSCs are thought to occupy a special microenvironment within the germ line called the niche (Schofield 1978). In 1978, Schofield first proposed the model that the niche exerts influence on occupying cells in order to maintain the stem cell population (Schofield 1978). According to this model, as long as a GSC resides within its niche, it will continue to proliferate and give rise to more GSCs. When a GSC moves away from its niche, it will begin to differentiate. Later, this model was experimentally validated in various organisms, including Drosophila, *C. elegans*, as well as the mouse testis (Kimble and White 1981; Xie and Spradling 2000; Kiger et al. 2001; Tulina and Matunis 2001; Brinster 2002). The niche is made up of specialized somatic cells that make contact with the GSCs and provide short-range signaling to maintain their "stemness." They can also receive and integrate cues from outside the niche and signal to the GSCs to modulate the amount of proliferation.

The niche's influence is thought to be dominant, in at least some systems, to the type of occupying cell (Schofield 1978; Scadden 2014). In the case of mammals and Drosophila, even a partially differentiated cell can revert to the GSC fate if the cell moves back into the area of the niche (Brawley and Matunis 2004; Kai and Spradling 2004). However, this ability may be specific to only some stem cell systems, as several other adult stem cell niches, such as the hair follicle, have not yet been demonstrated to have this ability (Hsu et al. 2011). That at least some germ line niches have this unique ability suggests that the balance between proliferation and differentiation in the germ line has additional layers of regulation in place to help respond to changing developmental needs and environmental cues when compared with other adult stem cells, which help ensure that subsequent generations are formed.

A GSC can divide into two general ways, symmetrically or asymmetrically (Fig. 2.1) (Morrison and Kimble 2006; Inaba and Yamashita 2012; Yadlapalli and Yamashita 2012). In asymmetric divisions, when a GSC divides mitotically to produce two daughter cells, only one retains a stem cell identity, while the other begins the path to differentiation, including gamete formation. These differentiating daughter cells move away from the influence of the niche and may undergo additional rounds of proliferation before forming fully differentiated cells. Asymmetric division has been well studied in Drosophila and mammals (Inaba and Yamashita 2012; Yadlapalli and Yamashita 2012; Januschke and Näthke 2014). Generally speaking, asymmetric cell division can take place in two ways (Gonczy 2008; Inaba and Yamashita 2012). First, the mitotic spindle can be oriented in such a way that one daughter cell is retained in the niche, while the other is positioned away from the influence of the niche. A second way to achieve asymmetric division is by the asymmetric distribution of cytoplasmic cell fate determinants among the resulting two daughter cells, such that only one cell is fated to become a GSC, while the other becomes a differentiating cell. In symmetric divisions, the two resulting

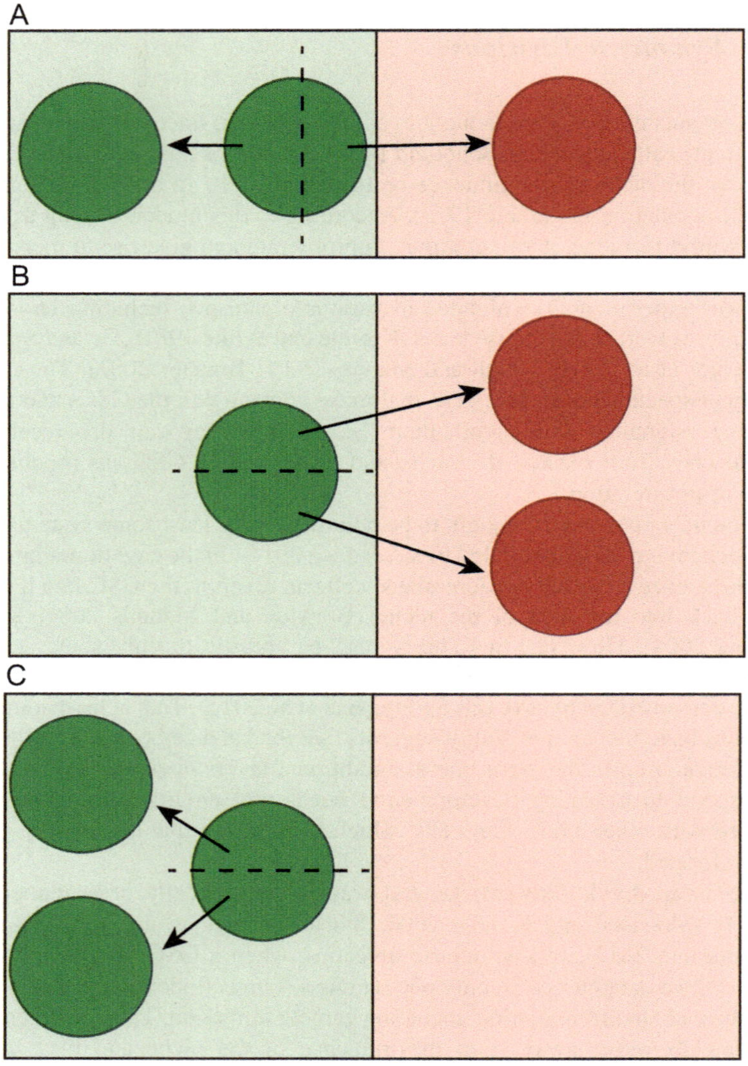

Fig. 2.1 Stem cells can divide symmetrically or asymmetrically. If the mitotic spindle of a dividing GSC is oriented perpendicular to the niche (**a**), one daughter cell will be retained in the niche as a GSC while the other is pushed away from the niche, thereby taking on the fate of a differentiating cell. If the mitotic spindle is oriented parallel to the niche, both daughter cells may take on the differentiating cell fate (**b**) or the GSC fate (**c**). *Green circle*—GSC, *green box*—niche area, *red circle*—differentiating cell, *red box*—non-niche area

daughter cells have the same fate, whether that is a stem cell fate or a differentiating fate. Symmetric divisions are thought to occur in the *C. elegans* hermaphrodite gonad (Crittenden et al. 2006). A given stem cell system need not exclusively utilize asymmetric or symmetric stem cell divisions. For example, a live imaging study in cultured Drosophila testis has shown that in a young healthy niche, GSCs divide asymmetrically about 80 % of the time, while other cells divide symmetrically, forming either two differentiating cells (10 %) or two self-renewing cells (10 %) (Sheng and Matunis 2011).

In the germ lines of certain organisms, the differentiating daughter cells undergo a fixed number of mitotic divisions before entering into a meiotic cell cycle to form gametes. These mitotically dividing cells are referred to as transit-amplifying cells or transient amplifying cells (Watt and Hogan 2000). Transit-amplifying (TA) cells are believed to serve as a way of amplifying the potential number of gametes formed, while limiting the potential of acquired mutations that are passed on by the precursor stem cell (Cairns 1975; Potten and Loeffler 1990). Transit-amplifying cells have been well studied in the male Drosophila germ line, where they are morphologically distinct from GSCs. They have also been identified in the mouse testis (Nakagawa et al. 2007). In *C. elegans*, it was earlier thought that transit-amplifying cells exist in the germ line (Maciejowski et al. 2006; Cinquin et al. 2010); however, recent evidence suggests that the worm hermaphroditic adult germ line might lack these cells altogether (Fox and Schedl 2015).

2.1.2 Adhesion and Competition Within the Niche Maintains GSCs Numbers

The number of GSCs maintained in the niche is tightly regulated. In Drosophila females, there are only 2–3 stem cells per niche, whereas in the *C. elegans* hermaphrodite there are likely 60–80 stem cells (Lin and Spradling 1993; Fox and Schedl 2015). If GSCs move away from the niche, they begin to lose their stem cell identity and ability to self-renew; therefore, there are mechanisms to maintain cells within the niche microenvironment. Somatic cells that make up the niche and the extracellular matrix surrounding it express adhesion molecules that maintain cells within the niche. The activity of these molecules may be regulated so as to control the number of GSCs in the niche (Chen et al. 2013). These adhesion molecules contribute to niche–cell adhesion as well as cell–cell adhesion (Chen et al. 2013). Within the Drosophila male and female GSC niches, cadherins and integrins are the major classes of adhesion molecules at play (Chen et al. 2013). For example, in the Drosophila ovary, GSCs are held in the niche by expression of E-cadherin at junctions between the GSCs and the niche cells (Song et al. 2002). The removal of E-cadherin from the junctions results in loss of GSCs from the niche (Song et al. 2002).

GSCs that move out of the niche environment must be replaced to maintain the stem cell population. Since the niche consists of a limited signaling environment and is often limited in physical space, GSCs are thought to be in constant competition to occupy the niche. The process of replacement of lost GSCs by division of neighboring GSCs in a stochastic manner is called neutral competition (Klein and Simons 2011). No one GSC has a long-term advantage over the other in this type of competition (Klein and Simons 2011). Lineage tracing experiments have shown that neutral competition of GSCs occurs in the mouse testis (Klein et al. 2010). However, a GSC may acquire mutations that could affect its ability to compete for the niche, which is referred to as non-neutral competition (Klein and Simons 2011). For example, in the Drosophila ovary, GSCs lacking the differentiation-promoting factor Bam (Bag of marbles) display increased levels of E-cadherin at the junctions with the niche cells and are preferentially retained in the niche (Jin et al. 2008). These cells also have increased proliferative capacity, suggesting that there is cross talk between factors that control adhesion, competition, and the balance between proliferation and differentiation.

2.1.3 Cell Signaling Maintains GSCs

In many germ line niches, a single key signaling program is necessary to maintain GSCs. In the Drosophila ovary, BMP signaling is the primary stem cell maintaining signal, while in the testis, the Janus kinase-signal transducer and activator of transcription (JAK/STAT) pathway performs this role (Xie and Spradling 1998; Kiger et al. 2001; Tulina and Matunis 2001). In the *C. elegans* hermaphrodite, GLP-1/Notch signaling is essential for GSC maintenance and proliferation (Austin and Kimble 1987). Similarly, in the mouse testis, Glial cell line-derived neurotrophic factor (GDNF), a member of the TGF-β family, is indispensible for self-renewal of GSCs (Kubota et al. 2004). Many other pathways have also been identified to be important players in maintaining GSCs or in promoting their differentiation and will be discussed in subsequent sections.

It is important to note that although the balance between proliferation and differentiation must be tightly regulated in the germ line, it also must allow for some flexibility in order to respond to environmental (nutritional state) and physiological changes (hormonal states and aging). This flexibility helps to ensure the survival of the species in a range of environmental conditions and during different life stages. For example, in *C. elegans*, specialized neurons in the pharynx sense population density and food abundance and are required for TGF-β-mediated proliferation of GSCs in the germ line niche (Dalfó et al. 2012). Likewise, during early larval development in the Drosophila female, ecdysone receptors repress precocious differentiation of the niche and the GSCs, but during later larval stages, once the niche has matured, ecdysone is required to initiate differentiation of the GSCs (Gancz et al. 2011). GSC function generally declines with age in adults resulting in decreased proliferation and impaired gametogenesis (Jones and Rando 2011; Gonen and Toledano 2014). GSCs from older mice, when transplanted into younger mice, maintain long past the normal lifespan of the donor animal, which suggests that younger niches can maintain older GSCs independent of their age (Ryu et al. 2006). Therefore, apart from the key signaling pathway that maintains the balance between proliferation and differentiation, there are often other cell signaling pathways that can modulate this balance under specific conditions.

In this chapter, we discuss three well-studied germ line model systems. We compare and contrast the known molecular and physical mechanisms behind the proliferation vs. differentiation balance in these systems. We discuss one female germ line, the Drosophila ovary, one male germ line, the mouse testis, and the hermaphrodite germ line of *C. elegans* for this comparison. We highlight conserved mechanisms that are essential to these systems as well as discuss other mechanisms that make each of these niche systems unique. We also briefly mention how this balance can be modulated in times of changing environmental/physiological conditions.

2.2 *C. elegans* Hermaphrodite

2.2.1 Anatomy of the Gonad

The gonad of the *C. elegans* hermaphrodite contains two U-shaped tubes or arms that meet at a common uterus (Hirsh et al. 1976; Kimble and White 1981). Germ cells are born at the distal end of each arm and mature as they move proximally towards the vulva. At the very distal end of each gonad arm is a large somatic cell called the distal tip cell (DTC), which caps the distal mitotic GSCs and serves as a niche for promoting GSC proliferation (Kimble and White 1981). The population of germ cells closest to the DTC is referred to as the proliferative zone (PZ) or mitotic zone (MZ), and cells within this region are mitotically dividing (Kimble and White 1981; Hansen et al. 2004a). The DTC forms the niche for GSC maintenance and polarizes the germ line in the distal–proximal axis (Hansen and Schedl 2013; Kershner et al. 2013). The DTC body intimately contacts the germ cells in the distal end of the gonad and also extends multiple cellular processes throughout the proliferative zone, with several smaller processes intercalating between germ cells, with the DTC and its extensions forming a DTC plexus (Fitzgerald and Greenwald 1995; Hall et al. 1999; Crittenden et al. 2006; Byrd et al. 2014). The PZ consists of about 200–250 cells that have similar nuclear morphology, as visualized by DAPI staining (Crittenden et al. 2006; Maciejowski et al. 2006; Fox et al. 2011). Most cells in the PZ in the adult are in the S or G2 phases of the cell cycle, with very few cells in the G1 or M (Hansen et al. 2004a; Crittenden et al. 2006; Maciejowski et al. 2006; Fox et al. 2011). According to the most recent model, the self-renewing GSC population in the PZ consists of the distal-most 60–80 cells, while the remaining cells within the PZ are either undergoing a final mitotic division or have entered premeiotic S phase (Fox and Schedl 2015), suggesting that the GSC niche only fully functions within the most distal region of the PZ. This is a relatively large number of stem cells as compared to other systems, including the Drosophila ovary (see below). The large pool of stem cells could be a means to ensure rapid production of gametes when environmental conditions are conducive for reproduction (Hubbard et al. 2013). In *C. elegans*, GSCs appear to divide by symmetric divisions as daughter cells are arranged in various orientations relative to the parental GSC upon mitotic division (Crittenden et al. 2006).

As cells move away from the DTC, towards the proximal end of the gonad, they enter into meiosis in an assembly-line like manner to produce gametes. This progression can be easily observed by comparing nuclear morphology as visualized by DAPI staining. The first change in nuclear morphology is seen at the transition zone, where the DNA takes on a crescent shaped appearance, which marks entry into the leptotene and zygotene stages of meiotic prophase (Fig. 2.2a) (Francis et al. 1995; Dernburg et al. 1998; MacQueen and Villeneuve 2001). A large pachytene region follows the transition zone and extends until approximately the loop region, where the gonad arm reflexes back towards the vulva. As cells progress beyond the loop region, they continue to progress through gametogenesis, with approximately the first 40 cells differentiating into sperm (forming ~140–160 sperm per gonad arm) and all subsequent cells differentiating into oocytes (Kimble and White 1981). The sperm are stored in the spermatheca and fertilize oocytes that pass through the spermatheca and into the uterus that is common to both gonad arms (Hirsh et al. 1976).

The germ cells in the distal gonad are syncytial in nature as germ cell nuclei are only partially covered by cell membranes. The central core of the distal arm, called the rachis, is largely cytoplasmic and devoid of cells, as most germ cells line the periphery. Small clusters of cells are seen in the rachis, but only in the proliferative zone (Hirsh et al. 1976). Recently, these cells in the rachis have been suggested to act as cytoplasmic bridges to slow diffusion of morphogens across the proliferative zone, which aids in maintaining the size of the proliferative zone (Cinquin et al. 2015). Besides the DTC, the germ line is lined by five pairs of somatic sheath cells (Hirsh et al. 1976; Kimble and Hirsh 1979; Hall et al. 1999), which interact with each other, and with germ cells, via gap junctions (Hall et al. 1999).

2.2.2 *Proliferation vs. Differentiation Signaling*

As mentioned, cells closest to the DTC are the self-renewing stem cells. As they move away from the DTC, they cease proliferating and enter into meiosis. If the DTC is ablated, all cells enter into meiosis prematurely and the stem cell population becomes depleted (Kimble and White 1981). Likewise, moving the DTC more proximally results in a population of proliferating cells being present in this new location (Kimble and White 1981), suggesting that the DTC is necessary and sufficient to maintain the stem cell population. Genetic analyses revealed that the Notch signaling pathway is the key signal emanating from the DTC to promote stem cell proliferation/self-renewal (Fig. 2.2b). The DTC expresses the Notch signaling ligands LAG-2 and APX-1 (Tax et al. 1994; Henderson et al. 1994; Nadarajan et al. 2009), which interact with the GLP-1/Notch receptor that is

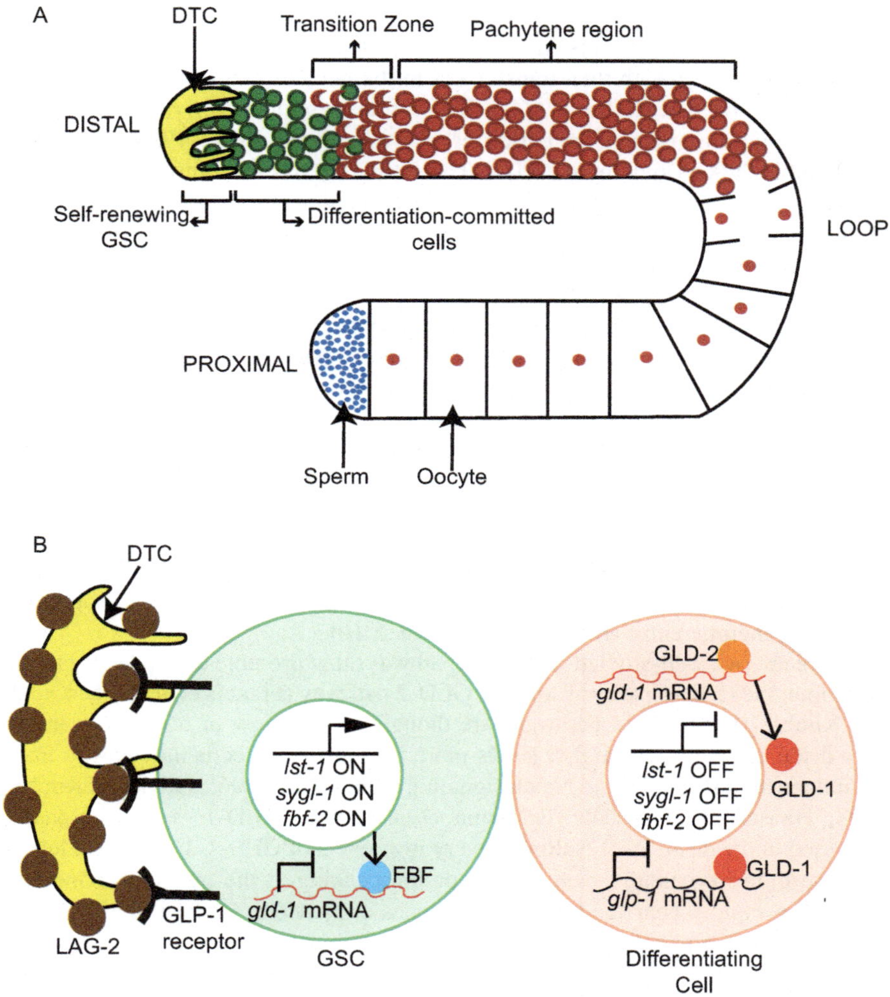

Fig. 2.2 (**a**) Schematic representation of the *C. elegans* adult hermaphrodite germ line. The somatic Distal Tip Cell (DTC) caps the distal end of the gonad and maintains a pool of self-renewing GSCs at the distal end followed by a region where cells are completing their last few rounds of mitosis or entering premeiotic S phase and are committed to differentiate. Proximal to this region, nuclei take on a crescent-shaped appearance in the Transition Zone. This is followed by a region where cells are in pachytene. At the loop region, oocytes begin to form. At the proximal end, sperm is contained in the spermatheca. (**b**) Key signals involved in the switch from proliferation to differentiation in the *C. elegans* hermaphrodite niche. The DTC expresses the GLP-1/Notch ligand, which makes contact with germ cells expressing the GLP-1 receptor. The activation of the receptor leads to transcription of the *lst-1*, *sygl-1*, and *fbf-2* genes. FBF protein binds *gld-1* mRNA and represses its translation. In germ cells away from the influence of the niche, GLD-1 and GLD-2 proteins are allowed to be expressed and GLD-2 in turn promotes the translation of GLD-1. The GLD-1 protein blocks translation of *glp-1* mRNA, thereby allowing the cell to enter the differentiation pathway

expressed on the surface of the germ cells in the proliferative zone (Crittenden et al. 1994). It is thought that this interaction results in cleavage of the intracellular domain of GLP-1, which then translocates to the nucleus and binds to the LAG-1 transcription factor and the LAG-3/SEL-8 transcription coactivator to form a transcriptional complex (Christensen et al. 1996; Roehl et al. 1996; Petcherski and Kimble 2000; Doyle et al. 2000). The formation of this complex leads to the transcription of downstream genes, including two redundant genes *lst-1* and *sygl-1*, in the distal-most cells of the PZ (Kershner et al. 2014). *lst-1* and *sygl-1* expression roughly corresponds to the region occupied by actively self-renewing GSCs and is therefore likely to represent the niche (Crittenden et al. 2006; Cinquin et al. 2010; Fox and Schedl 2015). It is not yet fully understood how *lst-1* and *sygl-1* expression is limited to the distal-most compartment of the PZ since GLP-1 receptor expression and LAG-2 expressing DTC projections are seen to extend beyond this region. Perhaps, Notch signaling levels are above a threshold only where the levels of GLP-1 and its ligands are at their highest, and/or there may be negative regulators of Notch that function more proximally in the PZ, mechanisms that remain to be elucidated. Until recently, it was unclear how LST-1 and SYGL-1 proteins function to maintain GSCs, as their protein sequences are not conserved outside of nematodes (Kershner et al. 2014). However, a new study shows that both LST-1 and SYGL-1 are responsible for inhibiting the expression of GLD-1, a key protein for promoting meiotic entry (Brenner and Schedl 2016).

There are two well-studied redundant pathways that promote meiotic entry in the germ line: the GLD-1 pathway and the GLD-2 pathway (Francis et al. 1995; Kadyk and Kimble 1998). These pathways are thought to have low or no activity in the very distal end, but have higher levels more proximally, as cells move away from the influence of the niche and Notch signaling (Jones et al. 1996; Kadyk and Kimble 1998; Hansen et al. 2004b; Eckmann et al. 2004). GLD-1 is translationally repressed in the distal PZ, resulting in very low levels of GLD-1 distally, but levels increase more proximally, with accumulation peaking at the transition zone and decreasing rapidly near the loop region (Jones et al. 1996; Hansen et al. 2004b). GLD-1 is itself a translational repressor and functions genetically with NOS-3, a Nanos homolog, to repress genes that promote proliferation and/or activate genes that promote meiotic entry (Jones and Schedl 1995; Jan et al. 1999; Kraemer et al. 1999; Hansen et al. 2004b; Eckmann et al. 2004). GLD-1 represses the translation of *glp-1* mRNA, thereby creating a negative feedback loop to reduce GLP-1 protein accumulation beyond the PZ (Marin and Evans 2003). GLD-2, on the other hand, encodes the catalytic subunit of cytoplasmic poly(A) polymerase (PAP) and functions as a translational activator (Wang et al. 2002; Kim et al. 2010b). GLD-2 binds the RNA-binding protein GLD-3 (a bicaudal-C homolog) to promote meiotic entry (Eckmann et al. 2002, 2004; Wang et al. 2002). GLD-2 is known to promote expression of *gld-1* mRNA (Hansen et al. 2004b; Suh et al. 2006); however, genetic analysis suggests that GLD-1 and GLD-2 pathways must regulate other mRNA targets, which are largely uncharacterized (Kadyk and Kimble 1998). A third pathway has been proposed to function redundantly with the GLD-1 and GLD-2 pathways to promote differentiation and may contain the GLD-4/GLS-1 cytoPAP complex (Hansen et al. 2004a; Millonigg et al. 2014).

The GLD pathways for promoting meiotic entry are regulated, in part, by the GLP-1 proliferation-promoting pathway through the FBF proteins, which are members of the PUF family of proteins (Zhang et al. 1997). GLP-1 signaling is known to activate *fbf-2* transcription (Lamont et al. 2004). FBF-2 works redundantly with its paralog, FBF-1, to maintain GSCs (Crittenden et al. 2002). In double mutant *fbf-1 fbf-2* worms, germ cells enter meiosis prematurely at lower temperatures (Crittenden et al. 2002). FBF represses translation of mRNAs in each branch of the GLD pathways (Crittenden et al. 2002; Eckmann et al. 2004; Suh et al. 2009).

In summary, cell-extrinsic signaling in the form of the GLP-1/Notch signal, initiated by the DTC niche cell, triggers a transcriptional response in germ cells instructing them to actively proliferate and maintain their stem cell identity. As cells move away from the DTC, the GLD-1, GLD-2, and third pathways for entry into meiosis are activated, allowing germ cells to enter the differentiation program. Recent evidence suggests that additional extrinsic signals mediated by gap junctions between the soma and the germ line also promote proliferation; however, the nature of these signals is unclear and could be a new avenue for research (McCarter et al. 1997; Starich et al. 2014).

Apart from the principal GLP-1 and GLD pathways, many other factors and pathways have been identified that can modulate the balance between proliferation and differentiation in the *C. elegans* germ line. These factors and pathways are thought to modulate or fine-tune the proliferation vs. differentiation decision and may help control the activities of the major pathways or run parallel to them. These additional regulatory controls include phosphorylation, RNA splicing, proteasomal degradation, and microRNA-mediated gene regulation (MacDonald et al. 2008; Mantina et al. 2009; Kerins et al. 2010; Bukhari et al. 2012; Wang et al. 2012, 2014; Gupta et al. 2015). Interestingly, another PUF protein, PUF-8, appears to have multiple roles in regulating the balance between proliferation and differentiation, much like its homolog Pumilio in Drosophila (see Sect. 2.3) (Datla et al. 2014). While PUF-8 works redundantly with MEX-3 to promote proliferation, it also works to suppress proliferation and/or promote differentiation, in part, by negatively regulating RAS/MAPK signaling (Ariz et al. 2009; Racher and Hansen 2012; Vaid et al. 2013). Recently, we have demonstrated that a chromodomain containing protein, MRG-1, may help to set a threshold, allowing the activity levels of the spatially controlled GLP-1/Notch and GLD pathways to properly control the activity of downstream genes. Furthermore, the activity of MRG-1 is controlled, at least in part, by proteasomal degradation (Gupta et al. 2015). Therefore, while the GLP-1/Notch and GLD signaling pathways are the key regulators of the proliferation vs. differentiation decision, there are multiple other inputs that help to ensure that this balance is precisely controlled and optimized for reproductive fitness.

Since the differentiation pathways are activated only when cells move away from the niche, a fundamental question that remains unanswered in *C. elegans* has to do with the mechanisms by which GSCs are retained in the niche. If GSCs exit the niche prematurely, the GSC pool will be rapidly drained and more differentiated cells would result. Conversely, if GSCs remain in the niche longer than appropriate, the balance would be tilted towards over-proliferation. We describe in Sects. 2.3 and 2.4 how adherens junctions and tight junctions help retain GSCs within the niche in the Drosophila ovary and mouse testis. Similar mechanisms have not yet been identified in the *C. elegans* hermaphrodite. It is speculated that the DTC plexus may help anchor the distal-most GSCs to the niche, but this has not yet been demonstrated (Crittenden et al. 2006; Byrd et al. 2014).

2.2.3 Plasticity

C. elegans is a free-living nematode that lives in the soil, and its germ line is likely exposed to large fluctuations in environmental conditions (Hubbard et al. 2013). As a consequence, the worm has developed unique adaptations to tide over stressful periods and reduce gamete production until conditions improve. Many of these adaptations have been well studied in early larval stages or in the transition from the last larval stage to an adult (Hubbard et al. 2013). However, a recent study sheds light into the adult niche's response to stressful conditions. In the absence of food, mitotically dividing germ cells become quiescent, i.e., enter a resting phase (Seidel and Kimble 2015). This is characterized by slower progression through S phase, arrest in G2 phase, disappearance of cells entering M phase, and rapid entry into M phase when food becomes available (Seidel and Kimble 2015). Interestingly, the ability of germ cells to become quiescent and recover from quiescence was independent of germ cell proximity to the DTC, i.e., it was independent of the GLP-1 pathway (Seidel and Kimble 2015). However, it is unknown what signals are instead responsible for induction and resumption from the quiescent state. Similar results were obtained when the worms were stressed under high NaCl conditions (Seidel and Kimble 2015). Therefore, *C. elegans* offers a tractable model system to study quiescence in adult stem cells.

2.3 Drosophila Ovariole

2.3.1 Organization of the Drosophila Ovariole

Adult female Drosophila have a pair of ovaries, each of which contain 16–23 (depending on the strain) tube-like structures called ovarioles (Fig. 2.3a) (Sarikaya et al. 2012). Each ovariole is a unit for egg production and is organized with a distinct polarity. An apical structure at the anterior end of the ovariole, called the germarium, houses stem cells and differentiating cells, which will become oocytes and supporting cells (King 1970; Telfer 1975; Mahowald and Kambysellis 1980; Spradling 1993). The germarium can be divided into distinct regions based on

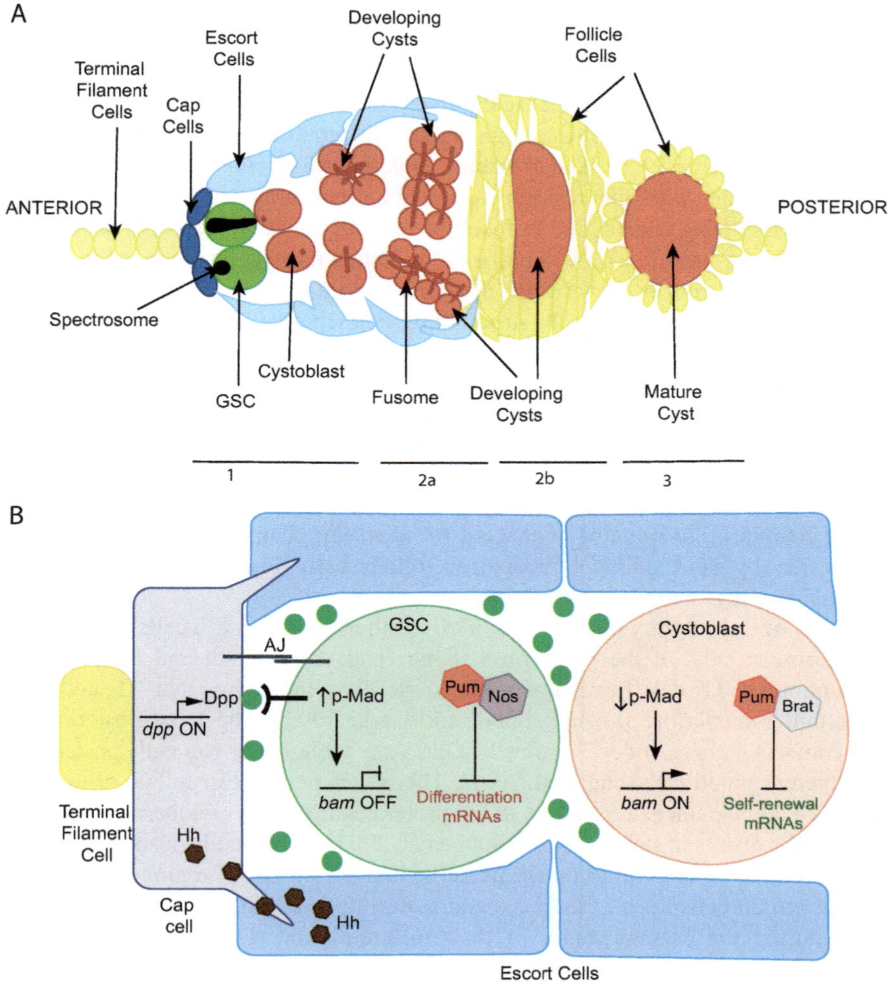

Fig. 2.3 (**a**) The Drosophila ovariole, depicting asymmetric GSC divisions and mitotic cyst divisions in region 1. In region 2a, 16 cell cysts are formed. By region 2b, the 16-cell developing cyst has flattened and occupies the entire width of the germ line. In region 3, the cyst becomes completely surrounded by follicle cells and is ready to bud. (**b**) Decapentapelagic (Dpp) is expressed by cap cells and escort cells. Cytoneme-mediated Hedgehog (Hh) delivery from cap cells to escort cells also activates Dpp transcription in escort cells. Neighboring GSCs transduce Dpp signal by a receptor and this results in high phosphorylated Mad (p-Mad) levels in the GSC. High p-Mad blocks Bam transcription, preventing the GSC from differentiating. At the same time, the Pum and Nanos complex prevents differentiating mRNAs from being translated. In the cystoblast, Dpp activation is low, which results in Bam transcription. This allows cystoblasts to enter the differentiation program. At the same time, Pum and Brat complex together to repress translation of genes responsive to Dpp. AJ—Adherens Junction

morphological differences—regions 1, 2a, 2b, and 3 (Fig. 2.3a). Region 1 comprises the anterior tip of the germarium and serves as a niche to maintain 2–3 GSCs (Wieschaus and Szabad 1979; Lin and Spradling 1993; Xie and Spradling 2000). Here, each GSC divides asymmetrically to produce a posterior daughter cell that is retained in the niche as a GSC and an anterior differentiating daughter cell called the cystoblast (King 1970). The cystoblast undergoes four subsequent synchronous mitotic divisions with the resulting cells remaining connected by cytoplasmic bridges, called ring canals, due to incomplete cytokinesis (King 1970; de Cuevas et al. 1997). As a result, a 16-cell ball of cells, called the cyst, is produced in region 2a. As the cyst moves posteriorly, the cells continue to develop and mature in regions 2a and 2b. In region 2b, surrounding epithelial cells, called follicle cells, envelop the 16-cell cyst, which flattens to take up the entire width of the germarium. Only one cell in the follicle develops into the oocyte, while the remaining 15 cells function as nurse cells (King 1970). Hence, the mitotic divisions that result in formation of a 16-cell cyst are not considered to be transit amplifying as they do not serve to amplify the number of oocytes produced (Gonen and Toledano 2014). By region 3, the enveloped cyst buds off from the germarium as an individual Stage 1 egg chamber, but remains connected to other egg chambers adjacent to it, on either side, by short stalks of specialized follicle cells (Fig. 2.3a) (Ruohola et al. 1991; Cummings and Cronmiller 1994).

A row of 8–10 tightly packed, disk-like, terminal filament (TF) cells make up the anterior-most cells of the germarium (King et al. 1968; Godt and Laski 1995). Disruption of TF cell formation leads to sterility, indicating that TF cells are important for ovariole function (Godt and Laski 1995). The most posterior TF cell contacts a cluster of 5–7 cap cells. GSCs are anchored to cap cells posteriorly by adherens junctions (Song et al. 2002). This attachment to cap cells is critical for stem cell maintenance, as removal of the protein components of adherens junctions from GSCs results in stem cell loss (Song et al. 2002). Escort cells (also called inner sheath cells or inner germarium sheath cells) line the surface of regions 1–2a of the germarium and envelope cystoblasts and maturing cysts with their cellular extensions during the earliest stages of their differentiation (Decotto and Spradling 2005). Escort cells project dynamic microtubule rich cellular processes to the GSC daughters and assist maturing cysts to move posteriorly through the germarium by passing the cyst from one escort cell to the other (Morris and Spradling 2011).

2.3.2 *Asymmetric Divisions Are Achieved by Spectrosome Orientation*

GSCs are the anterior-most cells within the niche and can be easily distinguished by their relatively large size and by the presence of a unique spherical cytoplasmic structure called the spectrosome, which is rich in cytoskeletal proteins and membrane vesicles, but excludes most ribosomes and mitochondria (Lin et al. 1994). During mitosis, the spectrosome is located on the anterior side of the GSC between the nucleus and the portion of the plasma membrane in contact with the cap cells and terminal filament cells (Deng and Lin 1997; Lin and Spradling 1997). At telophase, the spectrosome segregates asymmetrically during cystoblast differentiation, with the daughter cystoblast inheriting a reduced amount of spectrosomal material (Fig. 2.3a) (Deng and Lin 1997; Lin and Spradling 1997; de Cuevas and Spradling 1998). The spectrosome anchors one pole of the mitotic spindle so that the GSC divides along the anterior–posterior axis and only one cell is retained in the niche (Deng and Lin 1997). If GSCs are rapidly lost from the niche, the spectrosome aligns perpendicularly to the anterior–posterior axis so that both daughter cells are positioned in the niche to become GSCs, thereby replenishing the GSC population (Xie and Spradling 2000). Therefore, spectrosome orientation is an intracellular mechanism to determine whether a dividing GSC gives rise to daughter GSCs or a cystoblast and a GSC.

The spectrosomal material inherited by the cystoblast gradually develops into a large branched structure, called the fusome, which connects the ring canals in the cyst (Telfer 1975; de Cuevas and Spradling 1998). The fusome is asymmetrically distributed within the developing cyst and is important for cyst formation (Lin et al. 1994; de Cuevas and Spradling 1998). Disrupting the cytoskeletal components of the spectrosome results in failure to specify the oocyte in the developing follicle (Yue and Spradling 1992; Lin et al. 1994; de Cuevas et al. 1996). The asymmetrical distribution of the fusome within the cyst is also noteworthy, as the cell within the cyst that preferentially retains the most fusomal material will become the oocyte, while the remaining cells become nurse cells (Lin et al. 1994; de Cuevas and Spradling 1998; Huynh and Johnston 2004).

2.3.3 *Proliferation vs. Differentiation Signals*

Since the Drosophila ovariole has only 2–3 GSCs that give rise to many differentiating cells, in order to not deplete the stem cell population, the rates at which mitotic divisions occur is critical. Each GSC divides about once per day to give rise to 1–2 cystoblasts, but each cystoblast takes 4–5 days to exit the germarium (Margolis and Spradling 1995; Xie and Spradling 2000). Moreover, each GSC within the same germarium divides independently of the other (Deng and Lin 1997). Therefore, even though there are only 2–3 GSCs, the germarium can maintain multiple cysts at different stages of development without exhausting the GSC supply, ensuring that oocytes are continually produced. Additionally, for every GSC division, one cystoblast is produced, and although a cystoblast gives rise to a 16-cell cyst, only one of the cells within the cyst becomes an oocyte. In other words, for every GSC division, one GSC and one gamete are produced in this model. The primary mechanism that maintains this balance is the asymmetric division of the GSC, which ensures that only the daughter GSC is held in the niche, while the cystoblast is insulated from niche signals, thereby allowing it to enter the differentiation pathway. The principal molecular signal that determines this balance is *Decapentaplegic* (Dpp) signaling, assisted by additional signals, as discussed below (Fig. 2.3b).

The cap cells and escort cells express the Dpp ligand, a member of the *B*one *m*orphogenetic *p*rotein (BMP) superfamily of signaling molecules (Fig. 2.3b) (Xie and Spradling 1998, 2000; Wang et al. 2008). GSCs within the niche that are in contact with the cap and escort cells show high DPP signal activation, which causes increased levels of the downstream effector, phosphorylated Mad (p-Mad) (Song et al. 2004). Dpp signal activation is necessary and sufficient for the repression of *Ba*g-of-*m*arbles (Bam), a fusome protein that is a key differentiation-promoting factor, thereby allowing GSCs to maintain their stem cell identity (McKearin and Spradling 1990; McKearin and Ohlstein 1995; Ohlstein and McKearin 1997; Song et al. 2004). Cystoblasts are outside the niche and have low Dpp signal activation, which allows *bam* to be transcribed, thereby causing these cells to enter a differentiation program (Chen and McKearin 2003; Kai and Spradling 2003; Song et al. 2004). *G*lass *b*ottom *b*oat (Gbb), another BMP family ligand, is also necessary for GSC maintenance, but unlike Dpp, its overexpression does not result in complete repression of *bam* transcription or inhibit cystoblast differentiation (Wharton et al. 1991; Song et al. 2004).

In other developmental contexts Dpp can act as a long-range signal; however, in order for the balance between stem cells and differentiating cells to be maintained in the germarium, the range of Dpp activation is limited to just the 2–3 GSCs (Xie and Spradling 1998; Kai and Spradling 2003; Song et al. 2004). Several signaling pathways within the niche help restrict Dpp activity. The terminal filament (TF) cells secrete *Unpaired* (Upd) and related cytokines, which activate Jak/Stat signaling in the neighboring cap and escort cells to regulate *dpp* transcription in these cells (López-Onieva et al. 2008). Cap cells secrete the short-range signaling molecule *H*edge*h*og (Hh), which is delivered through cytonemes to escort cells, helping to maintain GSCs (Rojas-Ríos et al. 2012). These cytonemes are a recently discovered feature of cap cells and are reminiscent of the cytonemes of the *C. elegans* germ line niche cell, the DTC (see Sect. 2.2), although the cap cell cytonemes seem to be much shorter in comparison (Rojas-Ríos et al. 2012). Moreover, the germ cells themselves signal to the niche cells to limit Dpp movement and stability in order to repress Dpp activity in cells outside of the niche, thereby creating a negative feedback loop to limit Dpp activity (Liu et al. 2010). For a detailed review on restriction of proliferation-promoting signals in the germarium, please see Chen et al. (2011).

Several other signals emanating from the somatic cells of the niche help maintain GSCs. Piwi (*P*-element *i*nduced *w*impy test*i*s) is expressed in germ line stem cells and somatic cells of the niche and is required for spectrosome formation and GSC self-renewal (Lin and Spradling 1997; Cox et al. 1998, 2000). Piwi expression in the niche represses Bam expression in the GSCs (Szakmary et al. 2005). TF cells and cap cells also express Fs(1) Yb (Yb), which contributes to the maintenance of GSCs (King and Lin 1999; Szakmary et al. 2009). Yb is required for Piwi expression in the terminal filament and cap cells (King et al. 2001; Szakmary et al. 2009). Yb sequesters *Armi*tage (Armi), a putative RNA helicase involved in the piRNA pathway, to the Yb body, a cytoplasmic sphere in somatic cap cells to which Yb is exclusively localized, thereby allowing Piwi to enter the nucleus (Haase et al. 2010; Olivieri et al. 2010; Saito et al. 2010; Qi et al. 2011). However, Piwi's role in stem cell maintenance may be independent of its ability to enter the nucleus and remains to be elucidated (Klenov et al. 2011).

In addition to signaling supplied by the niche to the GSCs, factors within the germ cells are also important for maintaining their fate. For example, *Pum*ilio (Pum) is an RNA-binding protein that works with *N*an*os* (Nos) to maintain GSCs (Lin and Spradling 1997; Forbes and Lehmann 1998; Gilboa and Lehmann 2004; Wang and Lin 2004). Pum and Nos complex together and translationally represses differentiation genes, such as Mei-P26 and Brat (Harris et al. 2011; Joly et al. 2013). It is interesting that while Pum works in consort with Nos in the GSCs to repress differentiation and promote GSC maintenance, its role switches in cystoblasts where it partners with another RNA-binding protein, Brat, to repress genes that sensitize the cell to respond to Dpp (Sonoda and Wharton 2001; Harris et al. 2011; Newton et al. 2015). Brat is expressed exclusively in the cystoblasts and works with Pum to translationally repress proteins that contribute to Dpp responsiveness (Harris et al. 2011). In other words, the same protein, Pum, has opposing functions in two neighboring cells, based on context. Bgcn (*B*enign *g*onial *c*ell *n*eoplasm) is another cell-intrinsic factor required for cystoblast differentiation and controls the localization of Bam to the fusome (Lavoie et al. 1999). In cystoblasts, Bam directly interacts with Pum, with the help of Bgcn, to suppress Pum/Nos activity, thereby promoting differentiation (Kim et al. 2010a).

2.3.4 Plasticity

As stated in Sect. 2.1.1, germ line niches are able to respond to changing environments. With increasing age, there is a decline in reproductive capacity, which is due, in part, to decreased GSC proliferation and decreased GSC retention in the niche (Pan et al. 2007; Zhao et al. 2008). GSCs have a half-life of 4.6 weeks and are randomly lost by differentiation (Margolis and Spradling 1995; Xie and Spradling 1998). The age-related decrease in GSC function and number is associated with decreased BMP signaling and decreased E-cadherin accumulation between GSCs and their niche (Pan et al. 2007; Zhao et al. 2008). Increasing BMP signaling slows the aging of GSCs and promotes GSC proliferation (Pan et al. 2007; Zhao et al. 2008). However, despite the limited lifespan of a GSC, there is only a slight decline in the number of GSCs maintained in the niche of older germaria; GSCs lost with increasing age are actively replaced (Xie and Spradling 2000; Zhao et al. 2008). Cells within the developing cysts up to the 4-cell and 8-cell stages can revert into single stem cell-like cells if exposed to increased levels of Dpp (Kai and Spradling 2004). Recently, it was demonstrated that escort cells express low levels of Dpp, which helps maintain a pool of partially differentiated germ cells that can dedifferentiate into GSC-like cells if stem cells become depleted under normal or stress conditions (Liu et al. 2015). Therefore, in addition to the mechanism of asymmetric divisions that maintains the balance between proliferation and differentiation, dedifferentiation is employed by the Drosophila ovariole to restore equilibrium, when the GSC pool is perturbed.

The niche's response to nutritional changes is an active area of study (Shim et al. 2013). When supplied with a protein-rich diet, germ line stem cells and somatic stem cells have high division rates and their progeny also divide and expand quickly; this response is mediated by insulin signaling (Drummond-Barbosa and Spradling 2001). Furthermore, neural-derived Drosophila Insulin-Like Peptides (DILPs) directly regulate GSC division rate and the growth of cysts (LaFever and Drummond-Barbosa 2005). Under a poor diet, reduced insulin signaling slows the progression of GSCs through the G2 phase of cell division, while an insulin-independent pathway slows progression through the G1 phase (Hsu et al. 2008). Apart from insulin signaling, diet-induced changes in hedgehog signaling also control follicle stem cell proliferation (Hartman et al. 2013).

In summary, multiple cell signals emanating from the surrounding somatic cells of the germarium along with cell-intrinsic gene expression is essential to help regulate the balance between proliferation and differentiation in the Drosophila niche. While GSCs divide primarily by asymmetric divisions in this system, replacement of lost GSCs can also be achieved by symmetric divisions of GSCs or by dedifferentiation of cells from a pool of partially differentiated cells.

2.4 Mouse Testis

2.4.1 Organization of the Seminiferous Tubule

Each lobule of the mammalian testis contains a few tightly coiled structures called seminiferous tubules. Each tubule is a unit for continuous sperm production. The progression of gamete development in the mouse testis conforms loosely to a model of radial polarity (peripheral–central), instead of the strict linear polarity (distal–proximal/anterior–posterior) seen in the *C. elegans* hermaphrodite gonad or the Drosophila ovary. The outer circumference of the tubule supports immature germ cells called spermatogonia, and differentiation progresses towards the lumen producing spermatocytes, followed by spermatids (Fig. 2.4a). The most differentiated cells, the spermatozoa, are produced in the lumen of the seminiferous tubule.

Unlike other germ line models, in the mouse there does not appear to be a clearly defined anatomical region that corresponds to a niche. In fact, germ cells in various stages of differentiation maintain contact with the same somatic cell, the Sertoli cell. The Sertoli cell is a large cell that originates in the basement membrane of the tubule and projects towards the lumen. It also sends out cellular extensions to

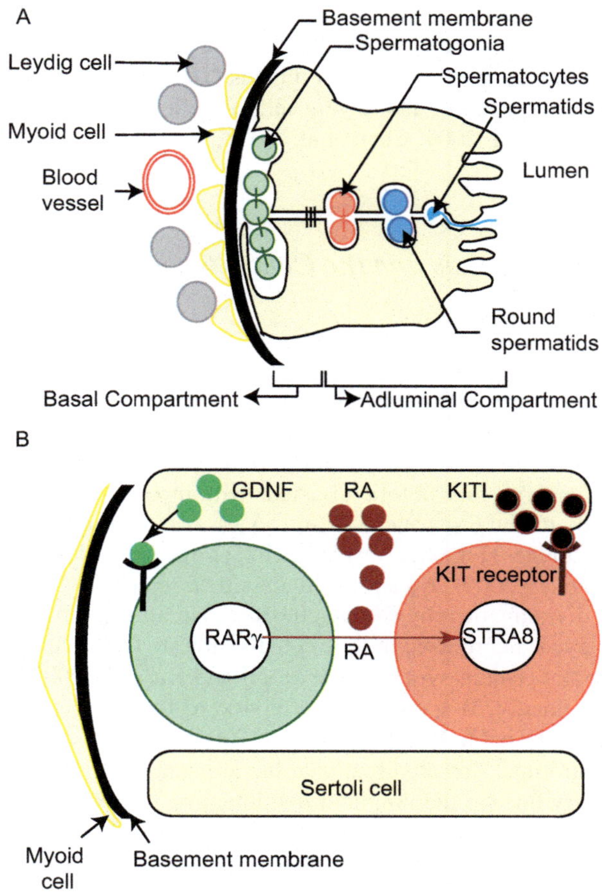

Fig. 2.4 (**a**) Sectional view of a seminiferous tubule. Germ cell development takes place in close association with somatic Sertoli cells. The Sertoli cells are held together by tight junctions, which divide the cell layer into an adluminal compartment and a basal compartment, the latter houses all of the spermatogonia. (**b**) A schematic representation of the key molecular signals involved in A to A_1 spermatogonia development. GDNF secreted by Sertoli cells binds to receptors on undifferentiated A spermatogonia to maintain them. Spermatogonia differentiate to A_1 spermatogonia upon RA stimulation by Sertoli cells. Nuclear receptor RARγ is essential for this. STRA8 and KIT receptor are then expressed in these cells. KIT receptor binds KITL ligand expressed by Sertoli cells

developing germ cells. Tight junctions hold adjacent Sertoli cells together, which results in the formation of a basal compartment and an adluminal compartment. The basal compartment houses the spermatogonia, while the adluminal compartment is home to maturing spermatocytes, spermatids, and spermatozoa. Therefore, in the mammalian testis the term niche is used to describe the close association of the undifferentiated spermatogonia with the epithelium of the seminiferous tubules (Oatley and Brinster 2012). The connective tissue surrounding the seminiferous tubules contains specialized cells called Leydig cells, which secrete testosterone, and the peritubular myoid cells, which line the outside of the basement membrane and provide rigidity to the seminiferous tubules. The adjacent Sertoli cells, interstitial peritubular myoid cells, and Leydig cells all influence the niche (Oatley et al. 2009, 2011; Welsh et al. 2009; Chen et al. 2014).

2.4.2 Spermatogenesis and the Commitment to Meiosis

Traditionally, the cells in different stages of early spermatogenesis have been described by a naming system based on mitotic hierarchy. Similar to cyst formation in the Drosophila ovary, each sequential mitotic division results in incomplete cell division and the daughter cells remain connected in 2, 4, 6, 8, 16, and, sometimes, 32 cell clusters. At the top of this hierarchy are the A_s cells. The single-celled (A_s) spermatogonia undergo four mitotic divisions with incomplete cytokinesis to produce sequentially paired (A_{pr}) and $A_{aligned\text{-}4}$, $A_{al\text{-}8}$, $A_{al\text{-}16}$ spermatogonia connected by intercellular bridges (Fig. 2.5) (Oakberg 1971; Tagelenbosch and de Rooij 1993; de Rooij and Russell 2000). A_{al} spermatogonia then mature to form A_1 spermatogonia, without dividing, thereby entering the differentiation pathway. A_1 spermatogonia undergo five synchronized mitotic cell divisions to form A_2, A_3, A_4, In, and B spermatogonia that remain syncytial (de Rooij and Russell 2000; de Rooij 2001; Yoshida 2012). Finally, B spermatogonia divide to form preleptotene spermatocytes that enter meiosis (de Rooij and Russell 2000).

However, with time, this unidirectional hierarchical description has undergone some revision. By lineage marking and live imaging, it was discovered that A_{pr}–$A_{al\text{-}8}$ spermatogonia can fragment and revert to forming shorter cysts or A_s cells (Nakagawa et al. 2007, 2010). While all of the cells from A_s to A_{al} are referred to as "undifferentiated" spermatogonia, in the prevailing model only some of A_s cells are considered to be actual stem cells (also called Spermatogonial Stem Cells or SSCs), the rest of the spermatogonia are thought to be transit-amplifying cells (Nakagawa et al. 2007, 2010). Both undifferentiated (A_s–A_{al}) and differentiating spermatogonia (A_1–B) are located in the basal compartment of the seminiferous tubule, while spermatocytes, spermatids, and spermatozoa are found in the adluminal compartment (Yoshida 2012). It remains unclear whether SSCs undergo symmetric or asymmetric divisions (de Rooij and Russell 2000; Oatley and Brinster 2012). Since A_{pr} and A_{al} spermatogonia can often be found in regions of the basal compartment adjacent to Leydig cells and blood vessels, one theory proposes that SSCs are positioned away from this area and divide asymmetrically to produce only one daughter cell that is retained in the niche (Shetty and Meistrich 2007; Yoshida et al. 2007). Despite this complexity, current understanding in the field recognizes that the transition to A_1 spermatogonia is the point where the irreversible commitment to meiosis is made and will be the focus of this section (Griswold 2016).

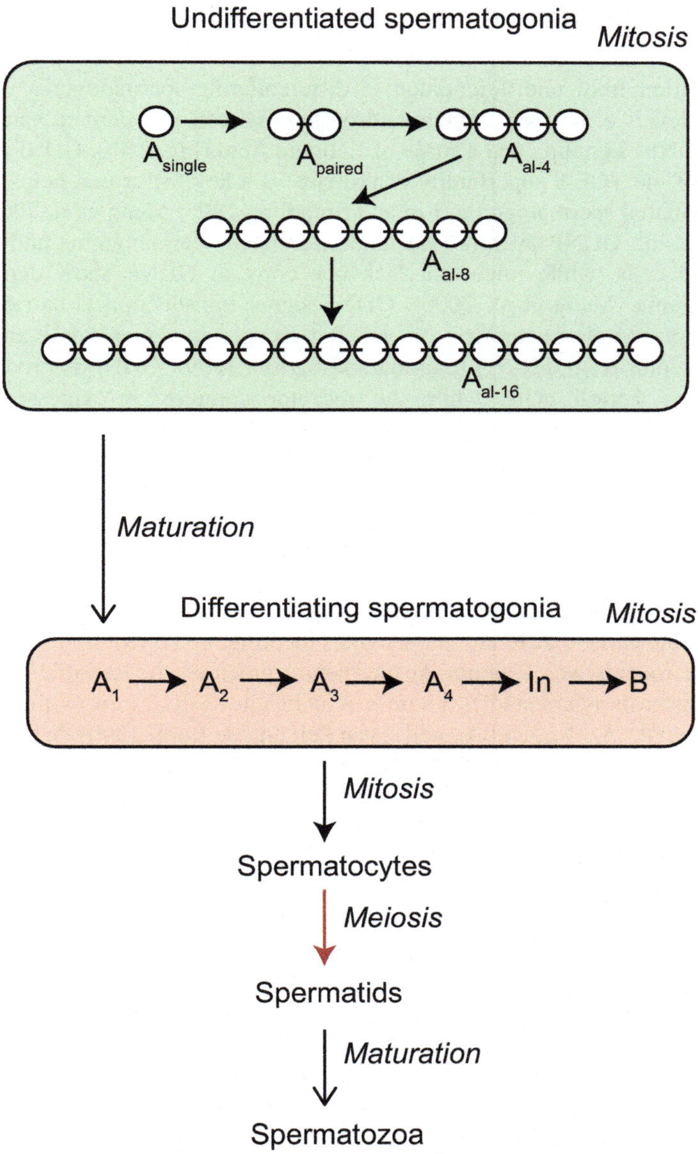

Fig. 2.5 The development of germ cells during spermatogenesis in the mouse testis. Spermatogonia are connected by intercellular bridges, and therefore spermatogonia exist as either syncytial or as A_{single} cells. Differentiating spermatogonia also exist as syncytial cell clusters, although they have not been shown in this figure. See text for details

2.4.3 Proliferation vs. Differentiation Signaling

The transition from undifferentiated to differentiating spermatogonia is orchestrated by levels of two key signals: a drop in *G*lial cell line-*d*erived *n*eurotrophic *f*actor (GDNF) signaling and a surge of Retinoic Acid (Fig. 2.4b). GDNF, a distant member of the TGF-β superfamily of proteins, is a key factor that helps maintain undifferentiated spermatogonia (Lin and Spradling 1993; Meng et al. 2000). Mice overexpressing GDNF overproduce undifferentiated spermatogonia and lack differentiated cells, while mice that lack one copy of *GDNF* show depletion of spermatogonia (Meng et al. 2000). GDNF signal transduction is carried out by the receptor complex comprising of the Ret receptor tyrosine kinase and *G*DNF *f*amily *r*eceptor-α1 (GFR-α1) (Saarma and Sariola 1999). GDNF is produced and secreted by Sertoli cells, while the receptor complex is expressed by the undifferentiated spermatogonia (Trupp et al. 1995; Golden et al. 1999; Meng et al. 2000; Yomogida et al. 2003). New data shows that GDNF production by peritubular myoid cells is also essential for spermatogonial development (Chen et al. 2016). Ret is also expressed by Sertoli cells; therefore, GDNF signaling works through both paracrine and autocrine mechanisms (Caires et al. 2012). GDNF induces the expression of several transcription factors that are important for SSC maintenance (Song and Wilkinson 2014).

One of the earliest differentiation signals identified to be required for spermatogenesis in rodents was Vitamin A, the dietary precursor of Retinoic Acid (RA). Spermatogenesis is arrested if Vitamin A deficiency is induced, with all tubules containing only A_s, A_{pr}, and A_{al} cells (van Pelt and de Rooij 1990). When Vitamin A is restored in the diet, spermatogenesis restarts synchronously (van Pelt and de Rooij 1990). Mice mutant for nuclear receptors of RA, such as RARα and RXRβ, are sterile (Lufkin et al. 1993; Kastner et al. 1996). Loss of another RA receptor, RARγ, results in impairment of the A_{al} to A_1 transition (Gely-Pernot et al. 2012). However, it remains unclear how many of these effects are due to RA's requirement by somatic cells, and how many are due to RA's direct action on germ cells.

Many of the mechanisms downstream of RA action that contribute to A_{al} to A_1 transition remain largely unknown (Busada and Geyer 2016). One of the best-studied downstream targets of RA is the gene Stra8 (STimulated by Retinoic Acid 8) (Oulad-Abdelghani et al. 1996). RA induces the expression of Stra8 in adult testis (Koubova et al. 2006), which is essential for initiation of spermatogenesis in the mouse testis and is required for preleptotene spermatocytes to transition into meiotic prophase (Baltus et al. 2006; Anderson et al. 2008). Furthermore, Stra8 promotes spermatogonial differentiation (Endo et al. 2015). It is expressed in the premeiotic spermatogonial cells of prepubertal and adult testes (Oulad-Abdelghani et al. 1996). However, although Stra8 is essential for initiation of spermatogenesis, it is not strictly required for differentiation of spermatogonia, as genetic ablation of Stra8 does not completely block differentiation, suggesting that RA acts through other targets in addition to Stra8 (Endo et al. 2015). It is found in both the cytoplasm and nucleus and may function as a transcription factor or a transcriptional co-regulator, but determination of its precise function and target genes has remained elusive (Oulad-Abdelghani et al. 1996; Tedesco et al. 2009). Recently, it was found that a periodic RA pulse within the seminiferous tubule triggers differentiation of spermatogonia (Sugimoto et al. 2012; Hogarth et al. 2015; Griswold 2016). The periodic RA-Stra8 signal plays an instructive role, and only some of the undifferentiated spermatogonia may be competent to respond to the signal and differentiate (Endo et al. 2015). This is supported by another study that found that undifferentiated spermatogonia exist as two subpopulations, one that is competent to respond to RA and differentiate and one that is not (Ikami et al. 2015). *D*ouble-sex and *mab-3* *r*elated *t*ranscription factor *1* (Dmrt1) also helps regulate Stra8 expression by restricting RA responsiveness (Matson et al. 2010). Spermatogonial expression of Dmrt1 suppresses precocious entry into meiosis by binding to the Stra8 promoter and blocking activation by the RA receptor transcriptional regulatory complex (Matson et al. 2010). Establishing differential competencies and differential ability to respond to signals are key mechanisms to help maintain hierarchy within the spermatogonial population, since spermatogonia at various stages of differentiation are all contained within the same basal compartment in close proximity.

In addition to RA, KIT, a receptor tyrosine kinase, is a well-established marker for differentiating spermatogonia and is expressed in A_1 to B spermatogonia, preleptotene spermatocytes, and Leydig cells (Manova et al. 1990; Yoshinaga et al. 1991; Schrans-stassen et al. 1999). KIT-mediated phosphatidylinositol 3′-kinase (PI3-kinase) activity is essential for spermatogonia to enter meiosis (Blume-Jensen et al. 2000; Kissel et al. 2000). Sertoli cells express the KIT ligand, KITL, also called the Stem Cell Factor (SCF) (Nakayama et al. 1988; Motro et al. 1991; Marziali et al. 1993). Activation of the KIT receptor in spermatogonia by neighboring KITL expressing Sertoli cells leads to activation of the PI3K pathway in spermatogonia, which is required for entry into meiosis (Blume-Jensen et al. 2000; Kissel et al. 2000). Interestingly, it was recently found that RA translationally activates KIT in neonatal testis, and it is also possible that RA is important for the translation of KIT in differentiating spermatogonia of the adult (Busada et al. 2015). Upon RA stimulation, nuclear receptor RAR/RXR heterodimers are activated and drive expression of the Sall4 transcription factor, which is known to regulate KIT expression (Hobbs et al. 2012; Gely-Pernot et al. 2015).

There are several other factors that help determine the fate of spermatogonia and have been recently reviewed in Griswold (2016). Moreover, there is evidence that both spermatogonia and Sertoli cells have an intrinsic cycle of gene expression (Timmons et al. 2002; Vernet et al. 2006; Yoshida et al. 2006). Surprisingly, and in contrast to the niche's dominance over the type of germ cell that occupies the niche in other systems (see Drosophila), in the mouse testis the spermatogonial cell-intrinsic program is dominant over the Sertoli cell cycle (França et al. 1998). This was perhaps best illustrated by a transplantation experiment in which rat germ cells transplanted to mouse testis always differentiated following the rat 12.9 day cycle of spermatogenesis rather than the mouse 8.6 day cycle (França et al. 1998). In summary, mammalian spermatogenesis involves a complex interplay of cell-intrinsic programs and extrinsic signals that help ensure the continual production of sperm.

2.4.4 Plasticity

In the mouse testis, age-related decline is observed in stem cell and niche function (Ryu et al. 2006; Zhang et al. 2006). At 1 year of age, the total number of SSCs declines, although they still retain their colony-forming abilities (Zhang et al. 2006). Between 1 and 2 years of age, SSC numbers continue to decline and spermatogenesis decreases (Ryu et al. 2006). Moreover, testis of 2-year-old mice cannot support spermatogenesis when transplanted with SSCs from younger mice, indicating that the somatic cell niche environment also declines with age (Zhang et al. 2006). This is supported by results from a reciprocal experiment; SSCs from older mice support spermatogenesis when transplanted to a younger niche (Ryu et al. 2006). During aging, GDNF expression changes in a biphasic mode. In 15–19-month-old aging-fertile animals, GDNF is increased by 60 %, when compared with levels in control 2–4-month-old young-fertile mice. However, in aging-infertile mice, GDNF levels are 73 % of the levels in control mice. The initial surge in GDNF, in response to an aging niche, is thought to be a potentially compensatory response to restore niche activity. However, with increasing age, GDNF expression goes down, likely due to the loss of Sertoli cell function, as they also show reduced expression of GFR-α1 receptor (Ryu et al. 2006).

To summarize, cell-extrinsic signals, such as GDNF from Sertoli cells, are essential for spermatogenesis in the mouse testis. At the same time, since spermatogonia of differing stages of meiosis all share the same physical space, a critical mechanism to help control the balance between proliferation and differentiation is the establishment of differing competencies to respond to extrinsic cues. The tight junctions that divide the Sertoli cell layer into two compartments help achieve further compartmentalization for developing germ cells. Several rounds of mitotic divisions amplify the proliferative capacity of the undifferentiated spermatogonia. The interplay of all of these mechanisms helps ensure that a proper balance between proliferation and differentiation is met.

2.5 Concluding Remarks

There are several common themes and intriguing differences between these different models of germ line stem cell proliferation, which we highlight in this section. Many germ lines seem to share some mechanisms to help maintain the balance between proliferation and differentiation. Distinct compartments (niche and non-niche) are produced by differential cell signaling and negative feedback loops that sharply demarcate the boundary between these two compartments. One main cell-signaling pathway (e.g., Notch, BMP, GDNF) appears to be essential for this balance, while several other pathways help modulate this balance. The key pathway is initiated by the adjacent niche cell and leads to transcriptional activation of genes essential for GSC maintenance in all three systems. Several other pathways feed into the key pathway, or function parallel to it, and exist to help modulate its effect. Physical barriers between these compartments, such as cell junctions in the case of Drosophila and mouse or cell bridges in *C. elegans*, are employed to help separate compartments, while adhesion also plays an important role in preferentially retaining GSCs in the niche. In the mouse, since both undifferentiated and differentiating spermatogonia are housed in the basal compartment, differential cell competencies and responsiveness to extrinsic cues is the cell-intrinsic mechanism by which two distinct populations of germ cells are maintained. Furthermore, niche cells make intimate contact with germ cells to limit their range of influence to a select group of cells. The DTC and cap cells extend processes or cytonemes to GSCs, while escort cells extend processes to differentiating cells, and Sertoli cells send cellular projections to developing germ cells. Another common feature in these systems is that germ cells are often partially syncytial, presumably to enable rapid cellular communication across developmentally similar cohorts of cells. While developing germ cells in *C. elegans* are partially covered by cell membranes, incomplete cell divisions take place in the developing cysts of Drosophila and the A_{pr} to A_{al} spermatogonia in mouse. RNA-binding proteins, such as PUF family proteins, also work similarly in *C. elegans* and Drosophila. Both Pumilio from Drosophila and PUF-8 from *C. elegans* have roles on both sides of this balance, regulating proliferation and differentiation through distinct mechanisms. With increasing age, niche function generally declines and a drop in number of GSCs is observed.

It may seem as though that there are vast differences in the number of GSCs maintained by the niche in these three systems. While Drosophila has only 2–3 GSCs per ovariole, the *C. elegans* hermaphrodite has 60–80 GSCs in each of the two gonad arms. However, the total number of GSCs per individual organism is less stark, when one accounts for the fact that there are 16–23 ovarioles per ovary and two such ovaries in the female fly. In the mouse testis, we find that there are multiple gamete-producing units in the form of many seminiferous tubules. Therefore, it is tempting to speculate that isolating GSCs into multiple niches may be an adaptation in relatively longer living organisms, such as Drosophila and the mouse, to preserve gametogenesis in case of injury/damage. The GSCs may be maintained by symmetric or asymmetric cell divisions, or a combination thereof. We have seen that TA divisions are not always required for success in these models. The female Drosophila and *C. elegans* hermaphrodite do not seem to have a pool of transit-amplifying cells. In the case of *C. elegans*, this may be attributed to its short reproductive life and the presence of a large pool of proliferative cells. The worm does not *need* the TA cells to help amplify GSC divisions or for limiting the effects of acquiring deleterious mutations because it is short-lived and the likelihood of acquiring mutations is likely to be less of a concern. While the Drosophila ovariole also does not have TA divisions, it does maintain a pool of partially differentiated cells that can replenish the niche. On the other hand, mouse spermatogonia undergo several rounds of TA divisions to help sustain continual spermatogenesis over a much longer reproductive life, while protecting germ cells from passing on any acquired mutations.

Understanding the fundamental principles of how the balance between proliferation and differentiation in these models is achieved is vital to furthering our knowledge of how adult stem cell niches are regulated. An imbalance in this decision, in either direction, can dramatically reduce the reproductive fitness of the species. These three models of the proliferation vs. differentiation decision have many similar mechanisms that govern this balance. While the precise signaling pathways that are utilized differ between these organisms, many aspects of their utilization are similar. The last few years have been particularly exciting with new research pointing towards the fact that the germ line niche is a highly dynamic and plastic tissue compartment.

References

Anderson EL, Baltus AE, Roepers-Gajadien HL et al (2008) *Stra8* and its inducer, retinoic acid, regulate meiotic initiation in both spermatogenesis and oogenesis in mice. Proc Natl Acad Sci U S A 105:14976–14980. doi:10.1073/pnas.0807297105

Ariz M, Mainpal R, Subramaniam K (2009) *C. elegans* RNA-binding proteins PUF-8 and MEX-3 function redundantly to promote germline stem cell mitosis. Dev Biol 326:295–304. doi:10.1016/j.ydbio.2008.11.024

Austin J, Kimble J (1987) *glp-1* is required in the germ line for regulation of the decision between mitosis and meiosis in C. elegans. Cell 51:589–599. doi:10.1016/0092-8674(87)90128-0

Baltus AE, Menke DB, Hu Y-C et al (2006) In germ cells of mouse embryonic ovaries, the decision to enter meiosis precedes premeiotic DNA replication. Nat Genet 38:1430–1434. doi:10.1038/ng1919

Berry LW, Westlund B, Schedl T (1997) Germ-line tumor formation caused by activation of *glp-1*, a *Caenorhabditis elegans* member of the *Notch* family of receptors. Development 124:925–936

Blume-Jensen P, Jiang G, Hyman R et al (2000) Kit/stem cell factor receptor-induced activation of phosphatidylinositol 3′-kinase is essential for male fertility. Nat Genet 24:157–162. doi:10.1038/72814

Brawley C, Matunis E (2004) Regeneration of male germline stem cells by spermatogonial dedifferentiation in vivo. Science 304:1331–1334. doi:10.1126/science.1097676

Brenner JL, Schedl T (2016) Germline stem cell differentiation entails regional control of cell fate Regulator GLD-1 in *Caenorhabditis elegans*. Genetics 202:1085–1103. doi:10.1534/genetics.115.185678

Brinster RL (2002) Germline stem cell transplantation and transgenesis. Science 296:2174–2176. doi:10.1126/science.1071607

Bukhari SIA, Vasquez-Rifo A, Gagné D et al (2012) The microRNA pathway controls germ cell proliferation and differentiation in C. elegans. Cell Res 22:1034–1045. doi:10.1038/cr.2012.31

Busada JT, Geyer CB (2016) The role of Retinoic Acid (RA) in spermatogonial differentiation. Biol Reprod 94:1–10. doi:10.1095/biolreprod.115.135145

Busada JT, Chappell VA, Niedenberger BA et al (2015) Retinoic acid regulates *Kit* translation during spermatogonial differentiation in the mouse. Dev Biol 397:140–149. doi:10.1016/j.ydbio.2014.10.020

Byrd DT, Knobel K, Affeldt K et al (2014) A DTC niche plexus surrounds the germline stem cell pool in *Caenorhabditis elegans*. PLoS One. doi:10.1371/journal.pone.0088372

Caires KC, de Avila J, McLean DJ (2012) Endocrine regulation of spermatogonial stem cells in the seminiferous epithelium of adult mice. Biores Open Access 1:222–230. doi:10.1089/biores.2012.0259

Cairns J (1975) Mutation selection and the natural history of cancer. Nature 255:197–200. doi:10.1038/255197a0
Chen D, McKearin D (2003) Dpp signaling silences *bam* transcription directly to establish asymmetric divisions of germline stem cells. Curr Biol 13:1786–1791. doi:10.1016/j.cub.2003.09.033
Chen S, Wang S, Xie T (2011) Restricting self-renewal signals within the stem cell niche: multiple levels of control. Curr Opin Genet Dev 21:684–689. doi:10.1016/j.gde.2011.07.008
Chen S, Lewallen M, Xie T (2013) Adhesion in the stem cell niche: biological roles and regulation. Development 140:255–265. doi:10.1242/dev.083139
Chen L-Y, Brown PR, Willis WB, Eddy EM (2014) Peritubular myoid cells participate in male mouse spermatogonial stem cell maintenance. Endocrinology 155:4964–4974. doi:10.1210/en.2014-1406
Chen L-Y, Willis WD, Eddy EM (2016) Targeting the *Gdnf* Gene in peritubular myoid cells disrupts undifferentiated spermatogonial cell development. Proc Natl Acad Sci 113:201517994. doi:10.1073/pnas.1517994113
Christensen S, Kodoyianni V, Bosenberg M et al (1996) *lag-1*, a gene required for *lin-12* and *glp-1* signaling in *Caenorhabditis elegans*, is homologous to human CBF1 and Drosophila Su(H). Development 122:1373–1383
Cinquin O, Crittenden SL, Morgan DE, Kimble J (2010) Progression from a stem cell-like state to early differentiation in the *C. elegans* germ line. Proc Natl Acad Sci U S A 107:2048–2053. doi:10.1073/pnas.0912704107
Cinquin A, Zheng L, Taylor PH et al (2015) Semi-permeable diffusion barriers enhance patterning robustness in the *C. elegans* germline. Dev Cell 35:405–417. doi:10.1016/j.devcel.2015.10.027
Cox DN, Chao A, Baker J et al (1998) A novel class of evolutionarily conserved genes defined by *piwi* are essential for stem cell self-renewal. Genes Dev 12:3715–3727. doi:10.1101/gad.12.23.3715
Cox DN, Chao A, Lin HF (2000) *piwi* encodes a nucleoplasmic factor whose activity modulates the number and division rate of germline stem cells. Development 127:503–514
Crittenden SL, Troemel ER, Evans TC, Kimble J (1994) GLP-1 is localized to the mitotic region of the *C. elegans* germ line. Development 120:2901–2911
Crittenden SL, Bernstein DS, Bachorik JL et al (2002) A conserved RNA-binding protein controls germline stem cells in *Caenorhabditis elegans*. Nature 417:660–663. doi:10.1038/nature754
Crittenden SL, Leonhard KA, Byrd DT, Kimble J (2006) Cellular analyses of the mitotic region in the *Caenorhabditis elegans* adult germ line. Mol Biol Cell 17:3051–3061. doi:10.1091/mbc.E06-03-0170
Cummings CA, Cronmiller C (1994) The *daughterless* gene functions together with *Notch* and *Delta* in the control of ovarian follicle development in *Drosophila*. Development 120:381–394
Dalfó D, Michaelson D, Hubbard EJA (2012) Sensory regulation of the *C. elegans* germline through TGF-β-dependent signaling in the niche. Curr Biol 22:712–719. doi:10.1016/j.cub.2012.02.064
Datla US, Scovill NC, Brokamp AJ et al (2014) Role of PUF-8/PUF protein in stem cell control, sperm-oocyte decision and cell fate reprogramming. J Cell Physiol 229:1306–1311. doi:10.1002/jcp.24618
de Cuevas M, Spradling AC (1998) Morphogenesis of the *Drosophila* fusome and its implications for oocyte specification. Development 125:2781–2789
de Cuevas M, Lee JK, Spradling AC (1996) α-spectrin is required for germline cell division and differentiation in the *Drosophila* ovary. Development 122:3959–3968
de Cuevas M, Lilly MA, Spradling AC (1997) Germline cyst formation in Drosophila. Annu Rev Genet 31:405–428. doi:10.1146/annurev.genet.31.1.405
de Rooij DG (2001) Proliferation and differentiation of spermatogonial stem cells. Reproduction 121:347–354. doi:10.1530/rep.0.1210347
de Rooij DG, Russell LD (2000) All you wanted to know about spermatogonia but were afraid to ask. J Androl 6512:776–798. doi:10.1002/j.1939-4640.2000.tb03408.x
Decotto E, Spradling AC (2005) The *Drosophila* ovarian and testis stem cell niches: similar somatic stem cells and signals. Dev Cell 9:501–510. doi:10.1016/j.devcel.2005.08.012

Deng W, Lin H (1997) Spectrosomes and fusomes anchor mitotic spindles during asymmetric germ cell divisions and facilitate the formation of a polarized microtubule array for oocyte specification in *Drosophila*. Dev Biol 189:79–94. doi: S0012-1606(97)98669-3 [pii]\n10.1006/dbio.1997.8669

Dernburg AF, McDonald K, Moulder G et al (1998) Meiotic recombination in *C. elegans* initiates by a conserved mechanism and is dispensable for homologous chromosome synapsis. Cell 94:387–398

Doyle TG, Wen C, Greenwald I (2000) SEL-8, a nuclear protein required for LIN-12 and GLP-1 signaling in *Caenorhabditis elegans*. Proc Natl Acad Sci U S A 97:7877–7881

Drummond-Barbosa D, Spradling AC (2001) Stem cells and their progeny respond to nutritional changes during Drosophila oogenesis. Dev Biol 231:265–278. doi:10.1006/dbio.2000.0135

Eckmann CR, Kraemer B, Wickens M, Kimble J (2002) GLD-3, a bicaudal-C homolog that inhibits FBF to control germline sex determination in *C. elegans*. Dev Cell 3:697–710. doi:10.1016/S1534-5807(02)00322-2

Eckmann CR, Crittenden SL, Suh N, Kimble J (2004) GLD-3 and control of the mitosis/meiosis decision in the germline of *Caenorhabditis elegans*. Genetics 168:147–160. doi:10.1534/genetics.104.029264

Endo T, Romer KA, Anderson EL et al (2015) Periodic retinoic acid—STRA8 signaling intersects with periodic germ-cell competencies to regulate spermatogenesis. Proc Natl Acad Sci 112: E2347–E2356. doi:10.1073/pnas.1505683112

Fitzgerald K, Greenwald I (1995) Interchangeability of *Caenorhabditis elegans* DSL proteins and intrinsic signalling activity of their extracellular domains in vivo. Development 121:4275–4282

Forbes A, Lehmann R (1998) Nanos and Pumilio have critical roles in the development and function of *Drosophila* germline stem cells. Development 125:679–690

Fox PM, Schedl T (2015) Analysis of germline stem cell differentiation following loss of GLP-1 Notch activity in *Caenorhabditis elegans*. Genetics 201:167–184. doi:10.1534/genetics.115.178061

Fox PM, Vought VE, Hanazawa M et al (2011) Cyclin E and CDK-2 regulate proliferative cell fate and cell cycle progression in the *C. elegans* germline. Development 138:2223–2234. doi:10.1242/dev.059535

França LR, Ogawa T, Avarbock MR et al (1998) Germ cell genotype controls cell cycle during spermatogenesis in the rat. Biol Reprod 59:1371–1377. doi:10.1095/biolreprod59.6.1371

Francis R, Maine E, Schedl T (1995) Analysis of the multiple roles of *gld-1* in germline development: interactions with the sex determination cascade and the *glp-1* signaling pathway. Genetics 139(2):607–630

Gancz D, Lengil T, Gilboa L (2011) Coordinated regulation of niche and stem cell precursors by hormonal signaling. PLoS Biol 9(11):e1001202. doi:10.1371/journal.pbio.1001202

Gely-Pernot A, Raverdeau M, Célébi C et al (2012) Spermatogonia differentiation requires retinoic acid receptor γ. Endocrinology 153:438–449. doi:10.1210/en.2011-1102

Gely-Pernot A, Raverdeau M, Teletin M et al (2015) Retinoic acid receptors control spermatogonia cell-fate and induce expression of the SALL4A transcription factor. PLoS Genet 11:1–23. doi:10.1371/journal.pgen.1005501

Gilboa L, Lehmann R (2004) Repression of primordial germ cell differentiation parallels germ line stem cell maintenance. Curr Biol 14:981–986. doi:10.1016/j

Godt D, Laski FA (1995) Mechanisms of cell rearrangement and cell recruitment in *Drosophila* ovary morphogenesis and the requirement of *bric à brac*. Development 121:173–187

Golden JP, DeMaro JA, Osborne PA et al (1999) Expression of neurturin, GDNF, and GDNF family-receptor mRNA in the developing and mature mouse. Exp Neurol 158:504–528. doi:10.1006/exnr.1999.7127

Gonczy P (2008) Mechanisms of asymmetric cell division: flies and worms pave the way. Nat Rev 9:355–366

Gonen O, Toledano H (2014) Why adult stem cell functionality declines with age? Studies from the fruit fly *Drosophila melanogaster* model organism. Curr Genomics 15:231–236. doi:10.2174/1389202915666140421213243

Griswold MD (2016) Spermatogenesis: the commitment to meiosis. Physiol Rev 96:1–17. doi:10.1152/physrev.00013.2015

Gupta P, Leahul L, Wang X et al (2015) Proteasome regulation of the chromodomain protein MRG-1 controls the balance between proliferative fate and differentiation in the *C. elegans* germ line. Development 142:291–302. doi:10.1242/dev.115147

Haase AD, Fenoglio S, Muerdter F et al (2010) Probing the initiation and effector phases of the somatic piRNA pathway in *Drosophila*. Genes Dev 24:2499–2504. doi:10.1101/gad.1968110

Hall DH, Winfrey VP, Blaeuer G et al (1999) Ultrastructural features of the adult hermaphrodite gonad of *Caenorhabditis elegans*: relations between the germ line and soma. Dev Biol 212:101–123

Hansen D, Schedl T (2013) Stem cell proliferation versus meiotic fate decision in *Caenorhabditis elegans*. In: Schedl T (ed) Germ cell development in *C. elegans*. Springer, New York, pp 71–99

Hansen D, Hubbard EJA, Schedl T (2004a) Multi-pathway control of the proliferation versus meiotic development decision in the *Caenorhabditis elegans* germline. Dev Biol 268:342–357. doi:10.1016/j.ydbio.2003.12.023

Hansen D, Wilson-berry L, Dang T, Schedl T (2004b) Control of the proliferation versus meiotic development decision in the *C. elegans* germline through regulation of GLD-1 protein accumulation. Development 131:93–104. doi:10.1242/dev.00916

Harris RE, Pargett M, Sutcliffe C et al (2011) Brat promotes stem cell differentiation via control of a bistable switch that restricts BMP signaling. Dev Cell 20:72–83. doi:10.1016/j.devcel.2010.11.019

Hartman TR, Strochlic TI, Ji Y et al (2013) Diet controls *Drosophila* follicle stem cell proliferation via Hedgehog sequestration and release. J Cell Biol 201:741–757. doi:10.1083/jcb.201212094

Henderson ST, Gao D, Lambie EJ, Kimble J (1994) *lag-2* may encode a signaling ligand for the GLP-1 and LIN-12 receptors of *C. elegans*. Development 120:2913–2924

Hirsh D, Oppenheim D, Klass M (1976) Development of the reproductive system of *Caenorhabditis elegans*. Dev Biol 49:200–219

Hobbs RM, Fagoonee S, Papa A et al (2012) Functional antagonism between Sall4 and Plzf defines germline progenitors. Cell Stem Cell 10:284–298. doi:10.1016/j.stem.2012.02.004

Hogarth CA, Arnold S, Kent T et al (2015) Processive pulses of retinoic acid propel asynchronous and continuous murine sperm production. Biol Reprod 92:37. doi:10.1095/biolreprod.114.126326

Hsu H-J, LaFever L, Drummond-Barbosa D (2008) Diet controls normal and tumorous germline stem cells via insulin-dependent and -independent mechanisms in *Drosophila*. Dev Biol 313:700–712

Hsu Y-C, Pasolli HA, Fuchs E (2011) Dynamics between stem cells, niche and progeny in the hair follicle. Cell 144:92–105. doi:10.1016/j.cell.2010.11.049.Dynamics

Hubbard EJA, Korta DZ, Dalfó D (2013) Physiological contol of germline development. In: Schedl T (ed) Germ cell development in *C. elegans*, Advances in Experimental Medicine and Biology. Springer, New York, pp 1–16

Huynh J-R, Johnston DS (2004) The origin of asymmetry: early polarisation of the *Drosophila* germline cyst and oocyte. Curr Biol 14:438–449. doi:10.1016/j.cub.2004.05.040

Ikami K, Tokue M, Sugimoto R et al (2015) Hierarchical differentiation competence in response to retinoic acid ensures stem cell maintenance during mouse spermatogenesis. Development 142:1582–1592. doi:10.1242/dev.118695

Inaba M, Yamashita YM (2012) Asymmetric stem cell division: precision for robustness. Cell Stem Cell 11:461–469. doi:10.1016/j.stem.2012.09.003

Jan E, Motzny CK, Graves LE, Goodwin EB (1999) The STAR protein, GLD-1, is a translational regulator of sexual identity in *Caenorhabditis elegans*. EMBO J 18:258–269. doi:10.1093/emboj/18.1.258

Januschke J, Näthke I (2014) Stem cell decisions: a twist of fate or a niche market? Semin Cell Dev Biol 34:116–123. doi:10.1016/j.semcdb.2014.02.014

Jin Z, Kirilly D, Weng C et al (2008) Differentiation-defective stem cells outcompete normal stem cells for niche occupancy in the *Drosophila* ovary. Cell Stem Cell 2:39–49. doi:10.1016/j.stem.2007.10.021

Joly W, Chartier A, Rojas-Rios P et al (2013) The CCR4 deadenylase acts with Nanos and Pumilio in the fine-tuning of Mei-P26 expression to promote germline stem cell self-renewal. Stem Cell Rep 1:411–424. doi:10.1016/j.stemcr.2013.09.007

Jones DL, Rando TA (2011) Emerging models and paradigms for stem cell ageing. Nat Cell Biol 13:506–512. doi:10.1038/ncb0511-506

Jones AR, Schedl T (1995) Mutations in *gld-1*, a female germ cell-specific tumor suppressor gene in *Caenorhabditis elegans*, affect a conserved domain also found in Src-associated protein Sam68. Genes Dev 9:1491–1504

Jones AR, Francis R, Schedl T (1996) GLD-1, a cytoplasmic protein essential for oocyte differentiation, shows stage- and sex-specific expression during *Caenorhabditis elegans* germline development. Dev Biol 180:165–183. doi:10.1006/dbio.1996.0293

Kadyk LC, Kimble J (1998) Genetic regulation of entry into meiosis in *Caenorhabditis elegans*. Development 125:1803–1813

Kai T, Spradling A (2003) An empty *Drosophila* stem cell niche reactivates the proliferation of ectopic cells. Proc Natl Acad Sci U S A 100:4633–4638. doi:10.1073/pnas.0830856100

Kai T, Spradling A (2004) Differentiating germ cells can revert into functional stem cells in *Drosophila melanogaster* ovaries. Nature 428:564–567. doi:10.1038/nature02463.1

Kastner P, Mark M, Leid M et al (1996) Abnormal spermatogenesis in RXRβ mutant mice. Genes Dev 10:80–92. doi:10.1101/gad.10.1.80

Kerins JA, Hanazawa M, Dorsett M, Schedl T (2010) PRP-17 and the pre-mRNA splicing pathway are preferentially required for the proliferation versus meiotic development decision and germline sex determination in *Caenorhabditis elegans*. Dev Dyn 239:1555–1572. doi:10.1002/dvdy.22274

Kershner A, Crittenden SL, Friend K et al (2013) Germline stem cells and their regulation in the nematode *Caenorhabditis elegans*. In: Hime G, Abud H (eds) Transcriptional and translational regulation of stem cells. Springer, Dordrecht, pp 307–328

Kershner AM, Shin H, Hansen TJ, Kimble J (2014) Discovery of two GLP-1/Notch target genes that account for the role of GLP-1/Notch signaling in stem cell maintenance. Proc Natl Acad Sci U S A 111:3739–3744. doi:10.1073/pnas.1401861111

Kiger AA, Jones DL, Schulz C et al (2001) Stem cell self-renewal specified by JAK-STAT activation in response to a support cell cue. Science 294:2542–2545. doi:10.1126/science.1066707

Kim JY, Lee YC, Kim C (2010a) Direct inhibition of Pumilo activity by Bam and Bgcn in *Drosophila* germ line stem cell differentiation. J Biol Chem 285:4741–4746. doi:10.1074/jbc.M109.002014

Kim KW, Wilson TL, Kimble J (2010b) GLD-2/RNP-8 cytoplasmic poly(A) polymerase is a broad-spectrum regulator of the oogenesis program. Proc Natl Acad Sci U S A 107:17445–17450. doi:10.1073/pnas.1012611107

Kimble J, Hirsh D (1979) The postembryonic cell lineages of the hermaphrodite and male gonads in *Caenorhabditis elegans*. Dev Biol 70:396–417

Kimble JE, White JG (1981) On the control of germ cell development in *Caenorhabditis elegans*. Dev Biol 81:208–219

King RC (1970) Ovarian development in *Drosophila melanogaster*. Academic, New York

King FJ, Lin H (1999) Somatic signaling mediated by *fs(1)Yb* is essential for germline stem cell maintenance during *Drosophila* oogenesis. Development 126:1833–1844

King RC, Aggarwal SK, Aggarwal U (1968) The development of the female *Drosophila* reproductive system. J Morphol 124:143–166. doi:10.1002/jmor.1051240203

King FJ, Szakmary A, Cox DN, Lin H (2001) *Yb* modulates the divisions of both germline and somatic stem cells through *piwi-* and *hh*-mediated mechanisms in the *Drosophila* ovary. Mol Cell 7:497–508. doi:10.1016/S1097-2765(01)00197-6

Kissel H, Timokhina I, Hardy MP et al (2000) Point mutation in Kit receptor tyrosine kinase reveals essential roles for Kit signaling in spermatogenesis and oogenesis without affecting other Kit responses. EMBO J 19:1312–1326. doi:10.1093/emboj/19.6.1312

Klein AM, Simons BD (2011) Universal patterns of stem cell fate in cycling adult tissues. Development 138:3103–3111. doi:10.1242/dev.060103

Klein AM, Nakagawa T, Ichikawa R et al (2010) Mouse germ line stem cells undergo rapid and stochastic turnover. Cell Stem Cell 7:214–224. doi:10.1016/j.stem.2010.05.017

Klenov MS, Sokolova OA, Yakushev EY et al (2011) Separation of stem cell maintenance and transposon silencing functions of Piwi protein. Proc Natl Acad Sci U S A 108:18760–18765. doi:10.1073/pnas.1106676108

Koubova J, Menke DB, Zhou Q et al (2006) Retinoic acid regulates sex-specific timing of meiotic initiation in mice. Proc Natl Acad Sci U S A 103:2474–2479. doi:10.1073/pnas.0510813103

Kraemer B, Crittenden S, Gallegos M et al (1999) NANOS-3 and FBF proteins physically interact to control the sperm-oocyte switch in *Caenorhabditis elegans*. Curr Biol 9:1009–1018. doi:10.1016/S0960-9822(99)80449-7

Kubota H, Avarbock MR, Brinster RL (2004) Growth factors essential for self-renewal and expansion of mouse spermatogonial stem cells. Proc Natl Acad Sci U S A 101:16489–16494. doi:10.1073/pnas.0407063101

LaFever L, Drummond-Barbosa D (2005) Direct control of germline stem cell division and cyst growth by neural insulin in *Drosophila*. Science 309:1071–1073. doi: 10.1126/science.1111410

Lamont LB, Crittenden SL, Bernstein D et al (2004) FBF-1 and FBF-2 regulate the size of the mitotic region in the *C. elegans* germline. Dev Cell 7:697–707. doi:10.1016/j.devcel.2004.09.013

Lavoie CA, Ohlstein B, McKearin DM (1999) Localization and function of Bam protein require the *benign gonial cell neoplasm* gene product. Dev Biol 212:405–413. doi:10.1006/dbio.1999.9346

Lin H, Spradling AC (1993) Germline stem cell division and egg chamber development in transplanted *Drosophila* germaria. Dev Biol 159:140–152

Lin H, Spradling A (1997) A novel group of *pumilio* mutations affects the asymmetric division of germline stem cells in the *Drosophila* ovary. Development 124:2463–2476

Lin H, Yue L, Spradling AC (1994) The *Drosophila* fusome, a germline-specific organelle, contains membrane skeletal proteins and functions in cyst formation. Development 120:947–956

Liu M, Lim TM, Cai Y (2010) The *Drosophila* female germline stem cell lineage acts to spatially restrict DPP function within the niche. Sci Signal 3:ra57. doi:10.1126/scisignal.2000740

Liu Z, Zhong G, Chai PC et al (2015) Coordinated niche-associated signals promote germline homeostasis in the *Drosophila* ovary. J Cell Biol 211:469–484. doi:10.1083/jcb.201503033

López-Onieva L, Fernández-Miñán A, González-Reyes A (2008) Jak/Stat signalling in niche support cells regulates *dpp* transcription to control germline stem cell maintenance in the *Drosophila* ovary. Development 135:533–540. doi:10.1242/dev.016121

Lufkin T, Lohnes D, Mark M et al (1993) High postnatal lethality and testis degeneration in retinoic acid receptor α mutant mice. Proc Natl Acad Sci U S A 90:7225–7229. doi:10.1073/pnas.90.15.7225

MacDonald LD, Knox A, Hansen D (2008) Proteasomal regulation of the proliferation *vs.* meiotic entry decision in the *Caenorhabditis elegans* germ line. Genetics 180:905–920. doi:10.1534/genetics.108.091553

Maciejowski J, Ugel N, Mishra B et al (2006) Quantitative analysis of germline mitosis in adult *C. elegans*. Dev Biol 292:142–151. doi:10.1016/j.ydbio.2005.12.046

MacQueen AJ, Villeneuve AM (2001) Nuclear reorganization and homologous chromosome pairing during meiotic prophase require *C. elegans chk-2*. Genes Dev 15:1674–1687. doi:10.1101/gad.902601

Mahowald AP, Kambysellis MP (1980) Oogenesis. In: Ashburner M, Wright TRF (eds) Genetics and biology of Drosophila. Academic Press, London, pp 141–224

Manova K, Nocka K, Besmer P, Bachvarova RF (1990) Gonadal expression of c-*kit* encoded at the *W* locus of the mouse. Development 110:1057–1069

Mantina P, MacDonald L, Kulaga A et al (2009) A mutation in *teg-4*, which encodes a protein homologous to the SAP130 pre-mRNA splicing factor, disrupts the balance between proliferation and differentiation in the *C. elegans* germ line. Mech Dev 126:417–429. doi:10.1016/j.mod.2009.01.006

Margolis J, Spradling A (1995) Identification and behavior of epithelial stem cells in the *Drosophila* ovary. Development 121:3797–3807

Marin VA, Evans TC (2003) Translational repression of a *C. elegans* Notch mRNA by the STAR/KH domain protein GLD-1. Development 130:2623–2632. doi:10.1242/dev.00486

Marziali G, Lazzaro D, Sorrentino V (1993) Binding of germ cells to mutant Sl^d Sertoli cells is defective and is rescued by expression of the transmembrane form the c-*kit* ligand. Dev Biol 157:182–190

Matson CK, Murphy MW, Griswold MD et al (2010) The mammalian doublesex homolog DMRT1 is a transcriptional gatekeeper that controls the mitosis versus meiosis decision in male germ cells. Dev Cell 19:612–624. doi:10.1016/j.devcel.2010.09.010

McCarter J, Bartlett B, Dang T, Schedl T (1997) Soma–germ cell interactions in *Caenorhabditis elegans*: multiple events of hermaphrodite germline development require the somatic sheath and spermathecal lineages. Dev Biol 181:121–143. doi:10.1006/dbio.1996.8429

McKearin D, Ohlstein B (1995) A role for the *Drosophila* Bag-of-marbles protein in the differentiation of cystoblasts from germline stem cells. Development 121:2937–2947

McKearin DM, Spradling AC (1990) *bag-of-marbles*: a *Drosophila* gene required to initiate both male and female gametogenesis. Genes Dev 4:2242–2251. doi:10.1101/gad.4.12b.2242

Meng X, Lindahl M, Hyvonen ME et al (2000) Regulation of cell fate decision of undifferentiated spermatogonia by GDNF. Science 287:1489–1493. doi:10.1126/science.287.5457.1489

Millonigg S, Minasaki R, Nousch M, Eckmann CR (2014) GLD-4-mediated translational activation regulates the size of the proliferative germ cell pool in the adult *C. elegans* germ line. PLoS Genet 10(9):e1004647. doi:10.1371/journal.pgen.1004647

Morris LX, Spradling AC (2011) Long-term live imaging provides new insight into stem cell regulation and germline-soma coordination in the *Drosophila* ovary. Development 138:2207–2215. doi:10.1242/dev.065508

Morrison SJ, Kimble J (2006) Asymmetric and symmetric stem-cell divisions in development and cancer. Nature 441:1068–1074. doi:10.1038/nature04956

Motro B, van der Kooy D, Rossant J et al (1991) Contiguous patterns of c-*kit* and *steel* expression: analysis of mutations at the *W* and *Sl* loci. Development 113:1207–1221

Nadarajan S, Govindan JA, McGovern M et al (2009) MSP and GLP-1/Notch signaling coordinately regulate actomyosin-dependent cytoplasmic streaming and oocyte growth in C. elegans. Development 136:2223–2234. doi:10.1242/dev.034603

Nakagawa T, Nabeshima Y, Yoshida S (2007) Functional Identification of the actual and potential stem cell compartments in mouse spermatogenesis. Dev Cell 12:195–206. doi:10.1016/j.devcel.2007.01.002

Nakagawa T, Sharma M, Nabeshima Y et al (2010) Functional heirarchy and reversibility within the murine spermatogenic stem cell compartment. Science 328:62–68

Nakayama H, Kuroda H, Onoue H et al (1988) Studies of Sl/Sld in equilibrium with +/+ mouse aggregation chimaeras. II. Effect of the steel locus on spermatogenesis. Development 102:117–126

Newton FG, Harris RE, Sutcliffe C, Ashe HL (2015) Coordinate post-transcriptional repression of Dpp-dependent transcription factors attenuates signal range during development. Development 142:3362–3373. doi:10.1242/dev.123273

Oakberg EF (1971) Spermatogonial stem-cell renewal in the mouse. Anat Rec 169:515–531. doi:10.1002/ar.1091690305

Oatley JM, Brinster RL (2012) The germline stem cell niche unit in mammalian testes. Physiol Rev 92:577–595. doi:10.1152/physrev.00025.2011

Oatley JM, Oatley MJ, Avarbock MR et al (2009) Colony stimulating factor 1 is an extrinsic stimulator of mouse spermatogonial stem cell self-renewal. Development 136:1191–1199. doi:10.1242/dev.032243

Oatley MJ, Racicot KE, Oatley JM (2011) Sertoli cells dictate spermatogonial stem cell niches in the mouse testis. Biol Reprod 84:639–645. doi:10.1095/biolreprod.110.087320

Ohlstein B, McKearin D (1997) Ectopic expression of the *Drosophila* Bam protein eliminates oogenic germline stem cells. Development 124:3651–3662

Olivieri D, Sykora MM, Sachidanandam R et al (2010) An *in vivo* RNAi assay identifies major genetic and cellular requirements for primary piRNA biogenesis in *Drosophila*. EMBO J 29:3301–3317. doi:10.1038/emboj.2010.212

Oulad-Abdelghani M, Bouillet P, Décimo D et al (1996) Characterization of a premeiotic germ cell-specific cytoplasmic protein encoded by Stra8, a novel retinoic acid-responsive gene. J Cell Biol 135:469–477. doi:10.1083/jcb.135.2.469

Pan L, Chen S, Weng C et al (2007) Stem cell aging is controlled both intrinsically and extrinsically in the *Drosophila* ovary. Cell Stem Cell 1:458–469. doi:10.1016/j.stem.2007.09.010

Petcherski AG, Kimble J (2000) LAG-3 is a putative transcriptional activator in the *C. elegans* Notch pathway. Nature 405:364–368. doi:10.1038/35012645

Potten CS, Loeffler M (1990) Stem cells: attributes, cycles, spirals, pitfalls and uncertainties. Lessons for and from the crypt. Development 110:1001–1020

Qi H, Watanabe T, Ku HY et al (2011) The Yb body, a major site for Piwi-associated RNA biogenesis and a gateway for Piwi expression and transport to the nucleus in somatic cells. J Biol Chem 286:3789–3797. doi:10.1074/jbc.M110.193888

Racher H, Hansen D (2012) PUF-8, a Pumilio homolog, inhibits the proliferative fate in the *Caenorhabditis elegans* germline. G3 (Bethesda) 2:1197–1205. doi:10.1534/g3.112.003350

Roehl H, Bosenberg M, Blelloch R, Kimble J (1996) Roles of the RAM and ANK domains in signaling by the *C. elegans* GLP-1 receptor. EMBO J 15:7002–7012

Rojas-Ríos P, Guerrero I, González-Reyes A (2012) Cytoneme-mediated delivery of hedgehog regulates the expression of bone morphogenetic proteins to maintain germline stem cells in *Drosophila*. PLoS Biol 10(4):e1001298. doi:10.1371/journal.pbio.1001298

Ruohola H, Bremer KA, Baker D et al (1991) Role of neurogenic genes in establishment of follicle cell fate and oocyte polarity during oogenesis in Drosophila. Cell 66:433–449. doi:10.1016/0092-8674(81)90008-8

Ryu B-Y, Orwig KE, Oatley JM et al (2006) Effects of aging and niche microenvironment on spermatogonial stem cell self-renewal. Stem Cells 24:1505–1511. doi:10.1634/stemcells.2005-0580

Saarma M, Sariola H (1999) Other neurotrophic factors: glial cell line-derived neurotrophic factor (GDNF). Microsc Res Tech 45:292–302. doi:10.1002/(SICI)1097-0029(19990515/01)45:4/5<292::AID-JEMT13>3.0.CO;2-8

Saito K, Ishizu H, Komai M et al (2010) Roles for the Yb body components Armitage and Yb in primary piRNA biogenesis in *Drosophila*. Genes Dev 24:2493–2498. doi:10.1101/gad.1989510

Sarikaya DP, Belay AA, Ahuja A et al (2012) The roles of cell size and cell number in determining ovariole number in *Drosophila*. Dev Biol 363:279–289. doi:10.1016/j.ydbio.2011.12.017

Scadden DT (2014) Nice neighborhood: emerging concepts of the stem cell niche. Cell 157:41–50. doi:10.1016/j.cell.2014.02.013

Schofield R (1978) The relationship between the spleen colony-forming cell and the haemopoietic stem cell. Blood Cells 4:7–25

Schrans-stassen BHGJ, van de Kant HJG, de Rooij DG, van Pelt AMM (1999) Differential expression of c-*kit* in mouse undifferentiated and differentiating type A spermatogonia. Endocrinology 140:5894–5900. doi:10.1210/endo.140.12.7172

Seidel HS, Kimble J (2015) Cell-cycle quiescence maintains *Caenorhabditis elegans* germline stem cells independent of GLP-1/Notch. Elife 4:1–28. doi:10.7554/eLife.10832

Sheng XR, Matunis E (2011) Live imaging of the *Drosophila* spermatogonial stem cell niche reveals novel mechanisms regulating germline stem cell output. Dev Stem Cells 138:3367–3376. doi:10.1242/dev.065797

Shetty G, Meistrich ML (2007) The missing niche for spermatogonial stem cells: do blood vessels point the way? Cell Stem Cell 1:361–363. doi:10.1016/j.stem.2007.09.013

Shim J, Gururaja-Rao S, Banerjee U (2013) Nutritional regulation of stem and progenitor cells in *Drosophila*. Development 140:4647–4656. doi:10.1242/dev.079087

Song HW, Wilkinson MF (2014) Transcriptional control of spermatogonial maintenance and differentiation. Semin Cell Dev Biol 30:14–26. doi:10.1016/j.semcdb.2014.02.005

Song X, Zhu C-H, Doan C, Xie T (2002) Germline stem cells anchored by adherens junctions in the *Drosophila* ovary niches. Science 296:1855–1857. doi:10.1126/science.1069871

Song X, Wong MD, Kawase E et al (2004) Bmp signals from niche cells directly repress transcription of a differentiation-promoting gene, *bag of marbles*, in germline stem cells in the *Drosophila* ovary. Development 131:1353–1364. doi:10.1242/dev.01026

Sonoda J, Wharton RP (2001) *Drosophila* brain tumor is a translational repressor. Genes Dev 15:762–773. doi:10.1101/gad.870801

Spradling A (1993) Developmental genetics of oogenesis. In: Bate M (ed) The development of Drosophila melanogaster. Cold Spring Harbor Laboratory Press, Cold Spring Harbor, NY

Starich TA, Hall DH, Greenstein D (2014) Two classes of gap junction channels mediate soma-germline interactions essential for germline proliferation and gametogenesis in *Caenorhabditis elegans*. Genetics 198:1127–1153. doi:10.1534/genetics.114.168815

Sugimoto R, Nabeshima Y i, Yoshida S (2012) Retinoic acid metabolism links the periodical differentiation of germ cells with the cycle of Sertoli cells in mouse seminiferous epithelium. Mech Dev 128:610–624. doi:10.1016/j.mod.2011.12.003

Suh N, Jedamzik B, Eckmann CR et al (2006) The GLD-2 poly(A) polymerase activates *gld-1* mRNA in the Caenorhabditis elegans germ line. Proc Natl Acad Sci U S A 103:15108–15112. doi:10.1073/pnas.0607050103

Suh N, Crittenden SL, Goldstrohm A et al (2009) FBF and its dual control of *gld-1* expression in the *Caenorhabditis elegans* germline. Genetics 181:1249–1260. doi:10.1534/genetics.108.099440

Szakmary A, Cox DN, Wang Z, Lin H (2005) Regulatory relationship among *piwi*, *pumilio*, and *bag-of-marbles* in *Drosophila* germline stem cell self-renewal and differentiation. Curr Biol 128:171–178. doi:10.1016/j

Szakmary A, Reedy M, Qi H, Lin H (2009) The Yb protein defines a novel organelle and regulates male germline stem cell self-renewal in *Drosophila melanogaster*. J Cell Biol 185:613–627. doi:10.1083/jcb.200903034

Tagelenbosch RAJ, de Rooij DG (1993) A quantitative study of spermatogonial multiplication and stem cell renewal in the C3H/101 F_1 hybrid mouse. Mutat Res Fundam Mol Mech Mutagen 290:193–200. doi:10.1016/0027-5107(93)90159-D

Tax FE, Yeargers JJ, Thomas JH (1994) Sequence of *C. elegans lag-2* reveals a cell-signalling domain shared with *Delta* and *Serrate* of *Drosophila*. Nature 368:150–153

Tedesco M, La Sala G, Barbagallo F et al (2009) STRA8 shuttles between nucleus and cytoplasm and displays transcriptional activity. J Biol Chem 284:35781–35793. doi:10.1074/jbc.M109.056481

Telfer WH (1975) Development and physiology of the oöcyte-nurse cell syncytium. Adv Insect Phys 11:223–319. doi:10.1016/S0065-2806(08)60164-2

Timmons PM, Rigby PWJ, Poirier F (2002) The murine seminiferous epithelial cycle is pre-figured in the Sertoli cells of the embryonic testis. Development 129:635–647

Trupp M, Rydén M, Jörnvall H et al (1995) Peripheral expression and biological activities of GDNF, a new neurotrophic factor for avian and mammalian peripheral neurons. J Cell Biol 130:137–148. doi:10.1083/jcb.130.1.137

Tulina N, Matunis E (2001) Control of stem cell self-renewal in *Drosophila* spermatogenesis by JAK-STAT signaling. Science 294:2546–2549. doi:10.1126/science.1066700

Vaid S, Ariz M, Chaturbedi A et al (2013) PUF-8 negatively regulates RAS/MAPK signalling to promote differentiation of *C. elegans* germ cells. Development 140:1645–1654. doi:10.1242/dev.088013

van Pelt AMM, de Rooij DG (1990) Synchronization of the seminiferous epithelium after vitamin A replacement in vitamin A-deficient mice. Biol Reprod 3:363–367

Vernet N, Dennefeld C, Guillou F et al (2006) Prepubertal testis development relies on retinoic acid but not rexinoid receptors in Sertoli cells. EMBO J 25:5816–5825. doi:10.1038/sj.emboj.7601447

Wang Z, Lin H (2004) *Nanos* maintains germline stem cell self-renewal by preventing differentiation. Science 303:2016–2019. doi:10.1126/science.1093983

Wang L, Eckmann CR, Kadyk LC et al (2002) A regulatory cytoplasmic poly(A) polymerase in *Caenorhabditis elegans*. Nature 419:312–316. doi:10.1038/nature01039

Wang L, Li Z, Cai Y (2008) The JAK/STAT pathway positively regulates DPP signaling in the *Drosophila* germline stem cell niche. J Cell Biol 180:721–728. doi:10.1083/jcb.200711022

Wang C, Wilson-Berry L, Schedl T, Hansen D (2012) TEG-1 CD2BP2 regulates stem cell proliferation and sex determination in the *C. elegans* germ line and physically interacts with the UAF-1 U2AF65 splicing factor. Dev Dyn 241:505–521. doi:10.1002/dvdy.23735

Wang X, Gupta P, Fairbanks J, Hansen D (2014) Protein kinase CK2 both promotes robust proliferation and inhibits the proliferative fate in the *C. elegans* germ line. Dev Biol 392:26–41. doi:10.1016/j.ydbio.2014.05.002

Watt FM, Hogan BL (2000) Out of Eden: stem cells and their niches. Science 287:1427–1430. doi:10.1126/science.287.5457.1427

Welsh M, Saunders PTK, Atanassova N et al (2009) Androgen action *via* testicular peritubular myoid cells is essential for male fertility. FASEB J 23:4218–4230. doi:10.1096/fj.09-138347

Wharton KA, Thomsen GH, Gelbart WM (1991) *Drosophila* 60A gene, another transforming growth factor β family member, is closely related to human bone morphogenetic proteins. Proc Natl Acad Sci U S A 88:9214–9218

Wieschaus E, Szabad J (1979) The development and function of the female germ line in *Drosophila melanogaster*: a cell lineage study. Dev Biol 68:29–46. doi:10.1016/0012-1606(79)90241-0

Xie T, Spradling AC (1998) *decapentaplegic* is essential for the maintenance and division of germline stem cells in the *Drosophila* ovary. Cell 94:251–260. doi:10.1016/S0092-8674(00)81424-5

Xie T, Spradling AC (2000) A niche maintaining germ line stem cells in the *Drosophila* ovary. Science 290:328–330. doi:10.1126/science.290.5490.328

Yadlapalli S, Yamashita YM (2012) Spindle positioning in the stem cell niche. WIREs Dev Biol 1:215–230. doi:10.1002/wdev.16

Yomogida K, Yagura Y, Tadokoro Y, Nishimune Y (2003) Dramatic expansion of germinal stem cells by ectopically expressed human glial cell line-derived neurotrophic factor in mouse Sertoli cells. Biol Reprod 1307:1303–1307. doi:10.1095/biolreprod.103.015958

Yoshida S (2012) Elucidating the identity and behavior of spermatogenic stem cells in the mouse testis. Reproduction 144:293–302. doi:10.1530/REP-11-0320

Yoshida S, Sukeno M, Nakagawa T et al (2006) The first round of mouse spermatogenesis is a distinctive program that lacks the self-renewing spermatogonia stage. Development 133:1495–1505. doi:10.1242/dev.02316

Yoshida S, Sukeno M, Nabeshima Y-I (2007) A vasculature-associated niche for undifferentiated spermatogonia in the mouse testis. Science 317:1722–1726. doi:10.1126/science.1144885

Yoshinaga K, Nishikawa S, Ogawa M et al (1991) Role of c-*kit* in mouse spermatogenesis: identification of spermatogonia as a specific site of c-*kit* expression and function. Development 113:689–699

Yue L, Spradling AC (1992) *hu-li tai shao*, a gene required for ring canal formation during *Drosophila* oogenesis, encodes a homolog of adducin. Genes Dev 6:2443–2454. doi:10.1101/gad.6.12b.2443

Zhang B, Gallegos M, Puoti A et al (1997) A conserved RNA-binding protein that regulates sexual fates in the *C. elegans* hermaphrodite germ line. Nature 390:477–484. doi:10.1038/37297

Zhang X, Ebata KT, Robaire B, Nagano MC (2006) Aging of male germ line stem cells in mice. Biol Reprod 124:119–124. doi:10.1095/biolreprod.105.045591

Zhao R, Xuan Y, Li X, Xi R (2008) Age-related changes of germline stem cell activity, niche signaling activity and egg production in *Drosophila*. Aging Cell 7:344–354. doi:10.1111/j.1474-9726.2008.00379.x

Chapter 3
Control of Germline Stem Cell Lineages by Diet and Physiology

Kaitlin M. Laws and Daniela Drummond-Barbosa

Abstract Tight coupling of reproduction to environmental factors and physiological status is key to long-term species survival. In particular, highly conserved pathways modulate germline stem cell lineages according to nutrient availability. This chapter focuses on recent in vivo studies in genetic model organisms that shed light on how diet-dependent signals control the proliferation, maintenance, and survival of adult germline stem cells and their progeny. These signaling pathways can operate intrinsically in the germ line, modulate the niche, or act through intermediate organs to influence stem cells and their differentiating progeny. In addition to illustrating the extent of dietary regulation of reproduction, findings from these studies have implications for fertility during aging or disease states.

3.1 Introduction

Organisms face frequent challenges to their homeostasis, and sensing and responding appropriately to these challenges is essential for their survival and successful reproduction. Diet and various stressors in the external environment help determine the levels of many circulating factors, including nutrients, metabolites, and hormones, which in turn can influence the germ line, a special lineage that gives rise to gametes and allows species propagation (Ables et al. 2012; Hubbard 2011). Constant evolutionary pressure on reproduction has therefore led

K.M. Laws
Department of Biochemistry and Molecular Biology, Johns Hopkins Bloomberg School of Public Health, Baltimore, MD 21205, USA

D. Drummond-Barbosa (✉)
Department of Biochemistry and Molecular Biology, Johns Hopkins Bloomberg School of Public Health, Baltimore, MD 21205, USA

Department of Environmental Health Sciences, Johns Hopkins Bloomberg School of Public Health, Baltimore, MD 21205, USA
e-mail: dbarbosa@jhu.edu

S. Arur (ed.), *Signaling-Mediated Control of Cell Division*, Results and Problems in Cell Differentiation 59, DOI 10.1007/978-3-319-44820-6_3

to very tight coupling of nutrient availability, metabolic status and other aspects of whole-body physiology to the biology of germ cells.

In many systems, germline stem cells (GSCs) support gametogenesis throughout most of adult life. Germ cell development from the stem cell stage to fully differentiated gametes is energetically costly and entails a large number of cellular processes that impose varying metabolic demands. It is therefore not surprising that multiple steps of gametogenesis are regulated by diet and other physiological factors (Ables et al. 2012; Hubbard 2011; Gracida and Eckmann 2013b; Busada and Geyer 2016).

Over the past 15 years, many studies have tackled the complex question of how whole-body physiology controls adult GSC lineages by taking advantage of in vivo model systems amenable to genetic manipulation. In this chapter, we summarize and discuss the progress in this field, with a special focus on diet-dependent mechanisms that modulate adult GSC lineages in *Drosophila melanogaster*, *Caenorhabditis elegans*, and mammals.

3.2 Responses of Germline Stem Cell Lineages to Diet

By coupling reproduction to nutrient availability, organisms avoid the costly metabolic investment of reproduction under suboptimal conditions and improve evolutionary fitness. Diet-dependent regulation of germ cells is therefore found in a wide range of systems regardless of life history strategy or germ line organization. For example, abalone produces fewer gametes under nutrient stress (Rogers-Bennett et al. 2010), starved zebrafish slow down egg production (Wang et al. 2006), and women with anorexia or excessively low body fat do not ovulate (Group 2006; Rojas et al. 2015). In cases where GSCs support gametogenesis, the germ line can be influenced by diet-dependent signaling in GSCs or their progeny, in the niche (a specialized microenvironment that maintains stem cells), or in intermediate organs that provide relay signals.

3.2.1 The Drosophila melanogaster *Ovary*

The *Drosophila* ovary has a well-described cell biology (Spradling 1993). Each ovary contains 15–20 ovarioles, composed of progressively more developed egg chambers (or follicles) formed in an anterior germarium, which houses GSCs and follicle stem cells (FSCs) (Fig. 3.1a). Two to three GSCs are closely associated with a group of somatic cap cells, which are the major cell type in the GSC niche. Cap cells produce bone morphogenetic protein (BMP) signals that maintain the GSC fate by repressing a differentiation factor, while the physical association between cap cells and GSCs requires E-cadherin. Anterior to cap cells, a row of terminal filament cells also contributes to the niche. GSCs typically divide asymmetrically to

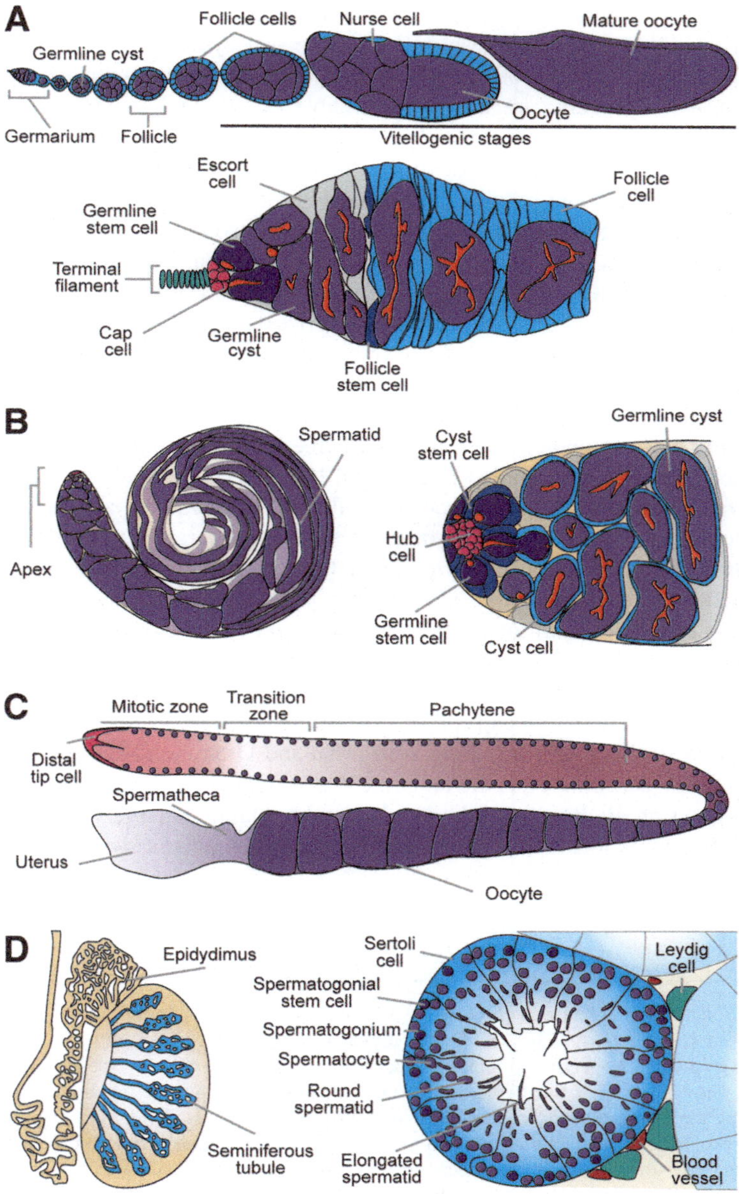

Fig. 3.1 GSC lineages. (**a**) Diagram of a *Drosophila* ovariole (*top*), which contains growing follicles. Each follicle is composed of a germline cyst surrounded by follicle cells and is produced from stem cell populations in the germarium (*bottom*). Germline stem cells (GSCs; *dark purple*) are juxtaposed to a somatic niche consisting primarily of cap cells (*pink*) and terminal filament cells (*teal*). GSCs divide asymmetrically, and their progeny generate 16-cell germline cysts (*light purple*) containing one oocyte and 15 nurse cells. The fusome (*orange*) becomes progressively more branched as cysts divide. Germline cysts initially associate with escort cells (*gray*) and are subsequently enveloped by follicle cells (*light blue*) generated by follicle stem cells (*dark blue*) to form follicles. (**b**) The *Drosophila* testis (*left*) is a blind-end tube. GSCs (*dark purple*) reside at its

self-renew and generate daughter cystoblasts. Cystoblasts divide four additional times with incomplete cytokinesis to form a 16-cell cyst: one of these cyst cells acquires an oocyte fate; the others support oocyte development as nurse cells. GSCs and their early progeny are easily identifiable based on the morphology of a specialized structure, the fusome. In GSCs, the fusome contacts the cap cell interface and remains round most of the time; as the cystoblast divides to form 16-cell cysts, the fusome becomes progressively more branched (Xie 2008). Early germ cells are closely associated with escort cells (also known as inner germarial sheath cells), which are required for the proper formation of 16-cell cysts (Kirilly et al. 2011). Two FSCs (abutting the posterior-most escort cells) give rise to follicle cells that envelop each 16-cell cyst to give rise to a follicle that buds off the germarium and proceeds through 14 developmental stages (Xie 2008).

Drosophila oogenesis is energetically demanding and highly regulated by diet (Ables et al. 2012). On a yeast-rich diet, each female lays an average of over 80 eggs per day, but upon shifting to a yeast-free (poor) diet, egg-laying rates drop to just one or two eggs daily (Drummond-Barbosa and Spradling 2001). This largely reversible response to diet occurs within 18–24 h and reflects the concerted regulation of multiple processes in oogenesis. The proliferation rates of GSCs and FSCs and the proliferation and growth of their progeny decrease, and follicles develop two- to threefold more slowly on a poor diet (Drummond-Barbosa and Spradling 2001). An additional effect of starvation in developing follicles is the accumulation of large aggregates of processing bodies and cortically enriched microtubules; this is a reversible response that requires microtubule motor proteins (Burn et al. 2015). GSC and cap cell loss over time is also accelerated by a poor diet (Drummond-Barbosa and Spradling 2001; Hsu and Drummond-Barbosa 2009). In addition, early germline cysts die at an increased frequency within the germarium, follicles entering vitellogenesis degenerate, and ovulation is largely blocked, causing an accumulation of mature stage 14 egg chambers within the ovary (Drummond-Barbosa and Spradling 2001).

Fig. 3.1 (continued) apical end in close association with hub cells (*pink*) and cyst stem cells (CySCs, *dark blue*) (*right*). GSCs and CySCs divide asymmetrically, and their progeny (germline cysts and cyst cells, respectively) remain associated with each other during spermatogenesis. (**c**) Diagram showing one of the two gonad arms of adult *C. elegans* hermaphrodites. A niche comprising the distal tip cell (DTC; *pink*) maintains progenitor cells in the mitotic, proliferative zone. As progenitor cells move away from the niche, they enter meiosis. Sperm produced during larval stages are stored in the spermatheca; oocytes (*purple*) generated later are fertilized by stored sperm (or sperm introduced by mating) before progressing to the uterus. (**d**) In the mouse testis (*left*), spermatogenesis takes place in seminiferous tubules. Cross-section of a seminiferous tubule (*right*) showing different stages of the lineage supported by basally located spermatogonial stem cells (SSCs, *dark purple*). SSCs divide to produce mitotically active differentiating progeny (spermatogonia) that undergo meiosis (spermatocytes) and spermiogenesis (spermatids) before they are released into the lumen of the tubule. Sperm undergoes further maturation in the epididymis, where they are eventually stored. Leydig cells (*teal*), blood vessels (*red*), and Sertoli cells (*blue*) play important roles in support of the SSC lineage

3.2.2 The Drosophila *Testis*

Drosophila male GSCs reside at the testis apex in a niche, the hub, composed of 10–15 somatic cells (Fig. 3.1b). Six to nine GSCs are closely associated with approximately twice as many somatic cyst stem cells (CySCs) at the hub. The cytokine unpaired (Upd) produced by the hub is required for adhesion of GSCs to the hub and for maintaining the CySC fate. Additional signals contribute to GSC and CySC maintenance, including BMPs and Hedgehog, respectively (Greenspan et al. 2015). GSCs divide asymmetrically to generate gonialblasts that form two-, four-, eight-, and 16-cell cysts, collectively referred to as spermatogonia. CySCs give rise to postmitotic cyst cells; a pair of cyst cells envelops the gonialblast and remains associated with the resulting germline cyst as it develops. Meiotic divisions produce a cyst with 64 spermatids, followed by their individualization and mature sperm generation (Fuller 1993).

Despite the obvious size and energy storage differences between oocytes and sperm, early stages of *Drosophila* spermatogenesis are also regulated by diet. Upon protein starvation, the number of GSCs and their division rates decrease, and these effects are reversible (McLeod et al. 2010). A reduction in protein and sugar levels also slows down GSC proliferation, at least in part as a result of increased rates of centrosome misorientation (Roth et al. 2012). A reduction in overall food intake without specific removal of nutrients has been reported to result in a slight delay in GSC loss with age (Mair et al. 2010), suggesting that specific dietary manipulations can impact GSC maintenance in unique ways. More recent studies examining the kinetics of the protein-starvation response show that although GSC proliferation and numbers decrease initially, after about a week, GSC numbers stabilize and proliferation rates return to normal levels. During this response, early spermatogonial cells die at the two- to four-cell stage, and their death requires caspase activity in surrounding cyst cells. When death of cysts cells is blocked by *dronc* caspase knockdown or DIAP1 overexpression, GSCs are no longer maintained during prolonged starvation, suggesting that elimination of cyst cells/spermatogonial units has a protective effect against protein starvation (Yang and Yamashita 2015).

3.2.3 The Caenorhabditis elegans *Gonad*

The hermaphrodite *C. elegans* has a gonad with two arms, each containing ~1000 germ cells and capped by a single somatic distal tip cell (DTC) at each end (Fig. 3.1c). Notch ligands produced by the DTC maintain a population of ~225 proliferating germ cells, including a stem cell pool of ~35–70 GSCs, in the distal mitotic zone (Kimble and Seidel 2008). As germ cells move proximally, they enter meiosis as a result of reduced Notch signaling and complex posttranscriptional regulation. Following intense oocyte growth, ovulation occurs, followed by fertilization (by sperm produced during larval development or introduced through mating) and egg laying (Hubbard and Greenstein 2005; L'Hernault 2009) (Fig. 3.1c).

Complete removal of food in adult hermaphrodites can lead to different phenotypes (Angelo and Van Gilst 2009). Maternal death may occur due to internal development of progeny (who will eat their way out of the mother) following reduced rates of egg laying. Animals that escape matricide undergo instead an adult reproductive diapause, where over the initial 10 days of starvation, the germ line is reduced to ~35 germ cells (presumably GSCs) that resist continued starvation for over 30 days and are able to reconstitute a fully functional germ line within 72 h of refeeding (Angelo and Van Gilst 2009). A more recent study showed that germ cells in the proliferative zone stop dividing and arrest in the G2 phase of the cell cycle, and meiotic entry is inhibited within a few hours of starvation of early adults, and that these effects are reversible upon refeeding. Interestingly, GSCs retain their stemness independently of Notch signaling during this starvation-induced arrest (Seidel and Kimble 2015).

3.2.4 The Mouse Testis

Mouse spermatogenesis takes place in the epithelial lining of seminiferous tubules in the testis (Fig. 3.1d) (de Rooij and Russell 2000). Large somatic Sertoli cells span the entire epithelium, thereby contacting all stages of germ cell development, and they secrete glial cell line-derived neurotrophic factor (GDNF), required for GSC self-renewal (Franca et al. 2016). GSCs, called spermatogonial stem cells (SSCs), represent a subset of undifferentiated A_{single} spermatogonia located basally in the epithelium. SSCs give rise to daughters that undergo four rounds of incomplete divisions remaining in clusters of two cells (A_{paired}) and four, eight, 16, and up to 32 cells (collectively called $A_{aligned}$ spermatogonia), subsequently differentiating into B spermatogonia (de Rooij and Russell 2000). B spermatogonia differentiate into spermatocytes, which undergo meiosis to form haploid spermatids that differentiate into sperm released into the lumen. A highly vascularized interstitium surrounding seminiferous tubules contains peritubular myoid cells, macrophages, and testosterone-producing Leydig cells, all of which support spermatogenesis (Oatley and Brinster 2012; DeFalco et al. 2015).

Although specific effects on SSCs have not been carefully analyzed, several studies point to possible connections with diet. For example, diet-induced obesity and vitamin D or zinc deficiency can lead to a decrease in male fertility in mice (Fan et al. 2015; Sun et al. 2015; Croxford et al. 2011). Adult mice deprived of dietary vitamin A show testicular degeneration as a result of increased apoptosis and sloughing of immature germ cells into the lumen in more severe cases (Boucheron-Houston et al. 2013). Several of these observations have parallels in humans: obese men are more likely to be infertile (Campbell et al. 2015), and there is a positive correlation between vitamin D and zinc and sperm quality in adult men (Blomberg Jensen 2014; Colagar et al. 2009).

3.3 Nutritional Control of GSC Lineages

GSC lineages sense and respond to nutritional inputs through multiple mechanisms integrated into a seamless physiological output. GSC lineages can directly receive dietary information through cellular energy sensors, or nutrient transport and sensing. These same inputs can act in the niche, indirectly influencing GSCs and their progeny. Finally, remote endocrine cells can produce hormones in a diet-dependent manner with broad physiological influence over the organism, including direct effects on the niche, GSCs, or their differentiating progeny, or indirect effects through one or more intermediate organs.

3.3.1 Nutrient Sensors Control GSC Lineages

Highly conserved nutrient-sensing pathways operate in a wide range of cells, including those in GSC lineages; yet, several examples illustrate how these pathways can control distinct processes depending on the cellular context. The energy sensor adenosine monophosphate (AMP)-activated protein kinase (AMPK) is a heterotrimeric protein composed of a catalytic α and regulatory β and γ subunits. When ATP levels are low (for example, due to low nutrients or prolonged exercise), AMP and ADP binding to the γ subunit leads to activation of AMPK, which also requires phosphorylation of the α subunit by liver kinase B1 (LKB1) or, in a few cases, by calmodulin-dependent protein kinase kinase β (CAMKKβ) (Hardie et al. 2016) (Fig. 3.2). AMPK activation inhibits anabolism and cell growth and promotes catabolic processes to restore cellular energetic balance (Hardie 2015; Hardie and Ashford 2014). AMPK inhibits growth in part through the inhibition of the kinase Target of Rapamycin (TOR, or mTOR in mammals). TOR exists as part of two distinct complexes, TORC1 and TORC2, which differ in their regulation and downstream roles (Bar-Peled and Sabatini 2014; Huang and Fingar 2014; Devreotes and Horwitz 2015). TORC1 is the best-understood complex and integrates diverse upstream inputs, including extracellular signals (e.g., insulin signaling) and intracellular cues (e.g., amino acid levels, AMPK activity), to control a variety of downstream cellular processes, including cell growth (Hindupur et al. 2015) (Fig. 3.2).

TOR signaling is required in *Drosophila* female GSCs. TOR activity promotes GSC proliferation via G2 independently of insulin signaling (see below). Intriguingly, maintenance of female *Drosophila* GSCs requires very precise regulation of TOR signaling levels. *Tor* mutation decreases GSC numbers (LaFever et al. 2010), and loss of *Tsc1* (which encodes an upstream inhibitor of TOR) function causes a significantly more severe GSC loss (LaFever et al. 2010; Sun et al. 2010) indicating that the either low or high TOR activity levels are detrimental to stem cell maintenance. *Tsc1* GSCs have low levels of BMP signaling, suggesting an impaired ability to respond to niche signals (Sun et al. 2010). Regulation of the *C. elegans*

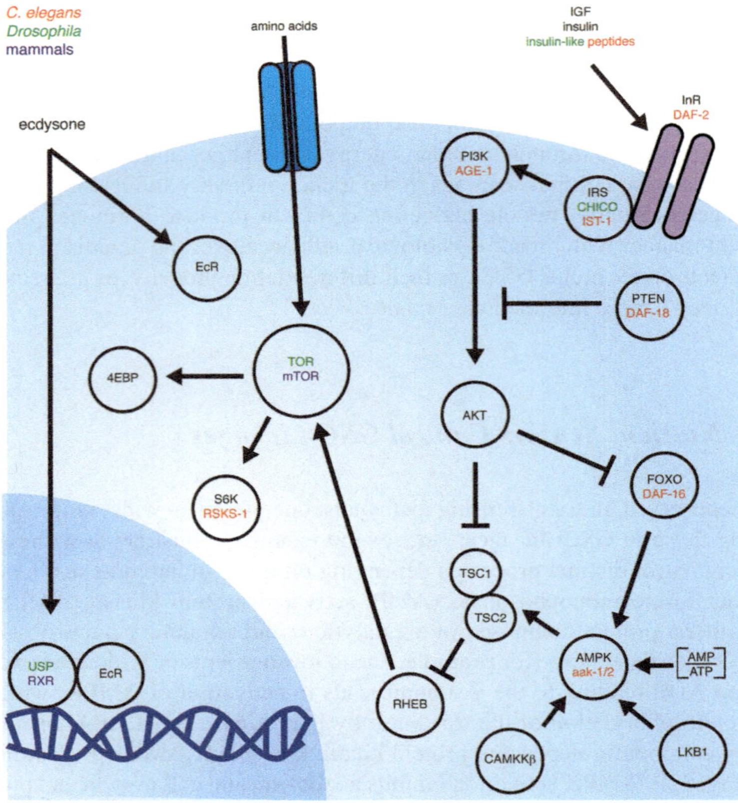

Fig. 3.2 Conserved nutrient sensing pathways. Cells respond to intrinsic energy and nutrient levels and external stimulation by hormones to activate an interdependent, conserved cellular response to diet. Species-specific protein names are indicated in *orange* (*C. elegans*), *green* (*Drosophila*), and *purple* (mammalian); common names are shown in *black*

germline progenitor pool by AMPK and TOR is mechanistically distinct in developing versus adult worms. During earlier nutrient-dependent developmental checkpoints, the homologs of mammalian AMPK, *aak-1* and *aak-2*, suppress germ line proliferation, and their simultaneous mutation leads to germline hyperplasia (Fukuyama et al. 2012; Narbonne and Roy 2006a). The *C. elegans* homolog of the TOR downstream effector ribosomal protein S6 kinase (S6K), *rsks-1*, is required for germ cell proliferation during the fourth larval instar (Korta et al. 2012). In stark contrast, *rsks-1* is dispensable for progenitor proliferation in adults (Korta et al. 2012), and AMPK mutants are still competent to undergo GSC quiescence in response to starvation (Seidel and Kimble 2015). The mechanisms controlling the response of adult *C. elegans* proliferative germ cells to diet remain largely unknown, and appear to be distinct from those controlling adult female *Drosophila* GSCs.

TOR is also required for the proliferation, growth, and survival of differentiating progeny of *Drosophila* female GSCs (LaFever et al. 2010). *Tor* mutant dividing germline cysts in mosaic germaria show increased death, and follicles containing *Tor* mutant cysts grow at a markedly decreased rate (LaFever et al. 2010). Conversely, follicles carrying homozygous mutant cysts for *Tsc1* grow significantly faster (LaFever et al. 2010). Under amino acid starvation, TORC1 inhibition by nitrogen permease regulator like 2 and 3 (Nprl2 and Nprl3) protects pre-vitellogenic follicles from apoptosis, and knockdown of *nprl2* and *nprl3* in the germ line prevents recovery of oogenesis following amino acid starvation (Wei and Lilly 2014). Mutations in follicle cells, in addition to affecting follicle cell growth itself, can also non-autonomously influence the growth of the underlying wild-type germ line, with *Tor* and *Tsc1* mutant follicle cells slowing down or accelerating follicle growth, respectively (LaFever et al. 2010). AMPK activity also controls follicle cells, and *ampk* mutant follicle cells are larger than wild-type cells (Haack et al. 2013). Follicles carrying *Tor* mutant cysts eventually arrest at or prior to vitellogenesis, depending on allele strength, and degenerate (Pritchett and McCall 2012; LaFever et al. 2010). In another dipteran species, the yellow fever mosquito *Aedes aegypti*, where ovarian follicles remain in a pre-vitellogenesis arrest until blood ingestion (Attardo et al. 2005), entry into vitellogenesis is also TOR dependent, suggesting evolutionary conservation. Specifically, ovarian TORC1 activity is stimulated by a blood meal (Hansen et al. 2005; Roy and Raikhel 2012), and global knockdown of S6K inhibits yolk deposition (Hansen et al. 2005). Finally, TORC1 activity also regulates meiotic entry in the *Drosophila* ovary, as *Tor* mutant germline clones enter meiosis prematurely before 16-cell cysts are formed, and the amino acid sensing GATOR1 complex promotes meiotic entry in the early germ line by inhibiting TOR activity. Since global amino acid deprivation is not a condition for meiotic entry in the *Drosophila* ovary, however, GATOR1 may be acting in a nutrient-independent role (Wei et al. 2014).

Optimal levels of mTOR activity are also required in the mouse testis. TOR inhibition by rapamycin in neonatal mice reduces testis size, likely due to decreased germ cell proliferation and a block to meiosis (Busada et al. 2015). On the other hand, activation of mTORC1 as a result of global mutation of *promyeolocytic leukemia zinc finger* (PLZF) leads to compromised GDNF signaling and progressive germ cell loss, and these defects are rescued by rapamycin feeding (Hobbs et al. 2010). Nutrient-sensing pathways in Sertoli cells also influence the germ line. For example, Sertoli cell-specific overactivation of TOR via *Tsc1* or *Tsc2* deletion or through decreased AMPK activity by conditional *Lkb1* knockout leads to progressive germ cell loss and loss of Sertoli cell quiescence (Tanwar et al. 2012). These studies suggest that tight regulation of TOR is a recurring theme in the regulation of GSC lineages. Intriguingly, men with Peutz–Jeghers syndrome, frequently associated with *LKB1* mutations (Hemminkl et al. 1998), are at risk for germ cell loss and Sertoli cell tumors, suggesting a conserved role for LKB1 in the human testis (Venara et al. 2001; Gourgari et al. 2012).

3.3.2 Diet-dependent Hormones Affect GSC Lineages

Long-range, diet-dependent signals are integral components of the control of GSC lineages. As discussed below, hormones can regulate GSC lineages through direct actions on the germ line itself or indirectly, via the niche or other somatic support tissues. Cross talk among different hormonal systems and their integration within specific cellular contexts can further refine downstream responses.

Activation of insulin/insulin-like signaling is a highly conserved hormonal response to nutrient availability, and it can also affect downstream production of other hormones, such as the *Drosophila* steroid hormone 20-hydroxyecdysone (20E). In mammals, insulin is secreted by pancreatic β cells in response to stimulation by glucose or amino acids, and signals through the insulin receptor (InR) (Fig. 3.2). Insulin-like growth factors 1 and 2 (IGF1 and IGF2) are primarily synthesized in the liver and act through their receptors IGFR-1 and IGFR-2 to control cell growth (Siddle 2011). Activation of InR/IGFR receptors leads to multiple downstream events, including increased phosphoinositide-3 kinase (PI3K) activity and downstream phosphorylation and inhibition of the transcriptional factor FOXO. The *Drosophila* genome encodes eight insulin-like peptides (ILPs), whereas *C. elegans* has as many as 40, but in both cases, single homologs of InR and other downstream effectors transduce these signals (Kannan and Fridell 2013; Murphy and Hu 2013). 20E, the major steroid hormone in *Drosophila*, is produced from dietary cholesterol or ergosterol. In females, late stage egg chambers produce 20E in a diet- and insulin-dependent manner, whereas the source of the much lower titers of 20E in male hemolymph remains unidentified (Schwedes and Carney 2012). 20E acts through a nuclear hormone receptor composed of the ecdysone receptor [EcR, the homolog of mammalian farnesoid X receptor and liver X receptor (King-Jones and Thummel 2005)] and Ultraspiracle (Usp, the homolog of retinoid X receptor, RXR) to induce a wide range of downstream direct and indirect targets, including the early response genes *E74*, *E75*, and *broad* (Uryu et al. 2015).

ILPs directly stimulate GSC proliferation in *Drosophila*. In females, neural ILPs act directly on the germ line to control the G2 phase of the GSC cell cycle via PI3K but independently of FOXO (LaFever and Drummond-Barbosa 2005; Hsu et al. 2008). While downregulation of the insulin pathway preferentially extends G2, both the G1 and G2 phases of the GSC division cycle are lengthened in females on a poor diet, suggesting the existence of a yet unknown diet-dependent signal that regulates G1 (Hsu et al. 2008). In male GSCs, germline-specific knockdown or inhibition of insulin signaling increases centrosome misorientation by disrupting localization of Apc2, a cortical protein required for centrosome anchoring (Roth et al. 2012). Conversely, constitutive activation of InR in the male germ line rescues diet-induced centrosome misorientation. Further, an intact centrosome orientation checkpoint is required for the slowdown of GSC division in males on a poor diet (Roth et al. 2012), suggesting a central role for this mechanism in the testis.

These studies in *Drosophila* are in contrast to findings in *C. elegans*. Although larval germ cells and adult germ cell tumors require insulin signaling for proliferation, normal adult germ cells proliferate in an insulin-independent manner (Michaelson et al. 2010; Hubbard 2011; Dillin et al. 2002). Nevertheless, in adult *C. elegans*, insulin signaling couples nutrient availability to oocyte development and progression through meiosis I through the activation of MAPK/ERK signaling (and, notably, independently of FOXO/DAF-16) (Lopez et al. 2013), indicating that germ cells in different developmental stages use distinct branches of insulin signaling.

Drosophila GSC maintenance also requires insulin signaling. In females, the mechanisms involved are clearly distinct from those controlling proliferation. Insulin signaling is not required in GSCs themselves for their maintenance. Instead, ILPs act directly on cap cells to promote Notch signaling, which is required for cap cell maintenance (Song et al. 2007), through FOXO inhibition; under low insulin signaling, FOXO induces high levels of the glycosyltransferase Fringe, leading to inhibition of the Notch receptor (Hsu and Drummond-Barbosa 2009, 2011; Yang et al. 2013). In addition, ILPs directly stimulate the physical association between cap cells and GSCs through E-cadherin, independently of Notch signaling (Hsu and Drummond-Barbosa 2009). By contrast, in males, insulin signaling is intrinsically required in GSCs for their maintenance, as indicated by the higher frequency of loss of homozygous mutant *InR* GSCs in mosaic testes. Full rescue of GSC loss induced by starvation, however, requires the constitutive activation of InR in both the germ line and hub cells, suggesting that insulin signaling in the niche may also contribute to GSC maintenance (McLeod et al. 2010). These studies imply that conserved signaling pathways regulating similar processes (e.g., GSC proliferation or maintenance) can evolve distinct mechanisms to achieve that goal even between the different sexes of a given species.

20E directly controls GSC proliferation and maintenance in the *Drosophila* ovary, and it also has additional roles in ovarian and testicular somatic cells. Temperature-sensitive *EcR* mutants show increased loss and reduced proliferation rates of GSCs, and both of these phenotypes reflect an intrinsic requirement for ecdysone signaling in GSCs based on genetic mosaic analyses (Ables and Drummond-Barbosa 2010). Ecdysone signaling controls GSCs independently of insulin signaling by modulating the responsiveness of GSCs to BMP ligands from the niche. This role requires the downstream target *E74* [a member of the *ets* proto-oncogene family (Burtis et al. 1990; Karim et al. 1990)] specifically, as GSCs mutant for *usp* or *E74*, but not *E75* or *broad*, show low levels of a BMP signaling reporter and are rapidly lost from the niche (Ables and Drummond-Barbosa 2010). *E75* [a homolog of mammalian *peroxisome proliferator-activated receptor* γ (*PPAR*γ) (King-Jones and Thummel 2005)] is also required in escort cells for GSC maintenance, and expression of a dominant negative form of EcR in escort cells disrupts female germ cell differentiation (Konig et al. 2011; Morris and Spradling 2012), underscoring the complexity of 20E regulation of the early female germ line. In later follicle development, *E78*, an *E75*-related gene that is also an early ecdysone target (Stone and Thummel 1993), is cell autonomously required for germ line survival and functionally interacts with ecdysone signaling

(Ables et al. 2015). Many additional targets of ecdysone contribute to its various roles in the female GSC lineage, although more in depth analysis is needed (Ables et al. 2016). In the testis, EcR knockdown in the somatic lineage leads to CySC and GSC loss and death of differentiating germ cells, although the specific mechanisms involved remain unclear (Li et al. 2014).

Progression through vitellogenesis is a major nutritional checkpoint in *Drosophila* and other insects. *Drosophila* temperature-sensitive *EcR* mutants and *E75* mutant germline clones fail to progress through vitellogenesis (Carney and Bender 2000; Buszczak et al. 1999), indicating a germ line requirement for ecdysone signaling. *Drosophila InR* global mutants are defective in ecdysteroid production and vitellogenesis, and these defects appear to be at least partially rescued by treatment with a juvenile hormone analog (Tu et al. 2002). Follicles containing germline clones that are mutant for *InR* or defective for PI3K signaling, however, have reduced rates of growth and blocked vitellogenesis, clearly indicating that insulin signaling through PI3K is required in the germ line itself for these processes (Hsu et al. 2008; LaFever and Drummond-Barbosa 2005). Unlike for GSC proliferation, however, insulin signaling does not involve FOXO and, instead, feeds into TOR signaling for the control of follicle growth (Hsu et al. 2008). Interestingly, insulin/TOR signaling in follicle cells is required for the processing body and microtubule changes that occur in the underlying germ line in response to diet (Burn et al. 2015). In *A. aegypti*, vitellogenesis also requires insulin signaling (Brown et al. 2008; Gulia-Nuss et al. 2011), and the ovary expresses EcR and shows induction of ecdysone response genes, including *E75*, upon blood feeding (Cho et al. 1995; Swevers and Iatrou 2009; Pierceall et al. 1999), suggesting conserved mechanisms of vitellogenesis control. After vitellogenesis in the *Drosophila* ovary, a reduction in insulin signaling induces a metabolic shift towards glycogen storage and mitochondrial quiescence, preparing the oocyte for fertilization (Sieber et al. 2016).

Multiple hormones also regulate mammalian GSC lineages, although the connection to diet is not always well understood. Retinoic acid (RA) is derived from dietary vitamin A and signals through retinoic acid receptors (RARs) and their partners, RXRs (Blomhoff and Blomhoff 2006). The multiple isoforms of RAR and RXR are differentially expressed, lending an additional level of specificity to RA signaling. Analogously to ecdysone signaling in *Drosophila*, RA signaling is required in both the germ line and somatic support cells of the mouse testis (Hogarth and Griswold 2013). Knockout of all three *RAR* or *RXR* isoforms specifically in the germ line suppresses spermatogonia proliferation, consistent with the role of vitamin A in maintaining spermatogenesis (Gely-Pernot et al. 2012; Ghyselinck et al. 2006). Furthermore, Sertoli cell-specific knockout of retinaldehyde dehydrogenases, which are required for RA synthesis, blocks sperm meiotic entry (Tong et al. 2013; Raverdeau et al. 2012). Testicular RA titers are also reduced in mice genetically depleted of macrophages, resulting in spermatogonial differentiation defects (DeFalco et al. 2015). Conversely, ectopic RA administration induces differentiation and meiotic entry in specific, poised subsets of spermatogonia (Endo et al. 2015). While spermatogonia can directly respond to RA in vivo (Zhou et al. 2008), it is unlikely that RAR/RXR are the sole mediators of its activity. For example, spermatogonia lacking all three isoforms of either RAR or RXR can still enter meiosis, albeit at a low frequency (Gely-Pernot et al. 2015), hinting that RA could act independently of the nuclear hormone receptor dimer or act indirectly through other testicular cell types (Gely-Pernot et al. 2015). Indeed, Sertoli cell-specific *RARα* knockout in *RARγ* mutant mice has a complete block to meiosis, suggesting that the RA-dependent meiosis signal requires paracrine signaling from Sertoli cells (Gely-Pernot et al. 2015). While spermatogonial RAR and RXR isoforms are dispensable for germ cell survival (Gely-Pernot et al. 2015), Sertoli cell RARα promotes germ cell survival in aging animals (Vernet et al. 2006). Sertoli cell-specific knockout of *RARα* leads to progressive testis deterioration, with death of spermatocytes and spermatids (Vernet et al. 2006). Intriguingly, this phenotype is not recapitulated in Sertoli cell knockout of all isoforms of RXR, invoking again an RXR-independent RA signaling mechanism. Bioinformatic analysis further suggests that several genes responsible for establishing tight junctions to create the blood–testis barrier contain retinoic acid response elements, suggesting that RAR activity in Sertoli cells may regulate the integrity of the blood–testis barrier (Chung et al. 2010).

The hypothalamic-pituitary-gonadal axis is a major regulator of mammalian spermatogenesis, and there is evidence to suggest that diet and obesity can impact its function. Gonadotropin-releasing hormone (GnRH) secreted from the brain induces secretion of follicle-stimulating hormone (FSH) and luteinizing hormone (LH) from the pituitary, which act on Sertoli and Leydig cells, respectively, to control SSC lineage activity (O'Shaughnessy 2014). LH induces testosterone production by Leydig cells and testosterone signals via the androgen receptor (AR), a nuclear hormone receptor predominantly expressed in Sertoli cells (Sar et al. 1990; reviewed in Walker and Cheng 2005), although recent evidence suggests the presence of a functional AR in human sperm (Aquila et al. 2007). Aromatase activity in the testis, and to a lesser extent in peripheral tissues, catalyzes the synthesis of estradiol, the most potent biological estrogen, from androgens (Marcus and Korenman 1976). Estrogens are integral regulators of spermatogenesis, in part because they feedback to the brain to regulate LH and FSH secretion (Simpson et al. 1999, 2000). While our current understanding of nutrient inputs to the hypothalamic-pituitary-gonadal axis is complicated by the tissue dysfunction associated with its disruption, studies in rats suggest that GnRH release decreases in fasted animals (Gruenewald and Matsumoto 1993). Administration of insulin to female volunteers produces a serum pulse of LH, consistent with insulin acting on the GnRH neurons (Moret et al. 2009). Conflicting reports in mouse models, however, debate whether insulin acts directly on GnRH neurons (Evans et al. 2014; DiVall et al. 2015), and the response to insulin could also be sexually dimorphic (Kovacs et al. 2002; Castellano et al. 2006). Furthermore, Leydig cell testosterone production is perturbed in mice with Sertoli cell-specific *InR* or *Igf1r* knockout, suggesting that diet-dependent signaling controls the hypothalamic-pituitary-gonadal axis at the testis level (Pitetti et al. 2013b). *Igf1* knockout mice also have reduced spermatogenesis, although it is unclear whether there is a developmental contribution to this phenotype (Baker et al. 1996). Metabolic state influences the hypothalamic-pituitary-gonadal axis in adult men and women. Women with diet-induced amenorrhea have reduced plasma levels of estrogen, FSH, and LH (Couzinet et al. 1999), and high aromatase expression in postmenopausal obese women leads to more circulating estrogen (Baglietto et al. 2009). Additionally, men with type 2 diabetes or obesity are more likely to be testosterone deficient, further suggesting a possible connection between the hypothalamic-pituitary-gonadal axis and diet (Kelly and Jones 2013; Wang et al. 2011).

The hypothalamic-pituitary-gonadal axis regulates both the germ line and somatic support tissues in the mammalian testis. Testosterone and FSH act semi-redundantly through their receptors (AR and FSHR, respectively) on Sertoli cells to promote germ cell survival (Walker and Cheng 2005). Estrogen signaling is also critical for germ line survival during spermatogenesis, as shown in a knock-in ER mutant mouse that does not respond to estrogen (Sinkevicius et al. 2008, 2009). In accordance, global knockout of aromatase in mice results in increased germ cell apoptosis in the testis (Robertson et al. 1999). In both estrogen-signaling incompetent and aromatase mutant mice, circulating LH and FSH levels are slightly higher and unchanged, respectively, suggesting that germ cell apoptosis is not simply a downstream effect of estrogens controlling LH and FSH secretion (Sinkevicius et al. 2009; Robertson et al. 1999; Simpson et al. 1999, 2000). FSH is required for testis growth in both mice and humans (Kumar et al. 1997; Phillip et al. 1998), and *FSHR* knockout mice have impaired spermatid elongation (Krishnamurthy et al. 2000). AR is absolutely required in Sertoli cells for completion of meiosis (Abel et al. 2008; De Gendt et al. 2004). Furthermore, mice with a conditional knockout of AR in peritubular myoid cells have significantly reduced sperm counts, suggesting multiple sites of action for testosterone (Zhang et al. 2006).

3.3.3 Additional Interorgan Communication Influences GSC Lineages

Beyond the more "traditional" hormone examples discussed above, the extent of GSC lineage regulation by signals originating in other organs, with distinct organs providing various types of information, is just beginning to be appreciated. The molecular mechanisms of many of these signaling axes remain unknown, but proteohormones, signaling lipids or metabolites, and mobilized nutrients from one organ to another are likely participants.

The nervous system regulates physiological circuits that marry inputs from the external environment, including diet, to whole-body physiology either through changes in organismal behavior or more directly. As discussed above, *Drosophila* ILPs are diet-dependent neuropeptides that directly regulate GSC lineages and nearby somatic support cells (Hsu and Drummond-Barbosa 2009; LaFever and Drummond-Barbosa 2005). Ecdysone produced by ovarian follicles acts on the brain to promote female-specific feeding behavior, increasing nutrient uptake and supporting oogenesis (Sieber and Spradling 2015). Octopaminergic neurons innervate the ovaries and reproductive tract and are essential for *Drosophila* ovulation (Lee et al. 2003; Monastirioti 2003; Deady and Sun 2015). Interestingly, a subset of octopaminergic neurons becomes hyperactive under starvation conditions (and promotes foraging behavior) (Yang et al. 2015), suggesting a potential molecular link between ovulation and nutrient availability. During *C. elegans* development, transforming growth factor β (TGFβ/DAF-7) expression in chemosensory neurons is activated by food availability and low dauer pheromone, which is involved in sensing population density (Ren et al. 1996; Schackwitz et al. 1996; Dalfo et al. 2012). *daf-7* mutant worms enter meiosis prematurely, thereby reducing progenitor number, and downstream signaling pathway components are required in the DTC, indicating a niche-mediated role for TGFβ in maintaining a large germline progenitor pool from which gametes can be generated (Dalfo et al. 2012). As mentioned earlier, GnRH released from the mammalian hypothalamus stimulates pituitary FSH and LH release, which in turn act on testis somatic cells to control the germ line (O'Shaughnessy 2014). In adult men, even short periods of fasting can suppress GnRH, leading ultimately to a fall in LH-induced testosterone production (Trumble et al. 2010). Thus, nutritional cues may be transmitted to the germ line via the brain in multiple organisms, although specific strategies can vary.

GSC lineages are also controlled by endocrine signals from other organs, including the adipose tissue. Accordingly, obese men are more likely to be subfertile (Martin 2014; Kawwass et al. 2015), and diet-induced obesity in rats is linked to decreased sperm motility (Fernandez et al. 2011; Palmer et al. 2012). Mammalian proteohormones secreted from adipocytes, or adipokines, modulate homeostasis by regulating multiple processes (Cao 2014). For example, adiponectin regulates the sensitivity of peripheral tissues to insulin (Yamauchi and Kadowaki 2013), while leptin signals satiety to the brain and controls metabolism in peripheral tissues (Moran and Phillip 2003). Several lines of evidence also suggest that adipokines might influence reproduction, although reports are conflicting (Kawwass et al. 2015). Adiponectin receptors 1 and 2 are expressed in the human and mouse hypothalamus (Dupont et al. 2014), and their transcripts are detected in Leydig cells and the testicular epithelium of rats (Caminos et al. 2008). Adiponectin knockout mice, however, are fertile, while mice lacking adiponectin receptor 2 do not produce sperm (Bjursell et al. 2007), suggesting possible adiponectin-independent roles for this receptor. The leptin receptor is also expressed in the mammalian testis (Landry et al. 2013), and leptin-deficient mice have increased germ cell death and fertility defects (Mounzih et al. 1997; Bhat et al. 2006). Conversely, spermatogenesis defects in men show an association with increased expression of leptin and its receptor in the testis (Ishikawa et al. 2007). Deletion of leptin receptor specifically in the hypothalamus recapitulates many of the phenotypes of leptin-deficient mice, however, suggesting that leptin acts via the brain to control fertility (Ahima et al. 2006). Although these studies are important steps towards our understanding of how mammalian adipokines influence the germ line

(Kawwass et al. 2015), much remains to be learned about how these and other adipocyte factors control reproduction.

Adipocyte factors also contribute to the control of GSC lineages in invertebrate model systems. In *Drosophila*, the fat body, an organ composed of adipocytes and hepatocyte-like oenocytes (Gutierrez et al. 2007), has endocrine roles that control development, metabolism, and behavior (Arrese and Soulages 2010). Although *C. elegans* lack a distinct fat storage organ, dedicated lipid storage cells are found in the intestine and epidermis (Mullaney and Ashrafi 2009). Adipokine signaling modules are conserved in *Drosophila* and *C. elegans*, and they influence the GSC lineages of these organisms (Rajan and Perrimon 2012; Laws et al. 2015; Kwak et al. 2013; Svensson et al. 2011). The sole adiponectin receptor in *Drosophila*, AdipoR, is intrinsically required by ovarian GSCs for their maintenance (Laws et al. 2015). In *C. elegans*, deletion of adiponectin receptor homologs *paqr-1*, *-2*, and *-3* causes extensive defects, including reduced brood size (Svensson et al. 2011). Adiponectin-like ligands have not yet been identified in *C. elegans* or *Drosophila*; nevertheless, ex vivo cultures of fly larval brains respond to stimulation by recombinant mammalian adiponectin (Kwak et al. 2013), suggesting that the *Drosophila* receptor recognizes an endogenous ligand with conserved three-dimensional structure. Additionally, Unpaired 2 (Upd2) is secreted from *Drosophila* adipocytes and stimulates brain ILP secretion (Rajan and Perrimon 2012), presumably indirectly affecting oogenesis. Human leptin transgenic expression or feeding rescues the *upd2* mutant phenotype, suggesting that Upd2 is the functional equivalent of leptin despite lack of primary sequence homology (Rajan and Perrimon 2012).

Studies in *Drosophila* also show that adipocytes play an important role in reproduction by transmitting nutritional information to the ovary. In *Drosophila* females, a slight decrease in amino acid levels within adult adipocytes through the knockdown of single amino acid transporters significantly increases the rate of GSC loss from the niche and partially blocks ovulation through distinct mechanisms (Armstrong et al. 2014). Low amino acid levels trigger the evolutionarily conserved amino acid response pathway through unloaded tRNA-mediated activation of the GCN2 kinase within adipocytes to cause GSC loss, whereas amino acids modulate TOR in adipocytes to regulate ovulation (Armstrong et al. 2014). Adipocyte factors acting downstream of the amino acid response pathway or TOR to control the GSC lineage, however, remain unidentified. In *A. aegypti*, global knockdown of amino acid transporters reduces egg laying (Carpenter et al. 2012), and TOR activity in the fat body is induced after blood feeding (Hansen et al. 2005) and required for ovarian follicle vitellogenesis (Hansen et al. 2004, 2005; Roy and Raikhel 2011, 2012; Carpenter et al. 2012). Fat body transcription of *vitellogenin*, which is physiologically triggered by 20E following a blood meal, can be induced by ex vivo treatment of cultured fat bodies with amino acids, but this effect is suppressed by TOR inhibition (Hansen et al. 2004, 2005). In vivo, knockdown of components of the TOR pathway reduces egg laying and can impair yolk uptake and egg viability (Hansen et al. 2004, 2005), although these effects likely reflect potential roles of TOR in multiple locations, as is the case in *Drosophila*.

Many additional organs are important regulators of physiology; therefore, it would be logical to explore their potential roles in contributing to the control of GSC lineages. For example, nutrients are absorbed at the intestine, and in *Drosophila* females, feeding conditions change the physiology of the midgut (Cognigni et al. 2011). Mating also leads to extensive remodeling of the midgut and increased lipid metabolism, and these changes are required for normal levels of fecundity (Reiff et al. 2015). Although some of these effects likely reflect changes in efficiency of digestion and nutrient absorption, it is conceivable that more active signaling occurs between the intestine and GSC lineages. It is also important to consider the effect of the gut microbiome in reproduction. The type of bacteria ingested by *C. elegans* influences its brood size (Yu et al. 2015). Also, in the absence of the nuclear hormone receptor *nhr-114* (the homolog of HNF4), worms are sterile when fed a specific strain of bacteria (Gracida and Eckmann 2013a). *nhr-114* activity is detected in both the germ line and gut, but it does not appear to be required in the germ line. Tryptophan supplementation rescues this phenotype, suggesting that *nhr-114* may help buffer dietary changes in the gut (Gracida and Eckmann 2013a). Interestingly, germ-free mice have impaired blood–testis barrier and lumen formation in seminiferous tubules, and reduced levels of serum LH, FSH, and testicular testosterone. Exposure of these mice to a strain of bacteria that secrete high levels of the short-chain fatty acid butyrate restores integrity of the blood–testis barrier (Al-Asmakh et al. 2014). Recent studies have also shown that muscles secrete peptide hormones, or myokines, in *Drosophila* and mice (Demontis et al. 2013, 2014; Demontis and Perrimon 2010), and genetic manipulations in muscles affect the physiology of the fly (Demontis and Perrimon 2010). In mice, osteocalcin secreted from bones modulates spermatogenesis by promoting testosterone production in Leydig cells (Oury et al. 2011). Further, in insects, sex peptides transferred during mating trigger a host of physiological changes, including many upstream of GSC lineage activity (Soller et al. 1997; Kubli 2003). In *C. elegans*, major sperm protein (MSP) released by sperm promotes oocyte growth, meiotic maturation, and ovulation in proximal oocytes via several mechanisms (Miller et al. 2001; Harris et al. 2006; Govindan et al. 2009; Kim et al. 2013). Future studies should consider many possible modes of action for various signals coming from multiple organs in regulating reproductive lineages.

3.4 Discussion

As reviewed here, highly conserved diet-dependent pathways control GSC lineages, revealing interesting similarities and differences in their specific roles in different contexts (Fig. 3.3) (Table 3.1). As the diet-dependent molecular, cellular, and physiological mechanisms controlling GSC lineages are further investigated in multiple models, common themes and more specific strategies shaped by evolution will become clearer. As this field advances, the continued use of tissue- and cell type-specific manipulations in an in vivo setting will be crucial to understand the

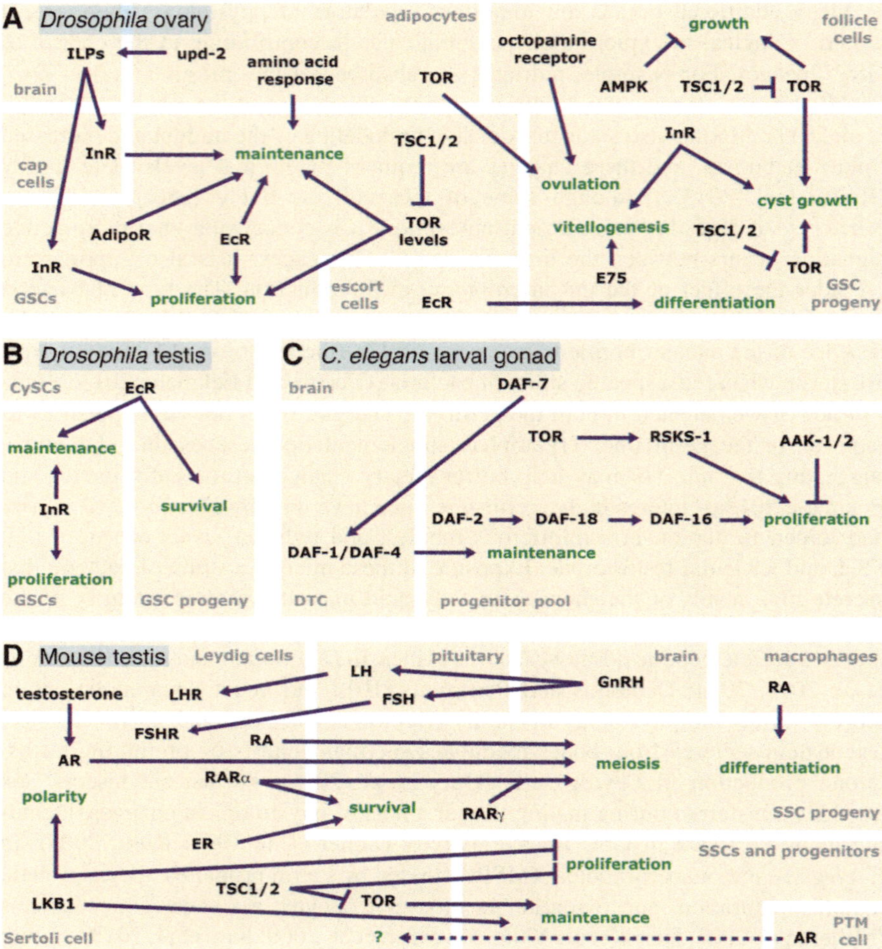

Fig. 3.3 Cell type-specific requirements for diet-dependent pathways involved in the control of GSCs and their progeny. (**a**) In the *Drosophila* ovary, intrinsic, local, and tissue non-autonomous signals coordinately regulate the GSC lineage. In addition to the niche, the brain, follicle cells, and adipocytes all communicate with the GSC lineage as part of the response to diet. (**b**) GSCs in the *Drosophila* testis require InR intrinsically for maintenance and proliferation. EcR in cyst stem cells (CySCs) promotes GSC maintenance and the survival of their progeny. (**c**) In the developing *C. elegans* gonad, but not in the adult, proliferation is regulated intrinsically by insulin, TOR, and AMPK (AAK-1/2). TGFβ ligand (DAF-7) promotes progenitor pool maintenance via the distal tip cell (DTC). (**d**) The mouse SSC pool is regulated by Leydig, Sertoli, and peritubular myeloid (PTM) cells, as well as macrophages. An additional layer of control is provided by the hypothalamic-pituitary-gonad axis, which controls the activity of these somatic support cells to influence spermatogenesis. Biological processes are indicated in *green* and cell types in *gray*. For details, see text and Table 3.1

Table 3.1 Physiological regulation of GSC lineages by diet-dependent pathways

Pathway	Organism	Role
AMPK	*C. elegans*	Inhibits germline proliferation during larval starvation (Fukuyama et al. 2012; Narbonne and Roy 2006b)
	Drosophila	Inhibits follicle cell growth in the ovary (Haack et al. 2013)
	Mouse	Sertoli cell LKB1 promotes SSC proliferation and maintenance (Tanwar et al. 2012)
TOR	*C. elegans*	Promotes larval progenitor proliferation (Korta et al. 2012)
	Drosophila	Controls GSC maintenance and proliferation (LaFever et al. 2010; Sun et al. 2010) and germline cyst survival (LaFever et al. 2010) in the ovary
		Regulates ovarian cyst growth intrinsically and via follicle cells (LaFever et al. 2010; Sun et al. 2010)
	A. aegypti	Elevated in ovary following blood meal (Hansen et al. 2005; Roy and Raikhel 2012)
		Required in the fat body for vitellogenesis (Hansen et al. 2004, 2005; Roy and Raikhel 2011, 2012; Carpenter et al. 2012)
	Mouse (*mTOR*)	Promotes germline proliferation and meiosis during development (Busada et al. 2015)
		Global hyperactivation inhibits SSC maintenance (Hobbs et al. 2010)
		Overactivation in Sertoli cells leads to intrinsic polarity defects, reduced SSC proliferation, and SSC loss (Tanwar et al. 2012)
Insulin	*C. elegans*	Promotes larval germ cell proliferation (Michaelson et al. 2010)
	Drosophila	Promotes ovarian GSC proliferation, cyst growth, and vitellogenesis (LaFever and Drummond-Barbosa 2005)
		Controls ovarian GSC maintenance via cap cells (Hsu and Drummond-Barbosa 2009)
		Promotes GSC proliferation and maintenance in testis (Roth et al. 2012; McLeod et al. 2010)
	A. aegypti	Required for progression through vitellogenesis (reviewed in Attardo et al. 2005)
	Mouse	Global requirement for *InR* and *Igf1r* for testicular development (Pitetti et al. 2013a)
		InR and *Igf1r* promote developmental Sertoli cell proliferation (Pitetti et al. 2013b)
		Global *IRS2* mice have small testes and progressive germ cell loss as adults (Griffeth et al. 2013)
		Global *Igf1* mice have reduced spermatogenesis (Baker et al. 1996)
Ecdysone	*Drosophila*	Promotes ovarian GSC maintenance, proliferation (Ables and Drummond-Barbosa 2010), and vitellogenesis (Buszczak et al. 1999)
		Required in escort cells for germline differentiation (Konig et al. 2011; Morris and Spradling 2012)
		Required in CySCs for GSC maintenance and progeny survival in the testis (Li et al. 2014)
	A. aegypti	After a blood meal, ecdysone response genes expressed in ovary (Pierceall et al. 1999)
		Required in the fat body for vitellogenesis (Martin et al. 2001)

(continued)

Table 3.1 (continued)

Pathway	Organism	Role
Retinoic acid	Mouse	Germ cell RARα and Sertoli cell RARγ are together required for meiosis (Gely-Pernot et al. 2015)
		Required in Sertoli cells for germ cell meiosis (Tong et al. 2013; Raverdeau et al. 2012)
Androgen receptor	Mouse	Required in Sertoli cells for cell survival and meiosis (Abel et al. 2008; Hobbs et al. 2010; De Gendt et al. 2004)
		Required intrinsically by PTMs for normal sperm counts (Zhang et al. 2006)
AdipoR	*C. elegans*	Global mutation reduces brood size (Svensson et al. 2011)
	Drosophila	Required for ovarian GSC maintenance (Laws et al. 2015)
	Mouse	*AdipoR2* global mutants are aspermic (Bjursell et al. 2007)
Leptin	*Drosophila*	Required in fat body to promote ILP secretion from the brain (Rajan and Perrimon 2012)
	Mouse	Promotes fertility (Mounzih et al. 1997) and germ cell survival (Bhat et al. 2006)

full range of contributions of any given hormone or other factor to the regulation of the germ line.

Not surprisingly, many other factors besides diet can impact organismal physiology. For example, changes in germ line activity accompany aging in multiple organisms (Tatar 2010; Oatley and Brinster 2012). *C. elegans* and *Drosophila* fecundity is impaired in older females (Herndon et al. 2002; Partridge and Fowler 1992), sperm aneuploidy increases as mice age (Lowe et al. 1995), and sperm count declines in older men (Eskenazi et al. 2003). The germ line ages intrinsically and is also affected by the aging of the soma. In *C. elegans*, oocytes deteriorate as mated hermaphrodites age, and naturally occurring cell death protects oocyte quality in younger animals, as cell death mutants *ced-3* and *ced-4* have a premature drop in oocyte quality. This is a germline-autonomous effect because mutations that block only somatic cell death do not impair oocyte quality (Andux and Ellis 2008). In contrast, there is a systemic effect of aging in mouse SSCs, as transplantation of SSC from old donors to young recipients restores youthful function to the older SSCs (Ryu et al. 2006). In *Drosophila* males and females, GSCs proliferate more slowly as they age (Wallenfang et al. 2006; Cheng et al. 2008; Pan et al. 2007). By contrast, long-lived *methuselah* mutant males do not experience a decrease in GSC proliferation as they age (Wallenfang et al. 2006). The number of niche cells and of GSCs declines with age in males and females (Wallenfang et al. 2006; Xie and Spradling 2000; Hsu and Drummond-Barbosa 2009; Zhao et al. 2008; Boyle et al. 2007). E-cadherin, and BMP and insulin signaling levels decline with age in the ovary (Pan et al. 2007; Hsu and Drummond-Barbosa 2009), and overexpression of E-cadherin or AdipoR in the germ line (Pan et al. 2007; Laws et al. 2015) or of BMP signals or ILPs in the soma (Pan et al. 2007; Hsu and Drummond-Barbosa 2009) can reverse age-related GSC loss. Somatic and germline overexpression of superoxide dismutase (SOD), an antioxidant enzyme, rescues the age-associated decline of GSC function and cap cell number (Pan et al. 2007), further emphasizing the complexity of effects of aging on the germ line.

Disease states, including cancer, also disrupt organismal homeostasis. Tumors are their own dynamic signaling centers (Karagiannis et al. 2010) and can both hijack normal metabolism to support their growth and secrete factors that have pleiotropic effects (Patel et al. 2014). For example, tumors induced in the midgut of adult *Drosophila* or transplanted into the fly hemocoel secrete the ILP-binding protein IMP-L2 (Kwon et al. 2015; Figueroa-Clarevega and Bilder 2015), causing systemic wasting, including degeneration of ovaries. In humans, this wasting, called cachexia, is a hallmark of end-stage cancer and is uncoupled from tumor burden (Petruzzelli and Wagner 2016; Fearon et al. 2012), underscoring the role of systemic factors in cancer pathologies. While our understanding of the effect of cachexia on mammalian GSC lineages is limited, cachectic patients are often insulin resistant (Honors and Kinzig 2012), and chronic inflammation, found in many cancers (Crusz and Balkwill 2015), can lead to Interleukin-6-mediated hyperactivation of the hypothalamic-pituitary-gonadal axis (Raber et al. 1997). Therefore, although it is clear that many cancer treatments can impair fertility (Suhag et al. 2015; Vakalopoulos et al. 2015), it is also possible that the physiological changes caused by the tumors themselves may have additional effects on the germ line of patients.

Finally, there is also evidence that diet can alter the epigenetic state of the germ line and thereby impact the next generation. For example, male mice fathered by fasted males have lower serum glucose concentrations than those fathered by normally fed animals (Anderson et al. 2006). Adult offspring of *Drosophila* males fed a high sugar diet have increased food intake, higher adiposity, and defects in lipid mobilization, a metabolic signature dependent on specific subtypes of heterochromatin (Ost et al. 2014). Notably, similar genomic derepression patterns are predictive for obesity in mice and humans, indicating a conserved pathway of diet-induced phenotypic variability (Ost et al. 2014). While the effects of paternal diet lasted a single generation in *Drosophila* (Ost et al. 2014), in certain mouse genetic backgrounds, paternal high fat diet causes offspring infertility in males and females for two generations (Fullston et al. 2012). Although the molecular mechanisms remain unknown, epidemiological studies have identified multigenerational ramifications for the children of males exposed to occupational and environmental toxins, including lead and pesticides, many of which could be mediated epigenetically (Soubry et al. 2014). As we learn more about how whole-body physiology controls GSC lineages, new light will be shed on how changes caused by diet, aging, diseases, infections, injuries, or other stressors affect fertility and, potentially, future generations.

3.5 Note Added in Proof

While this review article was in press, a study by Guo et al. (2016) was published showing that secreted paracrine factors from cultured oocytes act in surrounding cumulus cells to activate mTOR (by inhibiting an upstream negative regulator), and that mTOR signaling is important for oocyte-cumulus cell complex maturation and survival.

References

Abel MH, Baker PJ, Charlton HM, Monteiro A, Verhoeven G, De Gendt K, Guillou F, O'Shaughnessy PJ (2008) Spermatogenesis and sertoli cell activity in mice lacking sertoli cell receptors for follicle-stimulating hormone and androgen. Endocrinology 149 (7):3279–3285. doi:10.1210/en.2008-0086

Ables ET, Drummond-Barbosa D (2010) The steroid hormone ecdysone functions with intrinsic chromatin remodeling factors to control female germline stem cells in Drosophila. Cell Stem Cell 7(5):581–592. doi:10.1016/j.stem.2010.10.001

Ables ET, Laws KM, Drummond-Barbosa D (2012) Control of adult stem cells in vivo by a dynamic physiological environment: diet-dependent systemic factors in Drosophila and beyond. Wiley Interdiscip Rev Dev Biol 1(5):657–674. doi:10.1002/wdev.48

Ables ET, Bois KE, Garcia CA, Drummond-Barbosa D (2015) Ecdysone response gene E78 controls ovarian germline stem cell niche formation and follicle survival in Drosophila. Dev Biol 400(1):33–42. doi:10.1016/j.ydbio.2015.01.013

Ables ET, Hwang GH, Finger DS, Hinnant TD, Drummond-Barbosa D (2016) A genetic mosaic screen reveals ecdysone-responsive genes regulating Drosophila oogenesis. G3 (Bethesda) 6 (8):2629–2642. doi:10.1534/g3.116.028951

Ahima RS, Qi Y, Singhal NS (2006) Adipokines that link obesity and diabetes to the hypothalamus. Prog Brain Res 153:155–174. doi:10.1016/S0079-6123(06)53009-2

Al-Asmakh M, Stukenborg JB, Reda A, Anuar F, Strand ML, Hedin L, Pettersson S, Soder O (2014) The gut microbiota and developmental programming of the testis in mice. PLoS One 9 (8):e103809. doi:10.1371/journal.pone.0103809

Anderson LM, Riffle L, Wilson R, Travlos GS, Lubomirski MS, Alvord WG (2006) Preconceptional fasting of fathers alters serum glucose in offspring of mice. Nutrition 22 (3):327–331. doi:10.1016/j.nut.2005.09.006

Andux S, Ellis RE (2008) Apoptosis maintains oocyte quality in aging Caenorhabditis elegans females. PLoS Genet 4(12):e1000295. doi:10.1371/journal.pgen.1000295

Angelo G, Van Gilst MR (2009) Starvation protects germline stem cells and extends reproductive longevity in C. elegans. Science 326(5955):954–958. doi:10.1126/science.1178343

Aquila S, Middea E, Catalano S, Marsico S, Lanzino M, Casaburi I, Barone I, Bruno R, Zupo S, Ando S (2007) Human sperm express a functional androgen receptor: effects on PI3K/AKT pathway. Hum Reprod 22(10):2594–2605. doi:10.1093/humrep/dem243

Armstrong AR, Laws KM, Drummond-Barbosa D (2014) Adipocyte amino acid sensing controls adult germline stem cell number via the amino acid response pathway and independently of Target of Rapamycin signaling in Drosophila. Development 141(23):4479–4488. doi:10.1242/dev.116467

Arrese EL, Soulages JL (2010) Insect fat body: energy, metabolism, and regulation. Annu Rev Entomol 55:207–225. doi:10.1146/annurev-ento-112408-085356

Attardo GM, Hansen IA, Raikhel AS (2005) Nutritional regulation of vitellogenesis in mosquitoes: implications for anautogeny. Insect Biochem Mol Biol 35(7):661–675. doi:10.1016/j.ibmb.2005.02.013

Baglietto L, English DR, Hopper JL, MacInnis RJ, Morris HA, Tilley WD, Krishnan K, Giles GG (2009) Circulating steroid hormone concentrations in postmenopausal women in relation to body size and composition. Breast Cancer Res Treat 115(1):171–179. doi:10.1007/s10549-008-0069-3

Baker J, Hardy MP, Zhou J, Bondy C, Lupu F, Bellve AR, Efstratiadis A (1996) Effects of an Igf1 gene null mutation on mouse reproduction. Mol Endocrinol 10(7):903–918. doi:10.1210/mend.10.7.8813730

Bar-Peled L, Sabatini DM (2014) Regulation of mTORC1 by amino acids. Trends Cell Biol 24 (7):400–406. doi:10.1016/j.tcb.2014.03.003

Bhat GK, Sea TL, Olatinwo MO, Simorangkir D, Ford GD, Ford BD, Mann DR (2006) Influence of a leptin deficiency on testicular morphology, germ cell apoptosis, and expression levels of apoptosis-related genes in the mouse. J Androl 27(2):302–310. doi:10.2164/jandrol.05133

Bjursell M, Ahnmark A, Bohlooly YM, William-Olsson L, Rhedin M, Peng XR, Ploj K, Gerdin AK, Arnerup G, Elmgren A, Berg AL, Oscarsson J, Linden D (2007) Opposing effects of adiponectin receptors 1 and 2 on energy metabolism. Diabetes 56(3):583–593. doi:10.2337/db06-1432

Blomberg Jensen M (2014) Vitamin D and male reproduction. Nat Rev Endocrinol 10(3):175–186. doi:10.1038/nrendo.2013.262

Blomhoff R, Blomhoff HK (2006) Overview of retinoid metabolism and function. J Neurobiol 66 (7):606–630. doi:10.1002/neu.20242

Boucheron-Houston C, Canterel-Thouennon L, Lee TL, Baxendale V, Nagrani S, Chan WY, Rennert OM (2013) Long-term vitamin A deficiency induces alteration of adult mouse spermatogenesis and spermatogonial differentiation: direct effect on spermatogonial gene expression and indirect effects via somatic cells. J Nutr Biochem 24(6):1123–1135. doi:10.1016/j.jnutbio.2012.08.013

Boyle M, Wong C, Rocha M, Jones DL (2007) Decline in self-renewal factors contributes to aging of the stem cell niche in the Drosophila testis. Cell Stem Cell 1(4):470–478. doi:10.1016/j.stem.2007.08.002

Brown MR, Clark KD, Gulia M, Zhao Z, Garczynski SF, Crim JW, Suderman RJ, Strand MR (2008) An insulin-like peptide regulates egg maturation and metabolism in the mosquito Aedes aegypti. Proc Natl Acad Sci U S A 105(15):5716–5721. doi:10.1073/pnas.0800478105

Burn KM, Shimada Y, Ayers K, Lu F, Hudson AM, Cooley L (2015) Somatic insulin signaling regulates a germline starvation response in Drosophila egg chambers. Dev Biol 398 (2):206–217. doi:10.1016/j.ydbio.2014.11.021

Burtis KC, Thummel CS, Jones CW, Karim FD, Hogness DS (1990) The Drosophila 74EF early puff contains E74, a complex ecdysone-inducible gene that encodes two ets-related proteins. Cell 61(1):85–99

Busada JT, Geyer CB (2016) The role of Retinoic Acid (RA) in spermatogonial differentiation. Biol Reprod 94(1):10. doi:10.1095/biolreprod.115.135145

Busada JT, Niedenberger BA, Velte EK, Keiper BD, Geyer CB (2015) Mammalian target of rapamycin complex 1 (mTORC1) Is required for mouse spermatogonial differentiation in vivo. Dev Biol 407(1):90–102. doi:10.1016/j.ydbio.2015.08.004

Buszczak M, Freeman MR, Carlson JR, Bender M, Cooley L, Segraves WA (1999) Ecdysone response genes govern egg chamber development during mid-oogenesis in Drosophila. Development 126(20):4581–4589

Caminos JE, Nogueiras R, Gaytan F, Pineda R, Gonzalez CR, Barreiro ML, Castano JP, Malagon MM, Pinilla L, Toppari J, Dieguez C, Tena-Sempere M (2008) Novel expression and direct effects of adiponectin in the rat testis. Endocrinology 149(7):3390–3402. doi:10.1210/en.2007-1582

Campbell JM, Lane M, Owens JA, Bakos HW (2015) Paternal obesity negatively affects male fertility and assisted reproduction outcomes: a systematic review and meta-analysis. Reprod Biomed Online 31(5):593–604. doi:10.1016/j.rbmo.2015.07.012

Cao H (2014) Adipocytokines in obesity and metabolic disease. J Endocrinol 220(2):T47–T59. doi:10.1530/JOE-13-0339

Carney GE, Bender M (2000) The Drosophila ecdysone receptor (EcR) gene is required maternally for normal oogenesis. Genetics 154(3):1203–1211

Carpenter VK, Drake LL, Aguirre SE, Price DP, Rodriguez SD, Hansen IA (2012) SLC7 amino acid transporters of the yellow fever mosquito Aedes aegypti and their role in fat body TOR signaling and reproduction. J Insect Physiol 58(4):513–522. doi:10.1016/j.jinsphys.2012.01.005

Castellano JM, Navarro VM, Fernandez-Fernandez R, Roa J, Vigo E, Pineda R, Dieguez C, Aguilar E, Pinilla L, Tena-Sempere M (2006) Expression of hypothalamic KiSS-1 system and rescue of defective gonadotropic responses by kisspeptin in streptozotocin-induced diabetic male rats. Diabetes 55(9):2602–2610. doi:10.2337/db05-1584

Cheng J, Turkel N, Hemati N, Fuller MT, Hunt AJ, Yamashita YM (2008) Centrosome misorientation reduces stem cell division during ageing. Nature 456(7222):599–604. doi:10.1038/nature07386

Cho WL, Kapitskaya MZ, Raikhel AS (1995) Mosquito ecdysteroid receptor: analysis of the cDNA and expression during vitellogenesis. Insect Biochem Mol Biol 25(1):19–27

Chung SS, Choi C, Wang X, Hallock L, Wolgemuth DJ (2010) Aberrant distribution of junctional complex components in retinoic acid receptor alpha-deficient mice. Microsc Res Tech 73 (6):583–596. doi:10.1002/jemt.20797

Cognigni P, Bailey AP, Miguel-Aliaga I (2011) Enteric neurons and systemic signals couple nutritional and reproductive status with intestinal homeostasis. Cell Metab 13(1):92–104. doi:10.1016/j.cmet.2010.12.010

Colagar AH, Marzony ET, Chaichi MJ (2009) Zinc levels in seminal plasma are associated with sperm quality in fertile and infertile men. Nutr Res 29(2):82–88. doi:10.1016/j.nutres.2008.11.007

Couzinet B, Young J, Brailly S, Le Bouc Y, Chanson P, Schaison G (1999) Functional hypothalamic amenorrhoea: a partial and reversible gonadotrophin deficiency of nutritional origin. Clin Endocrinol (Oxf) 50(2):229–235

Croxford TP, McCormick NH, Kelleher SL (2011) Moderate zinc deficiency reduces testicular Zip6 and Zip10 abundance and impairs spermatogenesis in mice. J Nutr 141(3):359–365. doi:10.3945/jn.110.131318

Crusz SM, Balkwill FR (2015) Inflammation and cancer: advances and new agents. Nat Rev Clin Oncol 12(10):584–596. doi:10.1038/nrclinonc.2015.105

Dalfo D, Michaelson D, Hubbard EJ (2012) Sensory regulation of the C. elegans germline through TGF-beta-dependent signaling in the niche. Curr Biol 22(8):712–719. doi:10.1016/j.cub.2012.02.064

De Gendt K, Swinnen JV, Saunders PT, Schoonjans L, Dewerchin M, Devos A, Tan K, Atanassova N, Claessens F, Lecureuil C, Heyns W, Carmeliet P, Guillou F, Sharpe RM, Verhoeven G (2004) A Sertoli cell-selective knockout of the androgen receptor causes spermatogenic arrest in meiosis. Proc Natl Acad Sci U S A 101(5):1327–1332. doi:10.1073/pnas.0308114100

de Rooij DG, Russell LD (2000) All you wanted to know about spermatogonia but were afraid to ask. J Androl 21(6):776–798

Deady LD, Sun J (2015) A follicle rupture assay reveals an essential role for follicular adrenergic signaling in Drosophila ovulation. PLoS Genet 11(10):e1005604. doi:10.1371/journal.pgen.1005604

DeFalco T, Potter SJ, Williams AV, Waller B, Kan MJ, Capel B (2015) Macrophages contribute to the spermatogonial niche in the adult testis. Cell Rep 12(7):1107–1119. doi:10.1016/j.celrep.2015.07.015

Demontis F, Perrimon N (2010) FOXO/4E-BP signaling in Drosophila muscles regulates organism-wide proteostasis during aging. Cell 143(5):813–825. doi:10.1016/j.cell.2010.10.007

Demontis F, Piccirillo R, Goldberg AL, Perrimon N (2013) The influence of skeletal muscle on systemic aging and lifespan. Aging Cell 12(6):943–949. doi:10.1111/acel.12126

Demontis F, Patel VK, Swindell WR, Perrimon N (2014) Intertissue control of the nucleolus via a myokine-dependent longevity pathway. Cell Rep 7(5):1481–1494. doi:10.1016/j.celrep.2014.05.001

Devreotes P, Horwitz AR (2015) Signaling networks that regulate cell migration. Cold Spring Harb Perspect Biol 7(8):a005959. doi:10.1101/cshperspect.a005959

Dillin A, Crawford DK, Kenyon C (2002) Timing requirements for insulin/IGF-1 signaling in C. elegans. Science 298(5594):830–834. doi:10.1126/science.1074240

DiVall SA, Herrera D, Sklar B, Wu S, Wondisford F, Radovick S, Wolfe A (2015) Insulin receptor signaling in the GnRH neuron plays a role in the abnormal GnRH pulsatility of obese female mice. PLoS One 10(3):e0119995. doi:10.1371/journal.pone.0119995

Drummond-Barbosa D, Spradling AC (2001) Stem cells and their progeny respond to nutritional changes during Drosophila oogenesis. Dev Biol 231(1):265–278. doi:10.1006/dbio.2000.0135

Dupont J, Reverchon M, Bertoldo MJ, Froment P (2014) Nutritional signals and reproduction. Mol Cell Endocrinol 382(1):527–537. doi:10.1016/j.mce.2013.09.028

Endo T, Romer KA, Anderson EL, Baltus AE, de Rooij DG, Page DC (2015) Periodic retinoic acid-STRA8 signaling intersects with periodic germ-cell competencies to regulate spermatogenesis. Proc Natl Acad Sci U S A 112(18):E2347–E2356. doi:10.1073/pnas.1505683112

Eskenazi B, Wyrobek AJ, Sloter E, Kidd SA, Moore L, Young S, Moore D (2003) The association of age and semen quality in healthy men. Hum Reprod 18(2):447–454

Evans MC, Rizwan M, Mayer C, Boehm U, Anderson GM (2014) Evidence that insulin signalling in gonadotrophin-releasing hormone and kisspeptin neurones does not play an essential role in metabolic regulation of fertility in mice. J Neuroendocrinol 26(7):468–479. doi:10.1111/jne.12166

Fan Y, Liu Y, Xue K, Gu G, Fan W, Xu Y, Ding Z (2015) Diet-induced obesity in male C57BL/6 mice decreases fertility as a consequence of disrupted blood-testis barrier. PLoS One 10(4):e0120775. doi:10.1371/journal.pone.0120775

Fearon KC, Glass DJ, Guttridge DC (2012) Cancer cachexia: mediators, signaling, and metabolic pathways. Cell Metab 16(2):153–166. doi:10.1016/j.cmet.2012.06.011

Fernandez CD, Bellentani FF, Fernandes GS, Perobelli JE, Favareto AP, Nascimento AF, Cicogna AC, Kempinas WD (2011) Diet-induced obesity in rats leads to a decrease in sperm motility. Reprod Biol Endocrinol 9:32. doi:10.1186/1477-7827-9-32

Figueroa-Clarevega A, Bilder D (2015) Malignant Drosophila tumors interrupt insulin signaling to induce cachexia-like wasting. Dev Cell 33(1):47–55. doi:10.1016/j.devcel.2015.03.001

Franca LR, Hess RA, Dufour JM, Hofmann MC, Griswold MD (2016) The Sertoli cell: one hundred fifty years of beauty and plasticity. Andrology 4(2):189–212. doi:10.1111/andr.12165

Fukuyama M, Sakuma K, Park R, Kasuga H, Nagaya R, Atsumi Y, Shimomura Y, Takahashi S, Kajiho H, Rougvie A, Kontani K, Katada T (2012) C. elegans AMPKs promote survival and arrest germline development during nutrient stress. Biol Open 1(10):929–936. doi:10.1242/bio.2012836

Fuller M (1993) Spermatogenesis. In: Bate M, Arias AM (eds) The development of Drosophila melanogaster, vol I. Cold Spring Harbor Laboratory Press, New York, pp 71–147

Fullston T, Palmer NO, Owens JA, Mitchell M, Bakos HW, Lane M (2012) Diet-induced paternal obesity in the absence of diabetes diminishes the reproductive health of two subsequent generations of mice. Hum Reprod 27(5):1391–1400. doi:10.1093/humrep/des030

Gely-Pernot A, Raverdeau M, Celebi C, Dennefeld C, Feret B, Klopfenstein M, Yoshida S, Ghyselinck NB, Mark M (2012) Spermatogonia differentiation requires retinoic acid receptor gamma. Endocrinology 153(1):438–449. doi:10.1210/en.2011-1102

Gely-Pernot A, Raverdeau M, Teletin M, Vernet N, Feret B, Klopfenstein M, Dennefeld C, Davidson I, Benoit G, Mark M, Ghyselinck NB (2015) Retinoic acid receptors control spermatogonia cell-fate and induce expression of the SALL4A transcription factor. PLoS Genet 11(10):e1005501. doi:10.1371/journal.pgen.1005501

Ghyselinck NB, Vernet N, Dennefeld C, Giese N, Nau H, Chambon P, Viville S, Mark M (2006) Retinoids and spermatogenesis: lessons from mutant mice lacking the plasma retinol binding protein. Dev Dyn 235(6):1608–1622. doi:10.1002/dvdy.20795

Gourgari E, Saloustros E, Stratakis CA (2012) Large-cell calcifying Sertoli cell tumors of the testes in pediatrics. Curr Opin Pediatr 24(4):518–522. doi:10.1097/MOP.0b013e328355a279

Govindan JA, Nadarajan S, Kim S, Starich TA, Greenstein D (2009) Somatic cAMP signaling regulates MSP-dependent oocyte growth and meiotic maturation in C. elegans. Development 136(13):2211–2221. doi:10.1242/dev.034595

Gracida X, Eckmann CR (2013a) Fertility and germline stem cell maintenance under different diets requires nhr-114/HNF4 in C. elegans. Curr Biol 23(7):607–613. doi:10.1016/j.cub.2013.02.034

Gracida X, Eckmann CR (2013b) Mind the gut: dietary impact on germline stem cells and fertility. Commun Integr Biol 6(6):e26004. doi:10.4161/cib.26004

Greenspan LJ, de Cuevas M, Matunis E (2015) Genetics of gonadal stem cell renewal. Annu Rev Cell Dev Biol 31:291–315. doi:10.1146/annurev-cellbio-100913-013344

Griffeth RJ, Carretero J, Burks DJ (2013) Insulin receptor substrate 2 is required for testicular development. PLoS One 8(5):e62103. doi:10.1371/journal.pone.0062103

Group ECW (2006) Nutrition and reproduction in women. Hum Reprod Update 12(3):193–207. doi:10.1093/humupd/dmk003

Gruenewald DA, Matsumoto AM (1993) Reduced gonadotropin-releasing hormone gene expression with fasting in the male rat brain. Endocrinology 132(1):480–482. doi:10.1210/endo.132.1.8419144

Gulia-Nuss M, Robertson AE, Brown MR, Strand MR (2011) Insulin-like peptides and the target of rapamycin pathway coordinately regulate blood digestion and egg maturation in the mosquito Aedes aegypti. PLoS One 6(5):e20401. doi:10.1371/journal.pone.0020401

Guo J, Shi L, Gong X, Jiang M, Yin Y, Zhang X, Yin H, Li H, Emori C, Sugiura K, Eppig JJ, Su YQ (2016) Oocyte-dependent activation of MTOR in cumulus cells controls the development and survival of cumulus-oocyte complexes. J Cell Sci 129:3091–3103

Gutierrez E, Wiggins D, Fielding B, Gould AP (2007) Specialized hepatocyte-like cells regulate Drosophila lipid metabolism. Nature 445(7125):275–280. doi:10.1038/nature05382

Haack T, Bergstralh DT, St Johnston D (2013) Damage to the Drosophila follicle cell epithelium produces "false clones" with apparent polarity phenotypes. Biol Open 2(12):1313–1320. doi:10.1242/bio.20134671

Hansen IA, Attardo GM, Park JH, Peng Q, Raikhel AS (2004) Target of rapamycin-mediated amino acid signaling in mosquito anautogeny. Proc Natl Acad Sci U S A 101 (29):10626–10631. doi:10.1073/pnas.0403460101

Hansen IA, Attardo GM, Roy SG, Raikhel AS (2005) Target of rapamycin-dependent activation of S6 kinase is a central step in the transduction of nutritional signals during egg development in a mosquito. J Biol Chem 280(21):20565–20572. doi:10.1074/jbc.M500712200

Hardie DG (2015) AMPK: positive and negative regulation, and its role in whole-body energy homeostasis. Curr Opin Cell Biol 33:1–7. doi:10.1016/j.ceb.2014.09.004

Hardie DG, Ashford ML (2014) AMPK: regulating energy balance at the cellular and whole body levels. Physiology (Bethesda) 29(2):99–107. doi:10.1152/physiol.00050.2013

Hardie DG, Schaffer BE, Brunet A (2016) AMPK: an energy-sensing pathway with multiple inputs and outputs. Trends Cell Biol 26(3):190–201. doi:10.1016/j.tcb.2015.10.013

Harris JE, Govindan JA, Yamamoto I, Schwartz J, Kaverina I, Greenstein D (2006) Major sperm protein signaling promotes oocyte microtubule reorganization prior to fertilization in Caenorhabditis elegans. Dev Biol 299(1):105–121. doi:10.1016/j.ydbio.2006.07.013

Hemminkl A, Markie D, Tomlinson I, Avizlenyte E, Roth S, Loukola A, Bignell G, Warren W, Aminoff M, Hoglund P, Jarvinen H, Kristo P, Pelln K, Ridanpaa M, Salovaara R, Toro T,

Bodmer W, Olschwang S, Olsen AS, Stratton MR, de la Chapelle A, Aaltonen LA (1998) A serine/threonine kinase gene defective in Peutz-Jeghers syndrome. Nature 391:184–187

Herndon LA, Schmeissner PJ, Dudaronek JM, Brown PA, Listner KM, Sakano Y, Paupard MC, Hall DH, Driscoll M (2002) Stochastic and genetic factors influence tissue-specific decline in ageing C. elegans. Nature 419(6909):808–814. doi:10.1038/nature01135

Hindupur SK, Gonzalez A, Hall MN (2015) The opposing actions of target of rapamycin and AMP-activated protein kinase in cell growth control. Cold Spring Harb Perspect Med 5(7): a019141

Hobbs RM, Seandel M, Falciatori I, Rafii S, Pandolfi PP (2010) Plzf regulates germline progenitor self-renewal by opposing mTORC1. Cell 142(3):468–479. doi:10.1016/j.cell.2010.06.041

Hogarth CA, Griswold MD (2013) Retinoic acid regulation of male meiosis. Curr Opin Endocrinol Diabetes Obes 20(3):217–223. doi:10.1097/MED.0b013e32836067cf

Honors MA, Kinzig KP (2012) The role of insulin resistance in the development of muscle wasting during cancer cachexia. J Cachexia Sarcopenia Muscle 3(1):5–11. doi:10.1007/s13539-011-0051-5

Hsu HJ, Drummond-Barbosa D (2009) Insulin levels control female germline stem cell maintenance via the niche in Drosophila. Proc Natl Acad Sci U S A 106(4):1117–1121. doi:10.1073/pnas.0809144106

Hsu HJ, Drummond-Barbosa D (2011) Insulin signals control the competence of the Drosophila female germline stem cell niche to respond to Notch ligands. Dev Biol 350(2):290–300. doi:10.1016/j.ydbio.2010.11.032

Hsu HJ, LaFever L, Drummond-Barbosa D (2008) Diet controls normal and tumorous germline stem cells via insulin-dependent and -independent mechanisms in Drosophila. Dev Biol 313 (2):700–712. doi:10.1016/j.ydbio.2007.11.006

Huang K, Fingar DC (2014) Growing knowledge of the mTOR signaling network. Semin Cell Dev Biol 36:79–90. doi:10.1016/j.semcdb.2014.09.011

Hubbard EJ (2011) Insulin and germline proliferation in Caenorhabditis elegans. Vitam Horm 87:61–77. doi:10.1016/B978-0-12-386015-6.00024-X

Hubbard EJ, Greenstein D (2005) Introduction to the germ line. WormBook:1–4. doi:10.1895/wormbook.1.18.1

Ishikawa T, Fujioka H, Ishimura T, Takenaka A, Fujisawa M (2007) Expression of leptin and leptin receptor in the testis of fertile and infertile patients. Andrologia 39(1):22–27. doi:10.1111/j.1439-0272.2006.00754.x

Kannan K, Fridell YW (2013) Functional implications of Drosophila insulin-like peptides in metabolism, aging, and dietary restriction. Front Physiol 4:288. doi:10.3389/fphys.2013.00288

Karagiannis GS, Pavlou MP, Diamandis EP (2010) Cancer secretomics reveal pathophysiological pathways in cancer molecular oncology. Mol Oncol 4(6):496–510. doi:10.1016/j.molonc.2010.09.001

Karim FD, Urness LD, Thummel CS, Klemsz MJ, McKercher SR, Celada A, Van Beveren C, Maki RA, Gunther CV, Nye JA et al (1990) The ETS-domain: a new DNA-binding motif that recognizes a purine-rich core DNA sequence. Genes Dev 4(9):1451–1453

Kawwass JF, Summer R, Kallen CB (2015) Direct effects of leptin and adiponectin on peripheral reproductive tissues: a critical review. Mol Hum Reprod 21(8):617–632. doi:10.1093/molehr/gav025

Kelly DM, Jones TH (2013) Testosterone: a metabolic hormone in health and disease. J Endocrinol 217(3):R25–R45. doi:10.1530/JOE-12-0455

Kim S, Spike C, Greenstein D (2013) Control of oocyte growth and meiotic maturation in Caenorhabditis elegans. Adv Exp Med Biol 757:277–320. doi:10.1007/978-1-4614-4015-4_10

Kimble J, Seidel H (2008) C. elegans germline stem cells and their niche. StemBook, Cambridge, MA. doi:10.3824/stembook.1.95.1

King-Jones K, Thummel CS (2005) Nuclear receptors—a perspective from Drosophila. Nat Rev Genet 6(4):311–323. doi:10.1038/nrg1581

Kirilly D, Wang S, Xie T (2011) Self-maintained escort cells form a germline stem cell differentiation niche. Development 138(23):5087–5097. doi:10.1242/dev.067850

Konig A, Yatsenko AS, Weiss M, Shcherbata HR (2011) Ecdysteroids affect Drosophila ovarian stem cell niche formation and early germline differentiation. EMBO J 30(8):1549–1562. doi:10.1038/emboj.2011.73

Korta DZ, Tuck S, Hubbard EJ (2012) S6K links cell fate, cell cycle and nutrient response in C. elegans germline stem/progenitor cells. Development 139(5):859–870. doi:10.1242/dev.074047

Kovacs P, Parlow AF, Karkanias GB (2002) Effect of centrally administered insulin on gonadotropin-releasing hormone neuron activity and luteinizing hormone surge in the diabetic female rat. Neuroendocrinology 76(6):357–365, doi:67585

Krishnamurthy H, Danilovich N, Morales CR, Sairam MR (2000) Qualitative and quantitative decline in spermatogenesis of the follicle-stimulating hormone receptor knockout (FORKO) mouse. Biol Reprod 62(5):1146–1159

Kubli E (2003) Sex-peptides: seminal peptides of the Drosophila male. Cell Mol Life Sci 60 (8):1689–1704. doi:10.1007/s00018-003-3052

Kumar TR, Wang Y, Lu N, Matzuk MM (1997) Follicle stimulating hormone is required for ovarian follicle maturation but not male fertility. Nat Genet 15(2):201–204. doi:10.1038/ng0297-201

Kwak SJ, Hong SH, Bajracharya R, Yang SY, Lee KS, Yu K (2013) Drosophila adiponectin receptor in insulin producing cells regulates glucose and lipid metabolism by controlling insulin secretion. PLoS One 8(7):e68641. doi:10.1371/journal.pone.0068641

Kwon Y, Song W, Droujinine IA, Hu Y, Asara JM, Perrimon N (2015) Systemic organ wasting induced by localized expression of the secreted insulin/IGF antagonist ImpL2. Dev Cell 33 (1):36–46. doi:10.1016/j.devcel.2015.02.012

L'Hernault SW (2009) The genetics and cell biology of spermatogenesis in the nematode C. elegans. Mol Cell Endocrinol 306(1–2):59–65. doi:10.1016/j.mce.2009.01.008

LaFever L, Drummond-Barbosa D (2005) Direct control of germline stem cell division and cyst growth by neural insulin in Drosophila. Science 309(5737):1071–1073. doi:10.1126/science.1111410

LaFever L, Feoktistov A, Hsu HJ, Drummond-Barbosa D (2010) Specific roles of target of rapamycin in the control of stem cells and their progeny in the Drosophila ovary. Development 137(13):2117–2126. doi:10.1242/dev.050351

Landry D, Cloutier F, Martin LJ (2013) Implications of leptin in neuroendocrine regulation of male reproduction. Reprod Biol 13(1):1–14. doi:10.1016/j.repbio.2012.12.001

Laws KM, Sampson LL, Drummond-Barbosa D (2015) Insulin-independent role of adiponectin receptor signaling in Drosophila germline stem cell maintenance. Dev Biol 399(2):226–236. doi:10.1016/j.ydbio.2014.12.033

Lee HG, Seong CS, Kim YC, Davis RL, Han KA (2003) Octopamine receptor OAMB is required for ovulation in Drosophila melanogaster. Dev Biol 264(1):179–190

Li Y, Ma Q, Cherry CM, Matunis EL (2014) Steroid signaling promotes stem cell maintenance in the Drosophila testis. Dev Biol 394(1):129–141. doi:10.1016/j.ydbio.2014.07.016

Lopez AL 3rd, Chen J, Joo HJ, Drake M, Shidate M, Kseib C, Arur S (2013) DAF-2 and ERK couple nutrient availability to meiotic progression during Caenorhabditis elegans oogenesis. Dev Cell 27(2):227–240. doi:10.1016/j.devcel.2013.09.008

Lowe X, Collins B, Allen J, Titenko-Holland N, Breneman J, van Beek M, Bishop J, Wyrobek AJ (1995) Aneuploidies and micronuclei in the germ cells of male mice of advanced age. Mutat Res 338(1–6):59–76

Mair W, McLeod CJ, Wang L, Jones DL (2010) Dietary restriction enhances germline stem cell maintenance. Aging Cell 9(5):916–918. doi:10.1111/j.1474-9726.2010.00602.x

Marcus R, Korenman SG (1976) Estrogens and the human male. Annu Rev Med 27:357–370

Martin LJ (2014) Implications of adiponectin in linking metabolism to testicular function. Endocrine 46(1):16–28. doi:10.1007/s12020-013-0102-0

Martin D, Wang SF, Raikhel AS (2001) The vitellogenin gene of the mosquito Aedes aegypti is a direct target of ecdysteroid receptor. Mol Cell Endocrinol 173(1–2):75–86

McLeod CJ, Wang L, Wong C, Jones DL (2010) Stem cell dynamics in response to nutrient availability. Curr Biol 20(23):2100–2105. doi:10.1016/j.cub.2010.10.038

Michaelson D, Korta DZ, Capua Y, Hubbard EJ (2010) Insulin signaling promotes germline proliferation in C. elegans. Development 137(4):671–680. doi:10.1242/dev.042523

Miller MA, Nguyen VQ, Lee MH, Kosinski M, Schedl T, Caprioli RM, Greenstein D (2001) A sperm cytoskeletal protein that signals oocyte meiotic maturation and ovulation. Science 291 (5511):2144–2147. doi:10.1126/science.1057586

Monastirioti M (2003) Distinct octopamine cell population residing in the CNS abdominal ganglion controls ovulation in Drosophila melanogaster. Dev Biol 264(1):38–49

Moran O, Phillip M (2003) Leptin: obesity, diabetes and other peripheral effects—a review. Pediatr Diabetes 4(2):101–109. doi:10.1034/j.1399-5448.2003.00017.x

Moret M, Stettler R, Rodieux F, Gaillard RC, Waeber G, Wirthner D, Giusti V, Tappy L, Pralong FP (2009) Insulin modulation of luteinizing hormone secretion in normal female volunteers and lean polycystic ovary syndrome patients. Neuroendocrinology 89(2):131–139. doi:10.1159/000160911

Morris LX, Spradling AC (2012) Steroid signaling within Drosophila ovarian epithelial cells sex-specifically modulates early germ cell development and meiotic entry. PLoS One 7(10): e46109. doi:10.1371/journal.pone.0046109

Mounzih K, Lu R, Chehab FF (1997) Leptin treatment rescues the sterility of genetically obese ob/ob males. Endocrinology 138(3):1190–1193. doi:10.1210/endo.138.3.5024

Mullaney BC, Ashrafi K (2009) C. elegans fat storage and metabolic regulation. Biochim Biophys Acta 1791(6):474–478. doi:10.1016/j.bbalip.2008.12.013

Murphy CT, Hu PJ (2013) Insulin/insulin-like growth factor signaling in C. elegans. WormBook:1–43. doi:10.1895/wormbook.1.164.1

Narbonne P, Roy R (2006a) Inhibition of germline proliferation during C. elegans dauer development requires PTEN, LKB1 and AMPK signalling. Development 133(4):611–619. doi:10.1242/dev.02232

Narbonne P, Roy R (2006b) Regulation of germline stem cell proliferation downstream of nutrient sensing. Cell Div 1:29. doi:10.1186/1747-1028-1-29

O'Shaughnessy PJ (2014) Hormonal control of germ cell development and spermatogenesis. Semin Cell Dev Biol 29:55–65. doi:10.1016/j.semcdb.2014.02.010

Oatley JM, Brinster RL (2012) The germline stem cell niche unit in mammalian testes. Physiol Rev 92(2):577–595. doi:10.1152/physrev.00025.2011

Ost A, Lempradl A, Casas E, Weigert M, Tiko T, Deniz M, Pantano L, Boenisch U, Itskov PM, Stoeckius M, Ruf M, Rajewsky N, Reuter G, Iovino N, Ribeiro C, Alenius M, Heyne S, Vavouri T, Pospisilik JA (2014) Paternal diet defines offspring chromatin state and intergenerational obesity. Cell 159(6):1352–1364. doi:10.1016/j.cell.2014.11.005

Oury F, Sumara G, Sumara O, Ferron M, Chang H, Smith CE, Hermo L, Suarez S, Roth BL, Ducy P, Karsenty G (2011) Endocrine regulation of male fertility by the skeleton. Cell 144 (5):796–809. doi:10.1016/j.cell.2011.02.004

Palmer NO, Bakos HW, Fullston T, Lane M (2012) Impact of obesity on male fertility, sperm function and molecular composition. Spermatogenesis 2(4):253–263. doi:10.4161/spmg.21362

Pan L, Chen S, Weng C, Call G, Zhu D, Tang H, Zhang N, Xie T (2007) Stem cell aging is controlled both intrinsically and extrinsically in the Drosophila ovary. Cell Stem Cell 1 (4):458–469. doi:10.1016/j.stem.2007.09.010

Partridge L, Fowler K (1992) Direct and correlated responses to selection on age at reproduction in Drosophila melanogaster. Evolution 46(1):76–91. doi:10.2307/2409806

Patel S, Ngounou Wetie AG, Darie CC, Clarkson BD (2014) Cancer secretomes and their place in supplementing other hallmarks of cancer. Adv Exp Med Biol 806:409–442. doi:10.1007/978-3-319-06068-2_20

Petruzzelli M, Wagner EF (2016) Mechanisms of metabolic dysfunction in cancer-associated cachexia. Genes Dev 30(5):489–501. doi:10.1101/gad.276733.115

Phillip M, Arbelle JE, Segev Y, Parvari R (1998) Male hypogonadism due to a mutation in the gene for the beta-subunit of follicle-stimulating hormone. N Engl J Med 338(24):1729–1732. doi:10.1056/NEJM199806113382404

Pierceall WE, Li C, Biran A, Miura K, Raikhel AS, Segraves WA (1999) E75 expression in mosquito ovary and fat body suggests reiterative use of ecdysone-regulated hierarchies in development and reproduction. Mol Cell Endocrinol 150(1–2):73–89

Pitetti JL, Calvel P, Romero Y, Conne B, Truong V, Papaioannou MD, Schaad O, Docquier M, Herrera PL, Wilhelm D, Nef S (2013a) Insulin and IGF1 receptors are essential for XX and XY gonadal differentiation and adrenal development in mice. PLoS Genet 9(1):e1003160. doi:10.1371/journal.pgen.1003160

Pitetti JL, Calvel P, Zimmermann C, Conne B, Papaioannou MD, Aubry F, Cederroth CR, Urner F, Fumel B, Crausaz M, Docquier M, Herrera PL, Pralong F, Germond M, Guillou F, Jegou B, Nef S (2013b) An essential role for insulin and IGF1 receptors in regulating sertoli cell proliferation, testis size, and FSH action in mice. Mol Endocrinol 27(5):814–827. doi:10.1210/me.2012-1258

Pritchett TL, McCall K (2012) Role of the insulin/Tor signaling network in starvation-induced programmed cell death in Drosophila oogenesis. Cell Death Differ 19(6):1069–1079. doi:10.1038/cdd.2011.200

Raber J, O'Shea RD, Bloom FE, Campbell IL (1997) Modulation of hypothalamic-pituitary-adrenal function by transgenic expression of interleukin-6 in the CNS of mice. J Neurosci 17 (24):9473–9480

Rajan A, Perrimon N (2012) Drosophila cytokine unpaired 2 regulates physiological homeostasis by remotely controlling insulin secretion. Cell 151(1):123–137. doi:10.1016/j.cell.2012.08.019

Raverdeau M, Gely-Pernot A, Feret B, Dennefeld C, Benoit G, Davidson I, Chambon P, Mark M, Ghyselinck NB (2012) Retinoic acid induces Sertoli cell paracrine signals for spermatogonia differentiation but cell autonomously drives spermatocyte meiosis. Proc Natl Acad Sci U S A 109(41):16582–16587. doi:10.1073/pnas.1214936109

Reiff T, Jacobson J, Cognigni P, Antonello Z, Ballesta E, Tan KJ, Yew JY, Dominguez M, Miguel-Aliaga I (2015) Endocrine remodelling of the adult intestine sustains reproduction in Drosophila. Elife 4:e06930. doi:10.7554/eLife.06930

Ren P, Lim CS, Johnsen R, Albert PS, Pilgrim D, Riddle DL (1996) Control of C. elegans larval development by neuronal expression of a TGF-beta homolog. Science 274(5291):1389–1391

Robertson KM, O'Donnell L, Jones ME, Meachem SJ, Boon WC, Fisher CR, Graves KH, McLachlan RI, Simpson ER (1999) Impairment of spermatogenesis in mice lacking a functional aromatase (cyp 19) gene. Proc Natl Acad Sci U S A 96(14):7986–7991

Rogers-Bennett LD, Dondanville RF, Moore JD, Vilchis LI (2010) Response of red abalone reproduction to warm water, starvation, and disease stressors: implications of ocean warming. J Shellfish Res 29(3):599–611. doi:10.2983/035.029.0308

Rojas J, Chavez-Castillo M, Olivar LC, Calvo M, Mejias J, Rojas M, Morillo J, Bermudez V (2015) Physiologic course of female reproductive function: a molecular look into the prologue of life. J Pregnancy 2015:715735. doi:10.1155/2015/715735

Roth TM, Chiang CY, Inaba M, Yuan H, Salzmann V, Roth CE, Yamashita YM (2012) Centrosome misorientation mediates slowing of the cell cycle under limited nutrient conditions in Drosophila male germline stem cells. Mol Biol Cell 23(8):1524–1532. doi:10.1091/mbc.E11-12-0999

Roy SG, Raikhel AS (2011) The small GTPase Rheb is a key component linking amino acid signaling and TOR in the nutritional pathway that controls mosquito egg development. Insect Biochem Mol Biol 41(1):62–69. doi:10.1016/j.ibmb.2010.10.001

Roy SG, Raikhel AS (2012) Nutritional and hormonal regulation of the TOR effector 4E-binding protein (4E-BP) in the mosquito Aedes aegypti. FASEB J 26(3):1334–1342. doi:10.1096/fj.11-189969

Ryu BY, Orwig KE, Oatley JM, Avarbock MR, Brinster RL (2006) Effects of aging and niche microenvironment on spermatogonial stem cell self-renewal. Stem Cells 24(6):1505–1511. doi:10.1634/stemcells.2005-0580

Sar M, Lubahn DB, French FS, Wilson EM (1990) Immunohistochemical localization of the androgen receptor in rat and human tissues. Endocrinology 127(6):3180–3186. doi:10.1210/endo-127-6-3180

Schackwitz WS, Inoue T, Thomas JH (1996) Chemosensory neurons in parallel to mediate a pheromone response in C. elegans. Neuron 17:719–728

Schwedes CC, Carney GE (2012) Ecdysone signaling in adult Drosophila melanogaster. J Insect Physiol 58(3):293–302. doi:10.1016/j.jinsphys.2012.01.013

Seidel HS, Kimble J (2015) Cell-cycle quiescence maintains Caenorhabditis elegans germline stem cells independent of GLP-1/Notch. Elife 4. doi:10.7554/eLife.10832

Siddle K (2011) Signalling by insulin and IGF receptors: supporting acts and new players. J Mol Endocrinol 47(1):R1–R10. doi:10.1530/JME-11-0022

Sieber MH, Spradling AC (2015) Steroid signaling establishes a female metabolic state and regulates SREBP to control oocyte lipid accumulation. Curr Biol 25(8):993–1004. doi:10.1016/j.cub.2015.02.019

Sieber MH, Thomsen MB, Spradling AC (2016) Electron transport chain remodeling by GSK3 during oogenesis connects nutrient state to reproduction. Cell 164(3):420–432. doi:10.1016/j.cell.2015.12.020

Simpson E, Rubin G, Clyne C, Robertson K, O'Donnell L, Davis S, Jones M (1999) Local estrogen biosynthesis in males and females. Endocr Relat Cancer 6(2):131–137

Simpson E, Rubin G, Clyne C, Robertson K, O'Donnell L, Jones M, Davis S (2000) The role of local estrogen biosynthesis in males and females. Trends Endocrinol Metab 11(5):184–188

Sinkevicius KW, Burdette JE, Woloszyn K, Hewitt SC, Hamilton K, Sugg SL, Temple KA, Wondisford FE, Korach KS, Woodruff TK, Greene GL (2008) An estrogen receptor-alpha knock-in mutation provides evidence of ligand-independent signaling and allows modulation of ligand-induced pathways in vivo. Endocrinology 149(6):2970–2979. doi:10.1210/en.2007-1526

Sinkevicius KW, Laine M, Lotan TL, Woloszyn K, Richburg JH, Greene GL (2009) Estrogen-dependent and -independent estrogen receptor-alpha signaling separately regulate male fertility. Endocrinology 150(6):2898–2905. doi:10.1210/en.2008-1016

Soller M, Bownes M, Kubli E (1997) Mating and sex peptide stimulate the accumulation of yolk in oocytes of Drosophila melanogaster. Eur J Biochem 243(3):732–738

Song X, Call GB, Kirilly D, Xie T (2007) Notch signaling controls germline stem cell niche formation in the *Drosophila* ovary. Development 134:1071–1080. doi:10.1242/dev.003392

Soubry A, Hoyo C, Jirtle RL, Murphy SK (2014) A paternal environmental legacy: evidence for epigenetic inheritance through the male germ line. Bioessays 36(4):359–371. doi:10.1002/bies.201300113

Spradling AC (1993) Developmental genetics of oogenesis. In: Bate M (ed) The Development of *Drosophila melanogaster*. Cold Spring Harbor Laboratory Press, Plainview, NY

Stone BL, Thummel CS (1993) The Drosophila 78C early late puff contains E78, an ecdysone-inducible gene that encodes a novel member of the nuclear hormone receptor superfamily. Cell 75(2):307–320

Suhag V, Sunita BS, Sarin A, Singh AK, Dashottar S (2015) Fertility preservation in young patients with cancer. South Asian J Cancer 4(3):134–139. doi:10.4103/2278-330X.173175

Sun P, Quan Z, Zhang B, Wu T, Xi R (2010) TSC1/2 tumour suppressor complex maintains Drosophila germline stem cells by preventing differentiation. Development 137(15):2461–2469. doi:10.1242/dev.051466

Sun W, Chen L, Zhang W, Wang R, Goltzman D, Miao D (2015) Active vitamin D deficiency mediated by extracellular calcium and phosphorus results in male infertility in young mice. Am J Physiol Endocrinol Metab 308(1):E51–E62. doi:10.1152/ajpendo.00076.2014

Svensson E, Olsen L, Morck C, Brackmann C, Enejder A, Faergeman NJ, Pilon M (2011) The adiponectin receptor homologs in C. elegans promote energy utilization and homeostasis. PLoS One 6(6):e21343. doi:10.1371/journal.pone.0021343

Swevers L, Iatrou K. (2009) Ecdysteroids and ecdysteroid signaling pathways during insect oogenesis

Tanwar PS, Kaneko-Tarui T, Zhang L, Teixeira JM (2012) Altered LKB1/AMPK/TSC1/TSC2/mTOR signaling causes disruption of Sertoli cell polarity and spermatogenesis. Hum Mol Genet 21(20):4394–4405. doi:10.1093/hmg/dds272

Tatar M (2010) Reproductive aging in invertebrate genetic models. Ann N Y Acad Sci 1204:149–155. doi:10.1111/j.1749-6632.2010.05522.x

Tong MH, Yang QE, Davis JC, Griswold MD (2013) Retinol dehydrogenase 10 is indispensible for spermatogenesis in juvenile males. Proc Natl Acad Sci U S A 110(2):543–548. doi:10.1073/pnas.1214883110

Trumble BC, Brindle E, Kupsik M, O'Connor KA (2010) Responsiveness of the reproductive axis to a single missed evening meal in young adult males. Am J Hum Biol 22(6):775–781. doi:10.1002/ajhb.21079

Tu MP, Yin CM, Tatar M (2002) Impaired ovarian ecdysone synthesis of Drosophila melanogaster insulin receptor mutants. Aging Cell 1(2):158–160

Uryu O, Ameku T, Niwa R (2015) Recent progress in understanding the role of ecdysteroids in adult insects: germline development and circadian clock in the fruit fly Drosophila melanogaster. Zool Lett 1:32. doi:10.1186/s40851-015-0031-2

Vakalopoulos I, Dimou P, Anagnostou I, Zeginiadou T (2015) Impact of cancer and cancer treatment on male fertility. Hormones (Athens) 14(4):579–589. doi:10.14310/horm.2002.1620

Venara M, Rey R, Bergada I, Mendilaharzu H, Campo S, Chemes H (2001) Sertoli cell proliferations of the infantile testis: an intratubular form of Sertoli cell tumor? Am J Surg Pathol 25(10):1237–1244

Vernet N, Dennefeld C, Guillou F, Chambon P, Ghyselinck NB, Mark M (2006) Prepubertal testis development relies on retinoic acid but not rexinoid receptors in Sertoli cells. EMBO J 25 (24):5816–5825. doi:10.1038/sj.emboj.7601447

Walker WH, Cheng J (2005) FSH and testosterone signaling in Sertoli cells. Reproduction 130 (1):15–28. doi:10.1530/rep.1.00358

Wallenfang MR, Nayak R, DiNardo S (2006) Dynamics of the male germline stem cell population during aging of Drosophila melanogaster. Aging Cell 5(4):297–304. doi:10.1111/j.1474-9726.2006.00221.x

Wang T, Hung CC, Randall DJ (2006) The comparative physiology of food deprivation: from feast to famine. Annu Rev Physiol 68:223–251. doi:10.1146/annurev.physiol.68.040104.105739

Wang C, Jackson G, Jones TH, Matsumoto AM, Nehra A, Perelman MA, Swerdloff RS, Traish A, Zitzmann M, Cunningham G (2011) Low testosterone associated with obesity and the metabolic syndrome contributes to sexual dysfunction and cardiovascular disease risk in men with type 2 diabetes. Diabetes Care 34(7):1669–1675. doi:10.2337/dc10-2339

Wei Y, Lilly MA (2014) The TORC1 inhibitors Nprl2 and Nprl3 mediate an adaptive response to amino-acid starvation in Drosophila. Cell Death Differ 21(9):1460–1468. doi:10.1038/cdd.2014.63

Wei Y, Reveal B, Reich J, Laursen WJ, Senger S, Akbar T, Iida-Jones T, Cai W, Jarnik M, Lilly MA (2014) TORC1 regulators Iml1/GATOR1 and GATOR2 control meiotic entry and oocyte development in Drosophila. Proc Natl Acad Sci U S A 111(52):E5670–E5677. doi:10.1073/pnas.1419156112

Xie T (2008) Germline stem cell niches. StemBook, Cambridge, MA. doi:10.3824/stembook.1.23.1

Xie T, Spradling AC (2000) A niche maintaining germ line stem cells in the Drosophila ovary. Science 290(5490):328–330

Yamauchi T, Kadowaki T (2013) Adiponectin receptor as a key player in healthy longevity and obesity-related diseases. Cell Metab 17(2):185–196. doi:10.1016/j.cmet.2013.01.001

Yang H, Yamashita YM (2015) The regulated elimination of transit-amplifying cells preserves tissue homeostasis during protein starvation in Drosophila testis. Development 142 (10):1756–1766. doi:10.1242/dev.122663

Yang SA, Wang WD, Chen CT, Tseng CY, Chen YN, Hsu HJ (2013) FOXO/Fringe is necessary for maintenance of the germline stem cell niche in response to insulin insufficiency. Dev Biol 382(1):124–135. doi:10.1016/j.ydbio.2013.07.018

Yang Z, Yu Y, Zhang V, Tian Y, Qi W, Wang L (2015) Octopamine mediates starvation-induced hyperactivity in adult Drosophila. Proc Natl Acad Sci U S A 112(16):5219–5224. doi:10.1073/pnas.1417838112

Yu L, Yan X, Ye C, Zhao H, Chen X, Hu F, Li H (2015) Bacterial respiration and growth rates affect the feeding preferences, brood size and lifespan of Caenorhabditis elegans. PLoS One 10 (7):e0134401. doi:10.1371/journal.pone.0134401

Zhang C, Yeh S, Chen YT, Wu CC, Chuang KH, Lin HY, Wang RS, Chang YJ, Mendis-Handagama C, Hu L, Lardy H, Chang C (2006) Oligozoospermia with normal fertility in male mice lacking the androgen receptor in testis peritubular myoid cells. Proc Natl Acad Sci U S A 103(47):17718–17723. doi:10.1073/pnas.0608556103

Zhao R, Xuan Y, Li X, Xi R (2008) Age-related changes of germline stem cell activity, niche signaling activity and egg production in Drosophila. Aging Cell 7(3):344–354. doi:10.1111/j.1474-9726.2008.00379.x

Zhou Q, Nie R, Li Y, Friel P, Mitchell D, Hess RA, Small C, Griswold MD (2008) Expression of stimulated by retinoic acid gene 8 (Stra8) in spermatogenic cells induced by retinoic acid: an in vivo study in vitamin A-sufficient postnatal murine testes. Biol Reprod 79(1):35–42. doi:10.1095/biolreprod.107.066795

Chapter 4
Signaling-Mediated Regulation of Meiotic Prophase I and Transition During Oogenesis

Swathi Arur

Abstract Generation of healthy oocytes requires coordinated regulation of multiple cellular events and signaling pathways. Oocytes undergo a unique developmental growth and differentiation pattern interspersed with long periods of arrest. Oocytes from almost all species arrest in prophase I of oogenesis that allows for long period of growth and differentiation essential for normal oocyte development. Depending on species, oocytes that transit from prophase I to meiosis I also arrest at meiosis I for fairly long periods of time and then undergo a second arrest at meiosis II that is completed upon fertilization. While there are species-specific differences in *C. elegans*, *D. melanogaster*, and mammalian oocytes in stages of prophase I, meiosis I, or meiosis II arrest, in all cases cell signaling pathways coordinate the developmental events controlling oocyte growth and differentiation to regulate these crucial phases of transition. In particular, the ERK MAP kinase signaling pathway, cyclic AMP second messengers, and the cell cycle regulators CDK1/cyclin B are key signaling pathways that seem evolutionarily conserved in their control of oocyte growth and meiotic maturation across species. Here, I identify the common themes and differences in the regulation of key meiotic events during oocyte growth and maturation.

4.1 Introduction

In mammals, females are born with oocytes arrested at meiosis I. Meiosis I is composed of multiple interconnected differentiation programs, the key events being pairing and recombination of chromosomes and arrest of growing oocytes at the end of meiosis I. Due to the lack of germ line stem cells in vertebrate females, model systems like *C. elegans* and *Drosophila* have pioneered the studies on female meiotic I progression and the key signaling and molecular pathways that govern this progression. In this chapter, I highlight our current understanding on the

S. Arur, Ph.D. (✉)
Department of Genetics, UT M.D. Anderson Cancer Center, Houston, TX, USA
e-mail: sarur@mdanderson.org

S. Arur (ed.), *Signaling-Mediated Control of Cell Division*, Results and Problems in Cell Differentiation 59, DOI 10.1007/978-3-319-44820-6_4

environmental, hormonal, and signaling events that regulate the progression of meiosis I during female gametogenesis and compare and contrast the similarities and differences between worms, flies, and mammalian oocyte growth and meiotic maturation.

4.2 *C. elegans* Germ Line Development

C. elegans germ cells are specified in the embryo as Primordial Germ Cells (PGCs), but unlike in flies and mammals these cells do not undergo migration, but instead form the focal points that divide symmetrically along with the somatic gonad progenitors to populate the U-shaped gonad arms in a somatically female animal (Hubbard and Greenstein 2000; Strome and Updike 2015). During early stages of larval development, the germ line progenitors continue to divide symmetrically, until the end of Larval molt 3, at which point the progenitor stem cells in the hermaphroditic germ line differentiate into spermatocytes that progress through meiosis I and II eventually to form haploid mature sperm (reviewed in Hubbard and Greenstein 2000). The process of germ line stem cell proliferation and the balance and switch of a mitotic stem cell cycle to a meiotic cell cycle are discussed in Chap. 2. The end of the last larval molt-Larval Stage 4 completes the events of spermatogenesis that eventually results in mature sperm in a hermaphroditic germ line stored in a spermatheca. Upon the completion of spermatogenesis, coupled with the last molt into adulthood, a switch in sex determination of the germ cells results in onset of the female meiotic program in the hermaphroditic germ line. Through the remaining life of a hermaphroditic adult, the germ line produces oocytes (Hubbard and Greenstein 2000). A feminized germ line on the other hand only produces the female oogenic precursors and mature oocytes starting from larval stage 3. This chapter will focus on the oogenic events in the hermaphroditic germ line and highlight key differences (where relevant) from a feminized germ line.

4.2.1 Oogenic Meiotic Prophase Progression

The *C. elegans* germ line grows as a syncytial mass of nuclei that are attached to a common cytoplasm and appear very much like a growing stalk of brussel sprouts (Hall et al. 1999). In the syncytial germ line, plasma membranes do not fully surround each germ nucleus, and cytoplasmic bridge connects each germ nucleus to a common cytoplasm (Maddox et al. 2007). By convention, each nucleus surrounding cytoplasm and membranes are referred to as a germ cell. From the mitotic region that harbors the progenitor cells through the end of pachytene, germ cells are arranged on the surface of the gonadal tube with an interior cell/nucleus-free cytoplasmic region called the “rachis” or “core.” In oogenesis, the majority of pachytene cells appear to function as nurse cells, undergoing apoptosis after providing RNAs and proteins to the rachis (Gibert et al. 1984; Gumienny et al. 1999; Wolke et al. 2007).

The oogenic germ cells are in a very long meiotic prophase I, with the pachytene region being the longest, wherein each germ cell is thought to reside in pachytene for ~54–60 h (Jaramillo-Lambert et al. 2007). Events specific to meiotic pairing, recombination, and resolution of homologous sisters are covered in Chap. 4. This long stretch of pachytene can be divided into three major regions, early pachytene, mid-pachytene, and late pachytene, based on the distinct chromosomal behaviors (Phillips et al. 2009). Besides chromosomal events, the germ cells in this stage execute multiple cell biological events such as onset of meiosis (reviewed in Chap. 8, after the mitotic populations), apoptosis [wherein over 50–80 % of the female germ cells die and are either cleared from the germ line, by the nonprofessional phagocytic cells of the somatic gonad (Gumienny et al. 1999)], and plasma membrane synthesis. All of these chromosomal and cell biological behaviors are exquisitely coordinated in a normal growing germ line and occur in stereotypic spatial regions, within the context of the pachytene, suggesting that there must either be autonomous cues that the germ cells are executing as they progress through the long meiotic prophase I or that these nuclei are responding to spatially restricted molecular or environmental cues during the different stages of pachytene. At this time, not much is known regarding whether cues autonomous to the germ cells dictate normal meiotic progression in different regions of the germ line, but a clear understanding is beginning to emerge of the spatial cues at each of the stages of the germ cells that coordinate and drive these events.

Only a handful of signaling pathways have been identified in *C. elegans* that regulate progression through meiotic prophase, with the extracellular RAS/ERK MAPK pathway being one such key pathway. Activation of MPK-1 (*C. elegans* ERK) occurs in two distinct regions of the germ line: zone 1, or mid-pachytene, and zone 2 of growing mature oocytes (Lopez et al. 2013; Miller et al. 2001). In between these two zones active MPK-1 is downregulated at a stage where the germ cells are transiting from pachytene to diplotene (Lee et al. 2007; Lopez et al. 2013). Total MPK-1 is continuously expressed throughout the germ line. The two distinct zones of active MPK-1 signaling are triggered by (a) insulin receptor DAF-2 based activation in zone 1 (Lopez et al. 2013) in response to nutritional cues and (b) sperm-derived MSP signaling through the VAB-1 Ephrin (Eph) receptor in oocytes (Miller et al. 2001). In zone 1 or the pachytene region, MPK-1 is active for a sustained amount of time, with current estimates averaging at ~18 h/germ cell; in the proximal individual oocytes, MPK-1 is active for about 20 min each (Lee et al. 2007; Mattingly et al. 2015). During these stages of meiotic prophase, the germ line is transcriptionally silent and the meiotic events are controlled via translational or posttranslational events.

4.2.2 The LET-60 RAS/MPK-1 ERK Signaling Pathway as a Major Regulator of Meiotic Progression During Oogenesis

Abrogation of either DAF-2 (the Insulin receptor), LET-60 (ras), MEK-2 (mek), LIN-45 (raf), or MPK-1(erk) all result in female germ cells arrested in pachytene resulting in sterile adults that do not generate any gametes (Church et al. 1995; Lee et al. 2007; Lopez et al. 2013; Ohmachi et al. 2002). This leads to the model that the RAS/ERK pathway governs exit of oogenic germ cells from pachytene to the diplotene stage of meiotic prophase (Church et al. 1995). However, this observation is inconsistent with normal MPK-1 signaling in the germ line; MPK-1 is active from mid- to late pachytene and drops below detectable levels at the end of pachytene (Fig. 4.1a). This spatial activation pattern would suggest that active MPK-1 has to be downregulated to exit pachytene, rather than linger in pachytene, as the case seemed to be when *mpk-1* was genetically abrogated. Given that pachytene is the longest phase of meiotic prophase, a key question that arises is, does the rise of active MPK-1 in the mid-pachytene regulate any meiotic event, or is that rise important to exit pachytene, which given that the cells did not exit until the end of pachytene, and in fact remained in pachytene upon the rise of active MPK-1 was a flawed logic. Thus, the question that was posed was, is the rise in active MPK-1 correlative with progression of pachytene from early to mid-stage? And would the loss of *mpk-1* signaling cause an inability of these cells to exit, resulting in their arrest. To answer this question, Lee et al. (2007) used the RNA-binding proteins DAZ-1 and GLD-1 as molecular markers to distinguish between early, mid-, and late pachytene and assayed the stage of the arrested pachytene stage germ cells in the mpk-1 null mutant. DAZ-1 is high in the mitotic and transition zone and falls mid-way through pachytene in wild-type germ lines (Maruyama et al. 2005), with a reciprocal staining pattern with active MPK-1. Interestingly, in *mpk-1* null germ lines, DAZ-1 levels remain elevated throughout the arrested pachytene germ cells, indicating that these cells retain distal/mid-pachytene character (Lee et al. 2007). Additionally, GLD-1 (RNA-binding protein) and CEP-1 (the p53 homolog) levels were analyzed; GLD-1 falls at the end of pachytene where CEP-1 starts to rise in normal pachytene stage germ cells. In *mpk-1* null mutants, however, GLD-1 is continuously high in the arrested pachytene stage germ cells, while CEP-1 never accumulates (Lee et al. 2007). All of these data lead to a new model on the role of RAS/ERK pathway in controlling meiotic progression. The model proposes that active MPK-1 controls two distinct aspects of meiotic progression: first, the pathway enables progression from early to mid-pachytene, presumably, this is coupled by chromosomal behaviors some of which are now coming to light [(Nadarajan et al. 2016), also covered in Chap. 4] and second that it regulates pachytene progression, such that reduction of MPK-1 signaling (analyzed through the use of single point mutations in the MPK-1 gene, which reduces binding to MEK at nonpermissive temperatures and causes a reduction in MPK-1 signaling).

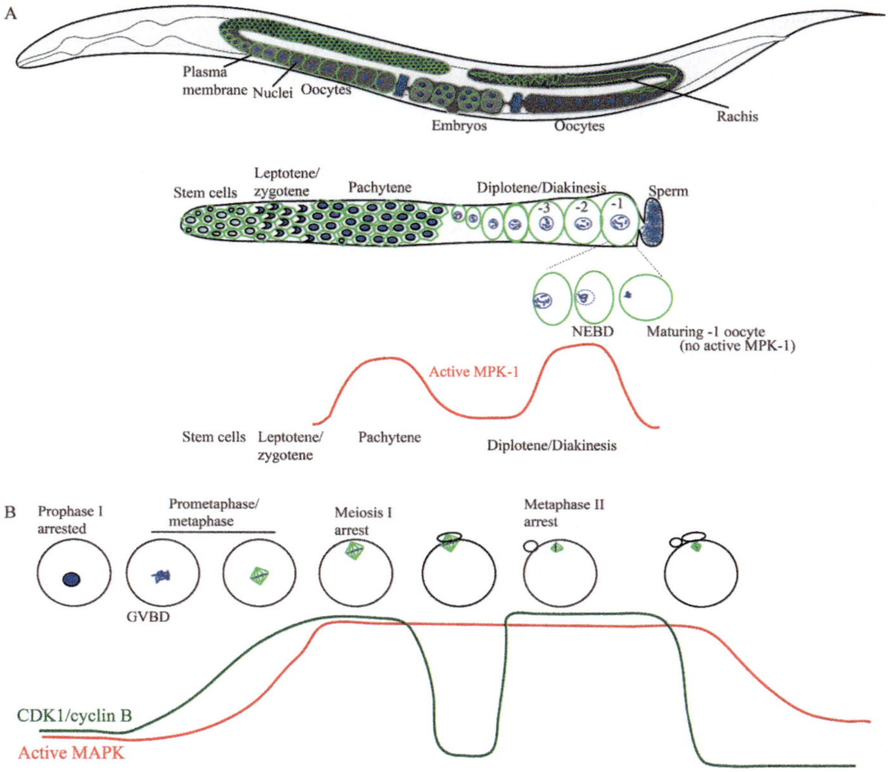

Fig. 4.1 MAP kinase signaling pathway regulates varied aspects of oocyte growth and maturation in worms and mammals. (**a**) (*Top*) An adult *C. elegans* hermaphroditic animal carries two U-shaped gonad arms. In the figure, the arm on the *left* represents the surface view of the gonad with plasma membranes in *green* and nuclei in *blue*. The *right* arm represents an internal view of the gonad, with plasma membranes that open into a common cytoplasm, rachis, and each growing oocyte is connected to the rachis via an opening. (*Middle*) Oogenic germ cells in one gonad arm of *C. elegans*, from stem cell pool on the *left* to growing oocytes on the *right*. In *C. elegans*, oocytes arrest in prophase I. (*Bottom*) MPK-1 ERK MAP kinase is active in two regions of the meiotic prophase in *C. elegans*: pachytene and growing oocytes. MPK-1 signal in pachytene is maintained by insulin like receptor DAF-2 and in growing oocytes by MSP signal from sperm. (Figure Adapted from Lee et al., Genetics, 2007. Express permission to reproduce the figure obtained from Tim Schedl and Genetics.) (**b**) Various stages of mammalian oocyte development and arrest periods. Active MAP kinase (*red*) and Maturation Promotion Factor (CDK1/cyclin B) are regulated dynamically during the various stages of mammalian oocyte development

In addition to triggering an entry of germ cells from early to mid-pachytene, MPK-1 signaling also regulates multiple aspects of cell biology during oogenesis, including control of cellular morphogenesis and membrane organization (Arur et al. 2009, 2011; Lee et al. 2007). Active MPK-1 is required for the gonadal tube arrangement of pachytene cells on the surface and an interior cytoplasmic rachis (a process termed "pachytene cellular organization"), forming a single row of cylindrically shaped oocytes in the proximal gonad (oocyte organization and differentiation), oocyte growth control, and migration of the oocyte nucleus to the distal surface. Genetic analysis as well as the pattern of activation indicates that these MPK-1 functions are largely or completely germ line autonomous. MPK-1 regulates these events that occur contemporaneously or simultaneously in the germ line via coordinated control of multiple distinct substrates that are phosphorylated and regulated in distinct regions, cell biological compartments, and times during germ line development. Currently, 31 novel MPK-1 substrates have been identified to regulate each of the eight processes (Arur et al. 2009). Analysis of each substrate in detail is shedding light into how MPK-1 regulates any given processes and the redundant mechanisms that drive the robustness in the germ line (Arur et al. 2009). However, it currently remains unknown, as to how MPK-1 mediates a switch from distal to proximal pachytene. The current model is that a number of MPK-1 substrates (currently unidentified) are likely involved in coordinating progression from distal to proximal pachytene.

4.2.3 Oocyte Growth and Development

Oocytes can achieve sizes >1000 times that of diploid somatic cells. Normal oocytes attain large sizes compared to their pachytene progenitors via influx of yolk protein and cytoplasmic constituents (Wolke et al. 2007). Clearly, sizes that oocytes achieve prior to undergoing a meiosis I arrest are stereotypical in all organisms. But mechanisms underlying their control remain unknown in most species. In *C. elegans*, large oocytes are observed under conditions of reduced MPK-1 activity (such as loss of *daf-2* insulin receptor signaling, *mpk-1* signaling, or loss of the phosphatase *ptp-2* null (Church et al. 1995; Gutch et al. 1998; Lopez et al. 2013). Conversely, small oocytes are generated under a condition where MPK-1 signaling is increased; *let-60 ras* gain of function alleles (Lee et al. 2007). Thus, the RAS/ERK signaling cascade appears to act as a rheostat to control oocyte size, with low MPK-1 activity promoting growth, high MPK-1 activity inhibiting growth, and presumably normal size oocytes produced at intermediate activity levels. Activated MPK-1 in diplotene, where it is normally downregulated appears to be involved in oocyte growth control. Multiple MPK-1 phosphorylation substrates have been identified that regulate the event of oocyte growth; however, mechanisms both cellular and mechanistic need to be delineated (Arur et al. 2011; Drake et al. 2014). It is likely that oocyte growth is controlled autonomously through each individual oocyte; however, a more likely possibility is that the flow of cytoplasmic contents from the central rachis of the gonad drives cytoplasmic constituents into each growing oocyte [connected with the rachis via a cytoskeletal bridge, much like a ring canal in *Drosophila melanogaster* oocytes (Robinson et al. 1994)], until the oocyte reaches a critical mass and completely buds off the central rachis. The signaling pathway thus likely regulates the growth of the oocyte through controlling either the cytoplasmic flow or the length of connection of the cytoplasmic bridge in each oocyte with the rachis. In case of the former, activation of MPK-1 in pachytene is likely critical for enabling the cytoplasmic flow; the latter likely relies on the activation of MPK-1 in the oocytes.

Besides the RAS/ERK signaling pathway, loss of proliferation in the germ line stem cell population (discussed in Chaps. 1–3) via loss of the GLP-1/Notch signaling also results in generation of large oocytes (Nadarajan et al. 2009). Phenomenologically, this could occur due to loss of the progenitor stem cell-based populations that can no longer sustain the development of the germ line, and thus likely result in misregulation of multiple signaling pathways due to loss of tissue architecture. It could, however, also be a direct effect of Notch signaling itself, wherein Notch pathway generates a specific signal in the distal stem cell region, which is later transported to the oogenic germ line, and directly affects oocyte size. Blocking the connection of the oogenic germ line with the progenitor population via injection of a droplet of oil results in the generation of large oocytes lending support to the model that a physical connection between the progenitor population and oogenic germ cells may be involved in regulating normal oocyte growth and development (Nadarajan et al. 2009). Molecules and cellular events that mediate these events, however, remain to be defined.

Interestingly, however, in a female germ line that lacks the sperm signal, and thus is unable to activate MPK-1 in oocytes via the Ephrin receptor, the oocytes are formed normally and arrest prior to oocyte maturation. The growth of these oocytes is equivalent to oocytes from hermaphroditic animals with robust sperm-dependent MPK-1 activation, suggesting that oocyte growth per se maybe regulated via events occurring more distally in the gonad, specifically in pachytene stage of gonadal development. These models remain to be tested.

4.2.4 Events Controlling Oocyte Meiotic Maturation

The oocytes of most sexually reproducing animals arrest in diplotene or diakinesis of meiotic prophase (Greenstein 2005). Species-specific hormonal signals trigger the resumption of meiosis termed meiotic maturation. Meiotic maturation is defined by the transition to metaphase I (M phase) and is accompanied by nuclear envelope breakdown (NEBD), rearrangement of the cortical cytoskeleton, and meiotic spindle assembly. In vertebrates, hormonal signaling activates cyclin-dependent kinase (CDK1) to promote M-phase entry (see below). Somatic cells of the gonad also function to maintain meiotic arrest in many species, see the section on mammalian oocyte arrest. Meiotic maturation of oocytes arrested in prophase I is under the control of signaling and translational events, since these oocytes are transcriptionally quiescent. Current studies are investigating the mechanisms that link signaling to translational control and how those regulate meiotic maturation.

In *C. elegans*, MPK-1 activation is also correlated with oocyte maturation/ovulation (Drake et al. 2014; Lee et al. 2007; Miller et al. 2001), but functionally, MPK-1 activity seems required mainly for initiation of the transition from prophase I, as timely activation of maturation/ovulation. MPK-1 activity is clearly required for the initiation of meiotic maturation such as nuclear translocation from the center of the cell to the anterior and cortical rearrangement of the oocyte (Harris et al. 2006; Kim et al. 2013; Lee et al. 2007). In accordance with this, in a female, where there is no sperm-dependent activation of MPK-1, oocytes do not initiate the process of meiotic maturation and remain arrest in prophase I (Harris et al. 2006; Kim et al. 2013; Lee et al. 2007). The sperm-dependent MPK-1 activation in zone 2, comprising of the most proximal ~five oocytes appears to be responsible for MPK-1's function in maturation/ovulation (Govindan et al. 2009; Greenstein 2005; Harris et al. 2006; Kim et al. 2013; Lee et al. 2007; Miller et al. 2001). However, because only the −1 oocyte undergoes maturation it suggests that MPK-1 activation alone is not sufficient and additional regulatory mechanisms, such as inhibition of CDK1 from proximal oocytes and/or inhibition from somatic gonadal sheath cells, must block maturation of the −2 through −5 oocytes (Govindan et al. 2006; McCarter et al. 1999).

The major sperm protein (MSP) that serves as a ligand to active MPK-1 signaling initiates the events of oocyte meiotic maturation (Miller et al. 2001). MSP likely acts both on the oocyte and the surrounding sheath cells; injection of purified MSP is sufficient to induce resumption of meiosis. Work from Govindan et al. established the role for Gαs-based activation of adenylate cyclase to elevate cAMP levels (Govindan et al. 2006, 2009). Blocking adenylate cyclase results in loss of cAMP signaling and blocks oocyte maturation, indicating a direct role of the adenylate cyclase and cAMP to regulate meiotic maturation in *C. elegans* oocytes. Mechanistically, activation of Gαs counteracts the inhibitory signals from Gαi/o that block meiotic maturation in the absence of MSP. Although MSP can act both on the oocyte and the sheath cells, the function of MSP directly on the oocytes remains to be determined. MSP regulates the functions described above via a communication between the somatic sheath cells and the oocyte. Interestingly, other regulators of oocyte meiotic maturation have been identified such as CEH-18 (A POU domain transcription factor) that also act in the soma to repress MPK-1 activation in the oocyte and mediate meiotic arrest (Govindan et al. 2009; Yamamoto et al. 2006). However, their exact mechanism of action remains to be defined.

Activated MPK-1 levels drop dramatically as the −1 oocyte undergoes maturation (Drake et al. 2014; Lee et al. 2007). This inactivation of MPK-1 is distinct from that in vertebrates where active ERK is present from maturation until release of the meiosis II arrest by fertilization (see below). Unlike in vertebrates though, in *C. elegans* sperm signal initiates ovulation and fertilization of a mature oocyte, and the events of both the events of meiosis are completed after fertilization. Thus, the evolutionary reasons for these differences are both intriguing and challenging to define.

Besides MSP, cyclin-dependent kinase CDK1 (Boxem et al. 1999) and the polo-like Kinase-1, PLK-1, dependent (Chase et al. 2000), each positively regulates events of oocyte meiotic maturation. The zinc finger domain-containing proteins OMA-1 and OMA-2 are redundantly required for oocyte maturation and ovulation

(Detwiler et al. 2001). In *oma-1* and *oma-2* double mutants, MPK-1 activation is not sustained and NEBD is suppressed. OMA-1 and OMA-2 may function upstream of two conserved cell cycle regulators, the MYT1-related kinase WEE-1.3 and CDK1 (Detwiler et al. 2001; Spike et al. 2014b). Additionally, recently the NHL1 domain protein LIN-41 has been identified to work antagonistically with the OMA-1 and OMA-2 proteins to control the prophase to metaphase transition in *C. elegans* (Spike et al. 2014a). Chapter 6 discusses the role of these proteins in mediating the events of oocyte meiotic maturation in further detail.

4.3 *Drosophila* Female Germ Line Development

Drosophila melanogaster germ cells, much like *C. elegans*, are specified in the embryo as primordial germ cells (PGCs) (Forbes and Lehmann 1999; Lehmann 1992). However, primordial germ cell specification from this point on is very different between these two species. In one cell embryo, right after fertilization of male and female pronuclei, nuclear divisions, in the absence of cytokinesis, give rise to a syncytial embryo. Of these early nuclear divisions, ~10 nuclei migrate to the posterior pole and become the first to be surrounded by a plasma membrane. These cells then undergo symmetric, but asynchronous divisions to give rise to a total of about 40 cells. These 40 cells are termed the pole cells and are the early primordial germ cells. Much like in *C. elegans*, these cells are transcriptionally quiescent (reviewed in Lehmann 1992).

During early embryogenesis, pole cells remain positioned at the posterior end of the embryo; however, as the germ layers form, and midgut starts to invaginate, the pole cells along with the midgut travel inside the embryo. Once inside the embryo, at day 10 of embryogenesis, the pole cells start to migrate through the midgut epithelium. These processes have been extensively reviewed in McLaughlin and Bratu (2015) and Williamson and Lehmann (1996). The migration and colonization of the embryonic gonad occur by the time the embryogenesis is complete. At the end of embryogenesis, another process starts to take place: the somatic gonadal cells begin to encapsulate the pole cells and form the embryonic gonads on either side of the embryo. The germ line stem cells continue to stay connected to the somatic gonadal cells through the rest of germ cell development (Forbes and Lehmann 1999; Williamson and Lehmann 1996).

These two cell types, the somatic gonadal cells and the pole cell precursors, form the major progenitors of the *D. melanogaster* female germ line or the ovary. The ovary comprises of two main stem cell populations, the germ line (GSC) and the follicle stem cells (FSC) (reviewed in Chaps. 1 and 3). These two cells give rise to the nurse cells, oocyte, and follicle cells of the mature egg chamber. Each stem cell population resides in a unique, specialized niche containing several types of somatic gonadal precursor cells (Fig. 4.2a). The somatic gonadal cells also harbor the germ line stem cell "niche" necessary for regulated balance between stem cell self-renewal and differentiation, reviewed in Chaps. 1–3.

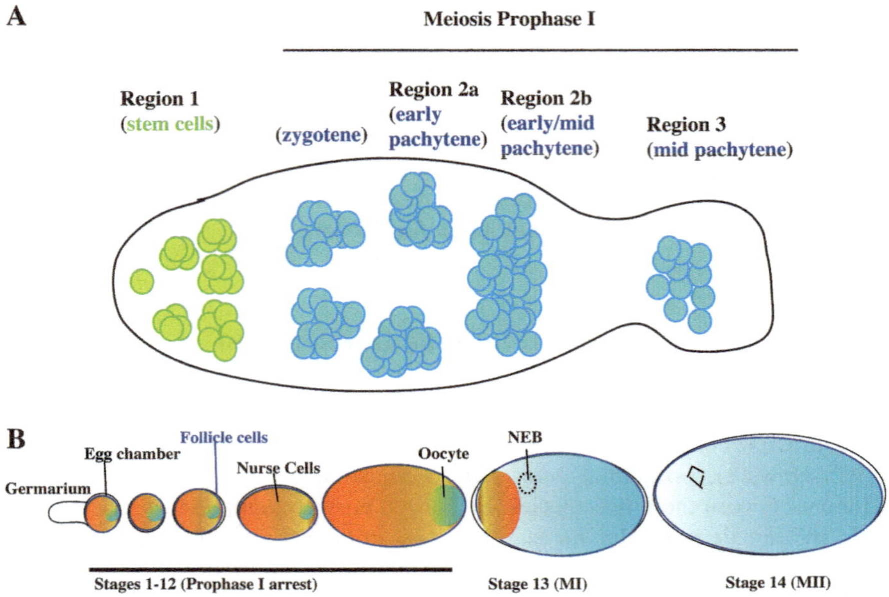

Fig. 4.2 Stages in *Drosophila* oogenesis. (**a**) Mature germarium with the various regions containing cells in differing stage of mitosis (*left*, *green*) or meiotic prophase I. (**b**) Once a germarium matures, it is surrounded by follicle cells (*blue line*, *orange*) that surround the maturing oocyte (*green*). A mature oocyte surrounded by the follicle cells makes an egg chamber. As newer oocytes are born, the older egg chambers progress through different stages of meiosis. At stage 13, the oocyte starts to undergo the process of meiotic maturation, marked by the nuclear envelope breakdown. The oocyte arrests in meiosis II at stage 14, until it passes through the oviduct, during which time the oocyte undergoes maturation in the absence of sperm

4.3.1 Meiotic Prophase I Progression

The basic unit of the ovary in *Drosophila* is the ovariole; there are 16–20 ovarioles per ovary, each being autonomous with its own stem cell populations and egg chambers at varying developmental stages (Fig. 4.2a). The ovariole can be divided into three principal regions (from anterior to posterior): the terminal filament, germarium, and vitellarium (also see Chap. 3). The germarium is the site of germ line stem cell (GSC) division, differentiation, and germ line cyst formation and is divided into four regions (1, 2a, 2b, and 3) (Fig. 4.2a). In germarium region 3, the germ line cyst, containing nurse cells and oocyte, is ensheathed by a somatic cell layer, together forming a structure named the "egg chamber" before being passed into the vitellarium. Here, I discuss the process of meiotic prophase progression and maturation, which primarily occurs in germarium region 2a (Fig. 4.2a).

The oocyte undergoes both developmental maturation and meiosis throughout the course of oogenesis, and these processes are intimately linked. The balance of oocyte differentiation and progression through meiosis is achieved by two major meiotic arrests during oogenesis. Prophase I of meiosis begins in region 2a of the germarium, where the events of synaptonemal complex can be visualized (reviewed in Lake and Hawley 2012; McLaughlin and Bratu 2015), and arrests in diplotene stage of prophase I at the beginning of stage 5 in the egg chamber. The remaining 15 cells develop into the nurse cells and enter an endocycle to become highly polyploid. Determination of the oocyte fate seems to be dictated by the accumulation of the cyclin-dependent kinase inhibitors p21 CIP/p27 KIP/p57 KIP2 that prevent endoreduplication via repressing the CDK2/cyclin E such that the oocyte maintains its meiotic state (Hong et al. 2003). While much is known about the signaling and molecular events underlying meiotic I arrest and progression in *C. elegans*, relatively little is understood in *D. melanogaster* regarding events that surround meiotic I arrest. Additionally, events such as gap junction-mediated germ line communications between somatic follicle cells and developing oocytes have not been established. While gap junction proteins have been identified in *D. melanogaster* oocytes (Stebbings et al. 2002), loss of gap junction proteins in the oocyte did not result in any developmental arrest phenotypes in the oocytes (Bohrmann and Zimmermann 2008), suggesting that the gap junction proteins do not play a major role in mediating meiotic I arrest in *D. melanogaster* unlike in *C. elegans* and mammals.

Likewise, unlike *C. elegans* where the MAPK pathway is central to meiotic I progression and meiotic maturation, in *D. melanogaster*, while the MAPK pathway is activated in the oocyte by a Mos (MAPK kinase kinase) homolog, loss of either MAPK or Mos does not result in meiotic arrest or any developmental defects in oocytes (Ivanovska et al. 2004), suggesting that MAPK activation plays a nonessential role in *D. melanogaster* development. Thus, currently much remains to be determined about the signaling and molecular events that regulate and sustain the prophase I arrest in flies (Fig. 4.3).

4.3.2 Oocyte Meiotic Maturation

The cyclin-dependent kinase signaling pathway CDK1/cyclin B regulates the process of meiotic maturation in the *D. melanogaster* oocyte at stage 13 of oogenesis (Von Stetina et al. 2008; Xiang et al. 2007). Loss of the active CDK1/cyclin B in *D. melanogaster* results in a delay in the process of meiotic maturation, including the process of NEBD (Von Stetina and Orr-Weaver 2011; Von Stetina et al. 2008). The Polo-like kinase, Polo, regulates the activity of CDK1/cyclin B. Polo binds to Matrimony (MTRM, a sterile alpha motif-containing protein that is expressed from the end of pachytene until the completion of meiosis I) in prophase. Phosphorylation of Matrimony, by a currently unknown kinase, results in generation of Polo binding site. Binding of Matrimony to Polo results in inactivation of Polo activity towards cell division cycle 25 (CDC25 or Twine). Activation of Polo results in activation of CDC25 phosphatase and thus activation of CDK1/cyclin B. Besides Matrimony, the Great Wall Kinase (Gwl) (Yu et al. 2004) also antagonizes Polo

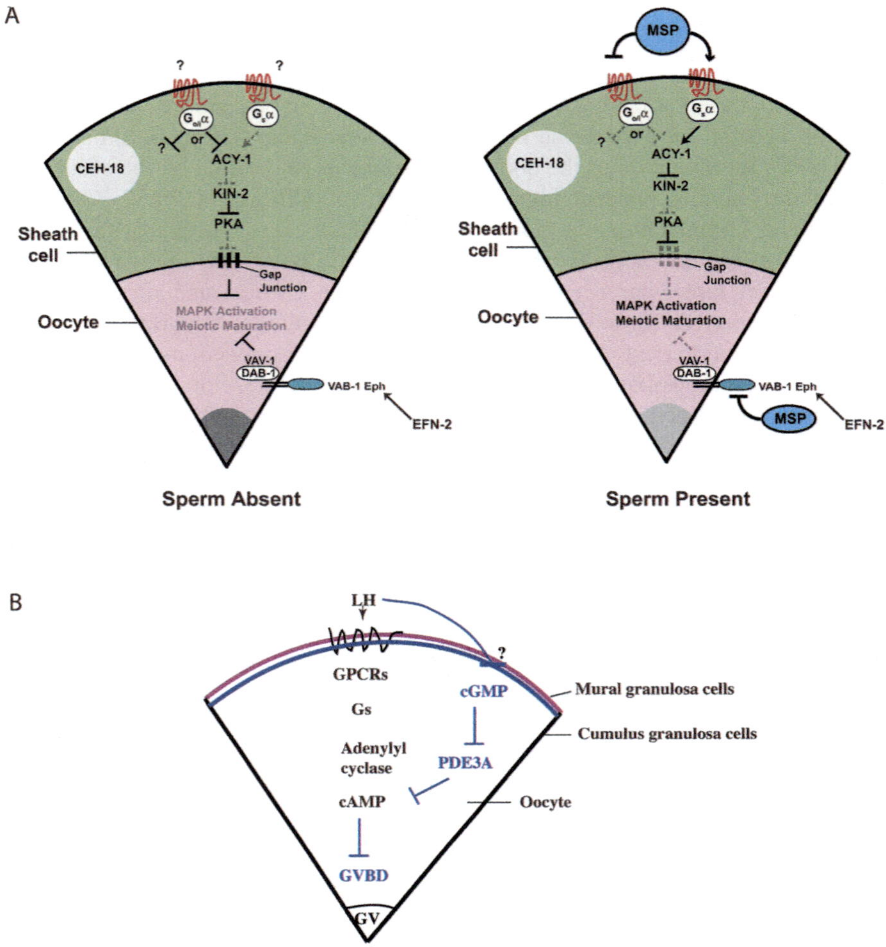

Fig. 4.3 Cell–cell communication and cyclic AMP coordinates oocyte meiotic maturation in worms and mammals. (**a**) In *C. elegans*, in the absence of the sperm signal, the Gαo/i pathway in somatic sheath cells surrounding the oocyte (*pink*) leads to the inactivation of adenylate cyclase 4 (ACY-4) and protein kinase A (PKA). This inhibition blocks MAP kinase (MAPK) activation and oocyte meiotic maturation. Additionally, the VAB-1/Ephrin Receptor in the oocyte also inhibits MAPK and oocyte maturation. Presence of the major sperm protein (MSP) antagonizes both the somatic Gαo/i and oocyte VAB-1 signaling pathways and activates Gαs pathway in the somatic sheath cells resulting in MAPK activation and meiotic maturation. (Adapted from Govindan et al. 2006 reprinted with express permission from David Greenstein and Current Biology.) (**b**) Cyclic AMP in the oocyte inhibits meiotic maturation in mammals. cGMP produced by the cumulus somatic cells enters the oocyte via the gap junctions (via currently unknown mechanisms) to inhibit PDE3A. PDE3A hydrolyses cAMP. Binding of luteinizing hormone (LH) to its G protein-coupled receptor (GPCR) (*blue*) activates the cAMp pathway and enables meiotic maturation via promoting GVBD

(Archambault et al. 2007) and maintains meiotic arrest. Activation of CDC25 allows cyclin-dependent kinase 1 (CDK1) to activate and promote germinal vesicular breakdown (GVBD) in prometaphase.

Besides, Polo and matrimony, α-endosulfine homolog is also thought to control CDK1/cyclin B activity and almost all aspects of meiotic maturation (Von Stetina et al. 2008). The endos encodes a conserved phosphoprotein; endos mutant oocytes display a severe delay in NEB, spindle formation, and chromosome congression defects very similar to twine mutants. The Endos protein appears to regulate CDK1/cyclin B activity, likely via novel substrates of CDK1/cyclin B (Von Stetina et al. 2008). On meiotic maturation at stage 13, Polo proteins overcome the inhibition of Matrimony, by sheer increase in their number, thus effectively inactivating CDC25 and activating CDK1/cyclin B and mediating meiotic resumption.

Interestingly, while there must be extrinsic signals that regulate oocyte maturation, not much is known about them. Some examples have, however, been defined. For example, the prostaglandin hormones or the steroid hormone ecdysone seem like likely stimulatory signals, since cyclooxygenase (COX), which results in prostaglandin synthesis, promotes early ovarian follicle maturation (Tootle and Spradling 2008). Similarly, ecdysone signaling is also required for progression of oogenesis and egg chamber maturation during mid-oogenesis (Buszczak et al. 1999). Because the follicles in mutants affecting either signaling pathways do not reach stage 13 (when meiotic maturation takes place), it is not known whether prostaglandins or ecdysone are required for meiotic maturation in flies.

4.4 Mammalian Female Germ Line Development

Mammalian females are born with differentiated oogenic germ cells arrested in meiosis I. Mammalian oocytes arrest in prophase during fetal development, an arrest that lasts for months in mice and years in women. In response to the leutenizing hormone signal from the pituitary at puberty, the oocyte transitions from prophase to metaphase II, where it remains arrested until fertilization causes the completion of meiosis. The prophase-to-metaphase transition is characterized by the breakdown of the nuclear envelope (NEBD, also known as germinal vesicle breakdown or GVBD) (Schuh and Ellenberg 2007). Because the oocytes are born under a condition of meiosis I arrest, not much is understood about earlier events leading up to the arrest in mammalian systems. This chapter will focus on signaling events that regulate meiosis I transition to metaphase and oocyte maturation. Events resulting in meiosis II resumption and fertilization are covered in Chaps. 7 and 8.

4.4.1 *Hormonal Control of Prophase I Arrest in Mammalian Oocytes*

4.4.1.1 cAMP and Luteinizing Hormone

Somatic cells surround mammalian oocytes, and these somatic cells are essential for maintaining prophase arrest in the oocyte. The somatic cells are made up of two layers, the outer layers or the mural granulosa and the inner cumulus layer. Oocytes or cumulus–oocyte complexes that are removed from the follicle resume meiosis spontaneously (Pincus and Enzmann 1935; Edwards 1965), suggesting that a meiosis inhibitory factor coming from the somatic cells helps maintain the arrest. Additionally, a physical continuity between the cellular layers connecting the somatic cells and the oocyte is essential for maintenance of prophase arrest. Gap junctions mediate the physical connection between the somatic cells and the oocyte; inhibition of gap junction proteins also results in resumption of meiosis, suggesting that gap junction-based communication is critical for oocyte meiotic arrest (Edry et al. 2006; Norris et al. 2008; Sela-Abramovich et al. 2006). The nature of the inhibitory signal that passes between the somatic cells and the oocyte currently remains undetermined.

Maintenance of prophase arrest in the fully grown mammalian oocyte requires high level of cyclic AMP (cAMP) in the oocyte. Inhibition of cAMP levels in the oocyte results in spontaneous resumption of oocyte progression to metaphase (Cho et al. 1974; Magnusson and Hillensjo 1977). cAMP is generated in the oocyte by the constitutively active Gs-linked receptor, GPR3 or GPR12, which act to stimulate adenylyl cyclase (Mehlmann et al. 2002 reviewed in Runft et al. 2002; Kalinowski et al. 2004; Mehlmann et al. 2004; Norris et al. 2007). Perturbing either the receptor, Gs, or adenylyl cyclase causes cAMP levels to decrease resulting in resumption of meiosis. Measurements of the cAMP concentration in isolated intact follicle-enclosed mouse oocytes reveal basal cAMP concentration of about ~700 nM (Norris et al. 2009). This cAMP concentration is sufficient to activate the cAMP-dependent kinase (PKA), PKAI and PKAII (Viste et al. 2005), both of which accumulate in the oocyte (Newhall et al. 2006). Cyclic AMP regulates meiotic prophase arrest by PKA-mediated phosphorylation of the phosphatase CDC25 (CDC25B in mouse) causing CDC25 inactive towards dephosphorylating CDK1; CDK1 remains phosphorylated and inactive (Lincoln et al. 2002; Oh et al. 2010; Zhang et al. 2008). Removal of the oocyte from its follicle (Vivarelli et al. 1983), or inhibition of the gap junction permeability within the follicle (Sela-Abramovich et al. 2006), causes cAMP to decrease in the oocyte, resulting in lack of PKA phosphorylation, which in turn activates CDC25, and CDC25 dephosphorylates CDK1 activating CDK1/cyclin B complex. Because inhibition of gap junctions results in initiation of meiosis, it is interesting to speculate that maybe the meiosis inhibitory signal that passes from the somatic cells to the oocyte through gap junctions is cAMP, which diffuses between the somatic cells and the oocytes and maintains itself at a high level in the oocyte. Interestingly, however, cAMP from the somatic cells (despite being high) is insufficient to maintain cAMP in the oocyte at an inhibitory (high) level (Mehlmann et al. 2002) suggesting that the simple diffusion-based model is inaccurate. One likely possibility is that the somatic cells maintain oocyte cAMP at ~700 nM level via a signal that inhibits the oocyte cAMP phosphodiesterase PDE3A. In accordance with this hypothesis, chemical or genetic inhibition of PDE3A in the oocyte (Masciarelli et al. 2004) helps maintain the oocyte arrest in prophase even when removed from the follicle. Further studies remain to be conducted, however, to understand the mechanisms underlying the control of PDE3A, the role of cAMP, and the mechanisms governing the control of cAMP in mediating prophase I arrest in oocytes.

Interestingly, high levels of cAMP are also necessary in the fetal ovary to enable the disassembly of synaptonemal complex during the development of mouse oocytes and for completion of prophase I until the arrest (Wang et al. 2015). These results suggest that cAMP needs to be high during early stages of oocyte development to enable normal oocyte growth and differentiation until arrest in prophase I. At which point it remains arrested for several weeks or years (if human), until a signal from LH results in downregulation of cAMP for resumption of meiosis.

4.4.1.2 Interplay Between cAMP and cGMP Regulates Prophase I Arrest

Besides cAMP, cyclic GMP is an additional key player in maintaining prophase I arrest in mammalian oocytes. cGMP is also thought to inhibit meiotic progression since the levels of cGMP in the oocyte decrease with time after removal from the follicle, and cGMP injection into an isolated oocyte delays spontaneous meiotic resumption (Tornell et al. 1990). Also, levels of cyclic GMP, much like cyclic AMP are also regulated by luteinizing hormone (Hubbard 1986). These observations together suggest that the prophase arrest is maintained by the coordinated function of cyclic nucleotide regulatory systems in both the somatic cells and oocyte, compartments that are connected by gap junctions. However, many unanswered questions remain. For example, it remains unknown as to exactly which guanylyl cyclases in the mural granulosa cells, and possibly in the oocyte, regulate cGMP, and in turn how they themselves are regulated. It also remains unknown whether like cAMP, cGMP has an additional role in oocyte growth and differentiation during meiosis prior to meiotic arrest.

4.4.2 The Role of Luteinizing Hormone in Mediating Progression of Meiosis I

Luteinizing hormone (LH) is released from the pituitary and is the primary stimulus for meiotic resumption in vertebrate oocytes. In mice, NEBD occurs ~2–4 h after exposure of isolated follicle containing oocytes to LH (Park et al. 2004). LH receptors are expressed both on the oocyte and the somatic cells suggesting that they likely act on one or the other cell type or both to mediate meiotic resumption (Amsterdam et al. 1975). The LH receptor is a G protein-coupled receptor that activates Gs, Gi, and Gq/11 (Herrlich et al. 1996; Rajagopalan-Gupta et al. 1998). Gs activation causes the production of cAMP in the granulosa cells, stimulating their PKA activity (Tsafriri et al. 1972). However, the effect of Gi or Gq stimulation on the function of the receptor remains undetermined.

How is LH conveyed from the somatic cells to the oocytes? In frogs and fish, unlike mice, LH stimulation of meiotic resumption is mediated by the synthesis of progesterone (Fortune 1983; Masui 1967; Nagahama and Yamashita 2008). However, while LH causes mammalian follicles to produce progesterone, which is essential for ovulation and subsequent implantation of the embryo, progesterone does not have a significant function in stimulating the prophase-to-metaphase transition in mammalian oocytes (reviewed by Tsafriri and Motola 2007). Thus, another process must account for how the LH signal is conveyed from the somatic cells to the oocyte.

4.4.3 LH-induced MAPK Activation and Meiotic Resumption

LH stimulates a key signaling pathway leading to meiotic resumption: MAP kinase (specifically the ERK MAPK) signaling pathway (Norris et al. 2008; Panigone et al. 2008). EGF receptor is expressed throughout the somatic cells (but not the oocyte), and current evidence suggests that synthesis and release of EGF receptor ligands from the somatic cells result in engagement of the EGF receptor and signaling to the oocyte. Within 30 min of exposure of the follicle to LH, EGF receptors are activated in the somatic cells in a manner dependent on phosphorylation and activation of PKA (Panigone et al. 2008). Application of EGF receptor ligands to follicles causes meiotic resumption (Dekel and Sherizly 1985; Nedachi and Conti 2004; Park et al. 2004), and loss of EGF receptor/Mos/MEK/ERK or any member of the EGFR/MAPK family results in loss of meiotic resumption, suggesting that LH signaling pathway leads to meiotic resumption via the EGFR pathway. Most importantly, studies of mutant mice with reduced EGFR activity and mice with genetic deletions of EGF ligands have shown that reducing EGFR activation largely, but not completely, inhibits meiotic resumption in response to LH (Hsieh et al. 2007).

EGF ligands that appear to be important for LH signaling are epiregulin and amphiregulin (Panigone et al. 2008). Granulosa cells from women as well as nonhuman primates also synthesize EGF ligands in response to LH (reviewed by Hsieh et al. 2009). EGF receptor activation results in the activation of MAP kinase in the cumulus cells, and this activation is necessary for activation in the oocyte. Interestingly, MAP kinase is also thought to be activated by an EGF receptor-independent pathway in the cumulus cells in a Src kinase-mediated manner in rat (Wayne et al. 2007). In this context MAP kinase activation may be a key link between LH stimulation and gap junction closure necessary for reinitiating meiosis. However, while very exciting, all of these models await rigorous experimental evidence.

4.5 Common Themes and Differences Across Evolution that Govern the Events of Meiotic Progression and Maturation

Despite the reproductive differences in the species reviewed in this chapter, varied signaling and molecular pathways function to regulate evolutionarily conserved events. The most striking of these is the communication between the somatic cells and the oocytes regulated via gap junctions in *C. elegans* and mammals. In both cases, high cyclic AMP levels along with adenylate cyclase activity in the somatic cells result in resumption of oocyte maturation. Although, in mammals, the high cAMP levels in the oocyte also assist in maintaining prophase I arrest. The signals that result in high cAMP levels in the somatic cells are the major sperm protein in *C. elegans* and luteinizing hormone in mammals, where they both engage the G protein-coupled receptors in the somatic cells. Major sperm protein, while produced by sperm, unlike LH produced by an endocrine system, can be thought of as a short range hormonal signal, since in essence the protein is secreted by the sperm very much like peptide hormones and elicits a robust signaling response even in the absence of physical sperm. Interestingly, in both *C. elegans* and mammals, meiotic maturation is associated with inactivation of the gap junctions between the oocyte and the somatic cells. It is therefore interesting to test whether in *D. melanogaster* gap junctions in the follicle cells play a role in regulating meiotic maturation. The source of a hormonal signal, however, maybe the key to understanding this regulation. In worms, MSP from the sperm seems to function as the hormonal signal to regulate the gap junctions and the cAMP levels; in flies it maybe a distinct mechanism that triggers this signaling event. Previously, these experiments have not been feasible; however, with the advent of genome editing technologies and continuous development of powerful imaging and genetic tools, in the future it will be interesting to test the function of gap junction proteins on the soma–oocyte communication in flies.

In vertebrates, ERK activation functions to promote oocyte meiotic maturation; events prior to this in prophase progression cannot be determined (Fan and Sun

2004; Liang et al. 2007). In *C. elegans*, ERK activation is necessary for meiotic prophase progression and is essential for onset of early events of meiotic maturation, such as NEBD and nuclear translocation. However, in *C. elegans*, active ERK levels are dramatically downregulated in the maturing oocyte before the completion of meiotic maturation. In fact, an inability to downregulate active ERK once meiotic maturation has begun results in early onset of mitosis and endocycles, rather a transition to metaphase I. This inactivation of ERK is distinct from that in mammals where active ERK is present from meiotic maturation until the release of the oocyte from meiosis II by fertilization. This difference is likely a reflection of the reproductive needs and architecture of the reproductive organs. In mammals, meiosis II arrest and meiosis I are spatially and temporally separated. Thus, a high active ERK signal in meiotic I maturation does not necessarily have an impact on meiosis II onset. However, in *C. elegans*, completion of meiosis I occurs only in the presence of the MSP signal from the sperm. The presence of the sperm ensures that meiosis I maturation is coupled with ovulation into the spermatheca, followed quickly by fertilization. This suggests that an inability to downregulate active ERK in the oocyte may result in an arrest in meiosis II, while in the spermatheca it results in aberrant fertilization and developmental defects. Currently, the phosphatase that mediates the downregulation of active ERK in the maturing oocyte in worms is unknown; identification of this phosphatase will enable us to directly test this model. It is also interesting to speculate that there maybe evolutionary intermediates where meiosis I arrest and transition while spatially and temporally separated from meiosis II may still occur in a relatively short time scale to enable cell biological and genetic dissection of the events. Interestingly, *D. melanogaster* would have served as such an intermediate; however, unlike worms and mammals, loss of Mos (the MAPK Kinase Kinase) or loss of ERK does not impact meiotic maturation or ovulation, although active ERK is robustly detected in the oocytes. This suggests that at least in flies ERK signaling and activity play a nonessential function during oocyte development.

While exciting, the regulatory events uncovered thus far suggest that we are barely scratching the surface in our understanding of the regulatory and signaling mechanisms that coordinate the various events underlying meiosis I progression, prophase arrest, and transition from prophase to metaphase. The parallels and differences between these model systems form a powerful foundation for uncovering common and novel principals underlying reproductive fitness in females.

Acknowledgments Arur Lab is funded by, NIH GM 98200, ACS RSG014-044-DDC and CPRIT RP160023.

References

Amsterdam A, Koch Y, Lieberman ME, Lindner HR (1975) Distribution of binding sites for human chorionic gonadotropin in the preovulatory follicle of the rat. J Cell Biol 67:894–900

Archambault V, Zhao X, White-Cooper H, Carpenter AT, Glover DM (2007) Mutations in Drosophila Greatwall/Scant reveal its roles in mitosis and meiosis and interdependence with Polo kinase. PLoS Genet 3, e200

Arur S, Ohmachi M, Nayak S, Hayes M, Miranda A, Hay A, Golden A, Schedl T (2009) Multiple ERK substrates execute single biological processes in Caenorhabditis elegans germ-line development. Proc Natl Acad Sci U S A 106:4776–4781

Arur S, Ohmachi M, Berkseth M, Nayak S, Hansen D, Zarkower D, Schedl T (2011) MPK-1 ERK controls membrane organization in C. elegans oogenesis via a sex-determination module. Dev Cell 20:677–688

Bohrmann J, Zimmermann J (2008) Gap junctions in the ovary of Drosophila melanogaster: localization of innexins 1, 2, 3 and 4 and evidence for intercellular communication via innexin-2 containing channels. BMC Dev Biol 8:111

Boxem M, Srinivasan DG, van den Heuvel S (1999) The Caenorhabditis elegans gene ncc-1 encodes a cdc2-related kinase required for M phase in meiotic and mitotic cell divisions, but not for S phase. Development 126:2227–2239

Buszczak M, Freeman MR, Carlson JR, Bender M, Cooley L, Segraves WA (1999) Ecdysone response genes govern egg chamber development during mid-oogenesis in Drosophila. Development 126:4581–4589

Chase D, Serafinas C, Ashcroft N, Kosinski M, Longo D, Ferris DK, Golden A (2000) The polo-like kinase PLK-1 is required for nuclear envelope breakdown and the completion of meiosis in Caenorhabditis elegans. Genesis 26:26–41

Cho WK, Stern S, Biggers JD (1974) Inhibitory effect of dibutyryl cAMP on mouse oocyte maturation in vitro. J Exp Zool 187:383–386

Church DL, Guan KL, Lambie EJ (1995) Three genes of the MAP kinase cascade, mek-2, mpk-1/sur-1 and let-60 ras, are required for meiotic cell cycle progression in Caenorhabditis elegans. Development 121:2525–2535

Dekel N, Sherizly I (1985) Epidermal growth factor induces maturation of rat follicle-enclosed oocytes. Endocrinology 116:406–409

Detwiler MR, Reuben M, Li X, Rogers E, Lin R (2001) Two zinc finger proteins, OMA-1 and OMA-2, are redundantly required for oocyte maturation in C. elegans. Dev Cell 1:187–199

Drake M, Furuta T, Suen KM, Gonzalez G, Liu B, Kalia A, Ladbury JE, Fire AZ, Skeath JB, Arur S (2014) A requirement for ERK-dependent Dicer phosphorylation in coordinating oocyte-to-embryo transition in C. elegans. Dev Cell 31:614–628

Edry I, Sela-Abramovich S, Dekel N (2006) Meiotic arrest of oocytes depends on cell-to-cell communication in the ovarian follicle. Mol Cell Endocrinol 252:102–106

Edwards RG (1965) Maturation in vitro of human ovarian oocytes. Lancet 2:926–929

Fan HY, Sun QY (2004) Involvement of mitogen-activated protein kinase cascade during oocyte maturation and fertilization in mammals. Biol Reprod 70(3):535–547

Forbes A, Lehmann R (1999) Cell migration in Drosophila. Curr Opin Genet Dev 9:473–478

Fortune JE (1983) Steroid production by Xenopus ovarian follicles at different developmental stages. Dev Biol 99:502–509

Gibert MA, Starck J, Beguet B (1984) Role of the gonad cytoplasmic core during oogenesis of the nematode Caenorhabditis elegans. Biol Cell 50:77–85

Govindan JA, Cheng H, Harris JE, Greenstein D (2006) Galphao/i and Galphas signaling function in parallel with the MSP/Eph receptor to control meiotic diapause in C. elegans. Curr Biol 16:1257–1268

Govindan JA, Nadarajan S, Kim S, Starich TA, Greenstein D (2009) Somatic cAMP signaling regulates MSP-dependent oocyte growth and meiotic maturation in C. elegans. Development 136:2211–2221

Greenstein, D. (2005). Control of oocyte meiotic maturation and fertilization. WormBook, 1–12

Gumienny TL, Lambie E, Hartwieg E, Horvitz HR, Hengartner MO (1999) Genetic control of programmed cell death in the Caenorhabditis elegans hermaphrodite germline. Development 126:1011–1022

Gutch MJ, Flint AJ, Keller J, Tonks NK, Hengartner MO (1998) The Caenorhabditis elegans SH2 domain-containing protein tyrosine phosphatase PTP-2 participates in signal transduction during oogenesis and vulval development. Genes Dev 12:571–585

Hall DH, Winfrey VP, Blaeuer G, Hoffman LH, Furuta T, Rose KL, Hobert O, Greenstein D (1999) Ultrastructural features of the adult hermaphrodite gonad of Caenorhabditis elegans: relations between the germ line and soma. Dev Biol 212:101–123

Harris JE, Govindan JA, Yamamoto I, Schwartz J, Kaverina I, Greenstein D (2006) Major sperm protein signaling promotes oocyte microtubule reorganization prior to fertilization in Caenorhabditis elegans. Dev Biol 299:105–121

Herrlich A, Kuhn B, Grosse R, Schmid A, Schultz G, Gudermann T (1996) Involvement of Gs and Gi proteins in dual coupling of the luteinizing hormone receptor to adenylyl cyclase and phospholipase C. J Biol Chem 271:16764–16772

Hong A, Lee-Kong S, Iida T, Sugimura I, Lilly MA (2003) The p27cip/kip ortholog dacapo maintains the Drosophila oocyte in prophase of meiosis I. Development 130:1235–1242

Hsieh M, Lee D, Panigone S, Horner K, Chen R, Theologis A, Lee DC, Threadgill DW, Conti M (2007) Luteinizing hormone-dependent activation of the epidermal growth factor network is essential for ovulation. Mol Cell Biol 27:1914–1924

Hsieh M, Zamah AM, Conti M (2009) Epidermal growth factor-like growth factors in the follicular fluid: role in oocyte development and maturation. Semin Reprod Med 27:52–61

Hubbard CJ (1986) Cyclic AMP changes in the component cells of Graafian follicles: possible influences on maturation in the follicle-enclosed oocytes of hamsters. Dev Biol 118:343–351

Hubbard EJ, Greenstein D (2000) The Caenorhabditis elegans gonad: a test tube for cell and developmental biology. Dev Dyn 218:2–22

Ivanovska I, Lee E, Kwan KM, Fenger DD, Orr-Weaver TL (2004) The Drosophila MOS ortholog is not essential for meiosis. Curr Biol 14:75–80

Jaramillo-Lambert A, Ellefson M, Villeneuve AM, Engebrecht J (2007) Differential timing of S phases, X chromosome replication, and meiotic prophase in the C. elegans germ line. Dev Biol 308:206–221

Kalinowski RR, Berlot CH, Jones TL, Ross LF, Jaffe LA, Mehlmann LM (2004) Maintenance of meiotic prophase arrest in vertebrate oocytes by a Gs protein-mediated pathway. Dev Biol 267:1–13

Kim S, Spike C, Greenstein D (2013) Control of oocyte growth and meiotic maturation in Caenorhabditis elegans. Adv Exp Med Biol 757:277–320

Lake CM, Hawley RS (2012) The molecular control of meiotic chromosomal behavior: events in early meiotic prophase in Drosophila oocytes. Annu Rev Physiol 74:425–451

Lee MH, Ohmachi M, Arur S, Nayak S, Francis R, Church D, Lambie E, Schedl T (2007) Multiple functions and dynamic activation of MPK-1 extracellular signal-regulated kinase signaling in Caenorhabditis elegans germline development. Genetics 177:2039–2062

Lehmann R (1992) Germ-plasm formation and germ-cell determination in Drosophila. Curr Opin Genet Dev 2:543–549

Liang CG, Su YQ, Fan HY, Schatten H, Sun QY (2007) Mechanisms regulating oocyte meiotic resumption: roles of mitogen-activated protein kinase. Mol Endocrinol 21(9):2037–2055

Lincoln AJ, Wickramasinghe D, Stein P, Schultz RM, Palko ME, De Miguel MP, Tessarollo L, Donovan PJ (2002) Cdc25b phosphatase is required for resumption of meiosis during oocyte maturation. Nat Genet 30:446–449

Lopez AL 3rd, Chen J, Joo HJ, Drake M, Shidate M, Kseib C, Arur S (2013) DAF-2 and ERK couple nutrient availability to meiotic progression during Caenorhabditis elegans oogenesis. Dev Cell 27:227–240

Maddox AS, Lewellyn L, Desai A, Oegema K (2007) Anillin and the septins promote asymmetric ingression of the cytokinetic furrow. Dev Cell 12:827–835

Magnusson C, Hillensjo T (1977) Inhibition of maturation and metabolism in rat oocytes by cyclic amp. J Exp Zool 201:139–147

Maruyama R, Endo S, Sugimoto A, Yamamoto M (2005) Caenorhabditis elegans DAZ-1 is expressed in proliferating germ cells and directs proper nuclear organization and cytoplasmic core formation during oogenesis. Dev Biol 277:142–154

Masciarelli S, Horner K, Liu C, Park SH, Hinckley M, Hockman S, Nedachi T, Jin C, Conti M, Manganiello V (2004) Cyclic nucleotide phosphodiesterase 3A-deficient mice as a model of female infertility. J Clin Invest 114:196–205

Masui Y (1967) Relative roles of the pituitary, follicle cells, and progesterone in the induction of oocyte maturation in Rana pipiens. J Exp Zool 166:365–375

Mattingly HH, Chen JJ, Arur S, Shvartsman SY (2015) A transport model for estimating the time course of ERK activation in the C. elegans germline. Biophys J 109:2436–2445

McCarter J, Bartlett B, Dang T, Schedl T (1999) On the control of oocyte meiotic maturation and ovulation in Caenorhabditis elegans. Dev Biol 205:111–128

McLaughlin JM, Bratu DP (2015) Drosophila melanogaster oogenesis: an overview. Methods Mol Biol 1328:1–20

Mehlmann LM, Jones TL, Jaffe LA (2002) Meiotic arrest in the mouse follicle maintained by a Gs protein in the oocyte. Science 297:1343–1345

Mehlmann LM, Saeki Y, Tanaka S, Brennan TJ, Evsikov AV, Pendola FL, Knowles BB, Eppig JJ, Jaffe LA (2004) The Gs-linked receptor GPR3 maintains meiotic arrest in mammalian oocytes. Science 306:1947–1950

Miller MA, Nguyen VQ, Lee MH, Kosinski M, Schedl T, Caprioli RM, Greenstein D (2001) A sperm cytoskeletal protein that signals oocyte meiotic maturation and ovulation. Science 291:2144–2147

Nadarajan S, Govindan JA, McGovern M, Hubbard EJ, Greenstein D (2009) MSP and GLP-1/Notch signaling coordinately regulate actomyosin-dependent cytoplasmic streaming and oocyte growth in C. elegans. Development 136:2223–2234

Nadarajan S, Mohideen F, Tzur YB, Ferrandiz N, Crawley O, Montoya A, Faull P, Snijders AP, Cutillas PR, Jambhekar A et al (2016) The MAP kinase pathway coordinates crossover designation with disassembly of synaptonemal complex proteins during meiosis. Elife 5:e12039

Nagahama Y, Yamashita M (2008) Regulation of oocyte maturation in fish. Dev Growth Differ 50 (Suppl 1):S195–S219

Nedachi T, Conti M (2004) Potential role of protein tyrosine phosphatase nonreceptor type 13 in the control of oocyte meiotic maturation. Development 131:4987–4998

Newhall KJ, Criniti AR, Cheah CS, Smith KC, Kafer KE, Burkart AD, McKnight GS (2006) Dynamic anchoring of PKA is essential during oocyte maturation. Curr Biol 16:321–327

Norris RP, Freudzon L, Freudzon M, Hand AR, Mehlmann LM, Jaffe LA (2007) A G(s)-linked receptor maintains meiotic arrest in mouse oocytes, but luteinizing hormone does not cause meiotic resumption by terminating receptor-G(s) signaling. Dev Biol 310:240–249

Norris RP, Freudzon M, Mehlmann LM, Cowan AE, Simon AM, Paul DL, Lampe PD, Jaffe LA (2008) Luteinizing hormone causes MAP kinase-dependent phosphorylation and closure of connexin 43 gap junctions in mouse ovarian follicles: one of two paths to meiotic resumption. Development 135:3229–3238

Norris RP, Ratzan WJ, Freudzon M, Mehlmann LM, Krall J, Movsesian MA, Wang H, Ke H, Nikolaev VO, Jaffe LA (2009) Cyclic GMP from the surrounding somatic cells regulates cyclic AMP and meiosis in the mouse oocyte. Development 136:1869–1878

Oh JS, Han SJ, Conti M (2010) Wee1B, Myt1, and Cdc25 function in distinct compartments of the mouse oocyte to control meiotic resumption. J Cell Biol 188:199–207

Ohmachi M, Rocheleau CE, Church D, Lambie E, Schedl T, Sundaram MV (2002) C. elegans ksr-1 and ksr-2 have both unique and redundant functions and are required for MPK-1 ERK phosphorylation. Curr Biol 12:427–433

Panigone S, Hsieh M, Fu M, Persani L, Conti M (2008) Luteinizing hormone signaling in preovulatory follicles involves early activation of the epidermal growth factor receptor pathway. Mol Endocrinol 22:924–936

Park JY, Su YQ, Ariga M, Law E, Jin SL, Conti M (2004) EGF-like growth factors as mediators of LH action in the ovulatory follicle. Science 303:682–684

Phillips CM, McDonald KL, Dernburg AF (2009) Cytological analysis of meiosis in Caenorhabditis elegans. Methods Mol Biol 558:171–195

Pincus G, Enzmann EV (1935) The comparative behavior of mammalian eggs in vivo and in vitro: I. The activation of ovarian eggs. J Exp Med 62:665–675

Rajagopalan-Gupta RM, Lamm ML, Mukherjee S, Rasenick MM, Hunzicker-Dunn M (1998) Luteinizing hormone/choriogonadotropin receptor-mediated activation of heterotrimeric guanine nucleotide binding proteins in ovarian follicular membranes. Endocrinology 139:4547–4555

Robinson DN, Cant K, Cooley L (1994) Morphogenesis of Drosophila ovarian ring canals. Development 120:2015–2025

Runft LL, Jaffe LA, Mehlmann LM (2002) Egg activation at fertilization: where it all begins. Dev Biol 245:237–254

Schuh M, Ellenberg J (2007) Self-organization of MTOCs replaces centrosome function during acentrosomal spindle assembly in live mouse oocytes. Cell 130:484–498

Sela-Abramovich S, Edry I, Galiani D, Nevo N, Dekel N (2006) Disruption of gap junctional communication within the ovarian follicle induces oocyte maturation. Endocrinology 147:2280–2286

Spike CA, Coetzee D, Eichten C, Wang X, Hansen D, Greenstein D (2014a) The TRIM-NHL protein LIN-41 and the OMA RNA-binding proteins antagonistically control the prophase-to-metaphase transition and growth of Caenorhabditis elegans oocytes. Genetics 198:1535–1558

Spike CA, Coetzee D, Nishi Y, Guven-Ozkan T, Oldenbroek M, Yamamoto I, Lin R, Greenstein D (2014b) Translational control of the oogenic program by components of OMA ribonucleoprotein particles in Caenorhabditis elegans. Genetics 198:1513–1533

Stebbings LA, Todman MG, Phillips R, Greer CE, Tam J, Phelan P, Jacobs K, Bacon JP, Davies JA (2002) Gap junctions in Drosophila: developmental expression of the entire innexin gene family. Mech Dev 113:197–205

Strome S, Updike D (2015) Specifying and protecting germ cell fate. Nat Rev Mol Cell Biol 16:406–416

Tootle TL, Spradling AC (2008) Drosophila Pxt: a cyclooxygenase-like facilitator of follicle maturation. Development 135:839–847

Tornell J, Billig H, Hillensjo T (1990) Resumption of rat oocyte meiosis is paralleled by a decrease in guanosine 3′,5′-cyclic monophosphate (cGMP) and is inhibited by microinjection of cGMP. Acta Physiol Scand 139:511–517

Tsafriri A, Motola S (2007) Are steroids dispensable for meiotic resumption in mammals? Trends Endocrinol Metab 18:321–327

Tsafriri A, Lindner HR, Zor U, Lamprecht SA (1972) In-vitro induction of meiotic division in follicle-enclosed rat oocytes by LH, cyclic AMP and prostaglandin E 2. J Reprod Fertil 31:39–50

Viste K, Kopperud RK, Christensen AE, Doskeland SO (2005) Substrate enhances the sensitivity of type I protein kinase a to cAMP. J Biol Chem 280:13279–13284

Vivarelli E, Conti M, De Felici M, Siracusa G (1983) Meiotic resumption and intracellular cAMP levels in mouse oocytes treated with compounds which act on cAMP metabolism. Cell Differ 12:271–276

Von Stetina JR, Orr-Weaver TL (2011) Developmental control of oocyte maturation and egg activation in metazoan models. Cold Spring Harb Perspect Biol 3:a005553

Von Stetina JR, Tranguch S, Dey SK, Lee LA, Cha B, Drummond-Barbosa D (2008) alpha-Endosulfine is a conserved protein required for oocyte meiotic maturation in Drosophila. Development 135:3697–3706

Wang Y, Teng Z, Li G, Mu X, Wang Z, Feng L, Niu W, Huang K, Xiang X, Wang C et al (2015) Cyclic AMP in oocytes controls meiotic prophase I and primordial folliculogenesis in the perinatal mouse ovary. Development 142:343–351

Wayne CM, Fan HY, Cheng X, Richards JS (2007) Follicle-stimulating hormone induces multiple signaling cascades: evidence that activation of Rous sarcoma oncogene, RAS, and the epidermal growth factor receptor are critical for granulosa cell differentiation. Mol Endocrinol 21:1940–1957

Williamson A, Lehmann R (1996) Germ cell development in Drosophila. Annu Rev Cell Dev Biol 12:365–391

Wolke U, Jezuit EA, Priess JR (2007) Actin-dependent cytoplasmic streaming in C. elegans oogenesis. Development 134:2227–2236

Xiang Y, Takeo S, Florens L, Hughes SE, Huo LJ, Gilliland WD, Swanson SK, Teeter K, Schwartz JW, Washburn MP et al (2007) The inhibition of polo kinase by matrimony maintains G2 arrest in the meiotic cell cycle. PLoS Biol 5, e323

Yamamoto I, Kosinski ME, Greenstein D (2006) Start me up: cell signaling and the journey from oocyte to embryo in C. elegans. Dev Dyn 235:571–585

Yu J, Fleming SL, Williams B, Williams EV, Li Z, Somma P, Rieder CL, Goldberg ML (2004) Greatwall kinase: a nuclear protein required for proper chromosome condensation and mitotic progression in Drosophila. J Cell Biol 164:487–492

Zhang Y, Zhang Z, Xu XY, Li XS, Yu M, Yu AM, Zong ZH, Yu BZ (2008) Protein kinase A modulates Cdc25B activity during meiotic resumption of mouse oocytes. Dev Dyn 237:3777–3786

Chapter 5
Prophase I: Preparing Chromosomes for Segregation in the Developing Oocyte

Rachel Reichman, Benjamin Alleva, and Sarit Smolikove

Abstract Formation of an oocyte involves a specialized cell division termed meiosis. In meiotic prophase I (the initial stage of meiosis), chromosomes undergo elaborate events to ensure the proper segregation of their chromosomes into gametes. These events include processes leading to the formation of a crossover that, along with sister chromatid cohesion, forms the physical link between homologous chromosomes. Crossovers are formed as an outcome of recombination. This process initiates with programmed double-strand breaks that are repaired through the use of homologous chromosomes as a repair template. The accurate repair to form crossovers takes place in the context of the synaptonemal complex, a protein complex that links homologous chromosomes in meiotic prophase I. To allow proper execution of meiotic prophase I events, signaling processes connect different steps in recombination and synapsis. The events occurring in meiotic prophase I are a prerequisite for proper chromosome segregation in the meiotic divisions. When these processes go awry, chromosomes missegregate. These meiotic errors are thought to increase with aging and may contribute to the increase in aneuploidy observed in advanced maternal age female oocytes.

5.1 Introduction

Oocyte formation involves a process termed meiosis. Meiosis is not unique to oocytes; it occurs in spermatocytes as well as other type of gametes (e.g., spores). Meiosis includes a reductional division, which is not found in mitotic cells. In meiosis, cells replicate their DNA once, followed by two subsequent rounds of division: a reductional division—meiosis I (MI)—and then an equational division—meiosis II (MII). When diploid cells undergo meiosis to form gametes such as sperm or spores (in yeast), it results in four identical haploid cells (e.g., Tamaki 1965). In oogenesis, a polar body is extruded in both meiotic divisions, and only one

R. Reichman • B. Alleva • S. Smolikove (✉)
Department of Biology, University of Iowa, Iowa City, IA 52242, USA
e-mail: sarit-smolikove@uiowa.edu

S. Arur (ed.), *Signaling-Mediated Control of Cell Division*, Results and Problems in Cell Differentiation 59, DOI 10.1007/978-3-319-44820-6_5

meiotic product will become the oocyte (Rodman 1971). Since the polar bodies do not divide further, meiosis in oogenesis creates three meiotic products, two of which are not fertilized and their DNA is not transmitted to the next generation.

Meiotic prophase I includes unique chromosomal processes that are not found in mitotic prophase (Fig. 5.1). These include programmed DNA damage in the form of DNA double-strand breaks (DSBs) that occur multiple times in different positions on each chromosome (Kolodkin et al. 1986). In meiosis, homologous chromosomes identify each other and come in close proximity (pairing), a process largely unique to meiotic cells. The tight association between homologous chromosomes is mediated by a meiotic-specific protein structure known as the synaptonemal complex through a process termed synapsis (Fawcett 1956; Moses 1956, 1958). These unique meiotic processes are targeted toward the formation of crossovers between each pair of homologous chromosomes. Completion of crossover formation is a prerequisite for successful meiotic divisions; without crossovers, chromosome segregation in MI is random, leading most frequently to inviable progeny.

Meiotic prophase is divided into five ordered stages: leptotene, zygotene, pachytene, diplotene, and diakinesis (Zickler and Kleckner 1999). DSB formation and the initiation of pairing interactions occur at the entrance to prophase I, whereas complete synapsis and stabilizing pairing interactions are observed in the pachytene stage in which DSBs are repaired to form crossovers and noncrossovers (gene conversions without crossover). In diplotene, the synaptonemal complex disassembles, and in diakinesis, chromosomes are condensed and chiasmata are typically observed. A chiasma is the physical/visual representation of a crossover event holding together two homologous chromosomes to form a bivalent.

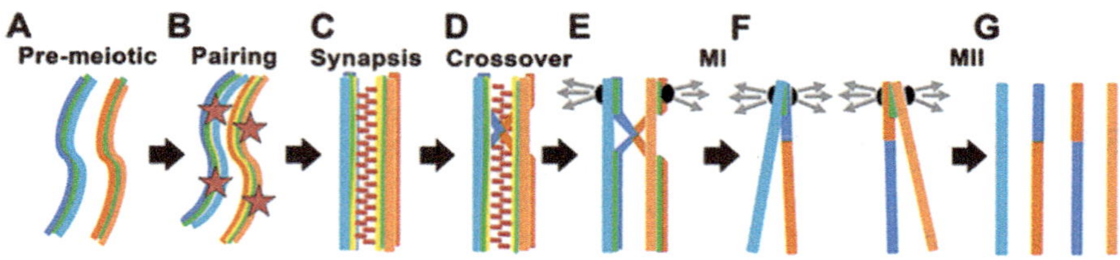

Fig. 5.1 Key meiotic events from chromosomal perspective. (A) In premeiotic S phase, each homolog is replicated once and resulting sister chromatids are held together by sister chromatid cohesions (*green*). Two replicated homologs are shown in *blue* and *orange*. (B) Pairing begins concurrently with axial elements (*yellow*) binding to the chromosomes. DNA double-stranded breaks are also made at this time (*red stars*), starting the process of homologous recombination. (C) Pairing interactions are stabilized when central and lateral components of the synaptonemal complex (*red zipper-like structure*) are assembled along the chromosomes. (D) Homologous recombination results in crossovers between homologs (one crossover shown here). (E) SC disassembles (some residual SC protein localization to chromosome is not shown here) and centromeres (*black circles*) of homologs are pulled to opposite directions (*gray arrows*). (F) In MI (transition from E to F), homologs separate, yet sister chromatids are still held together by cohesion that is maintained at the centromeric regions. (G) In meiosis II (transition from F to G), sister chromatids separate when the cohesions are removed from chromosomes. Meiosis results in four distinct haploid daughter gametes that are allelically different than the parent organism

Here, we will review the processes that chromosomes undergo to form DSBs and repair them. We will focus on the process of homologous recombination as well as on key structural features unique to meiosis that also contribute to this repair. The progression of meiosis requires communication between different steps in the meiotic program, which is inter- and intra-connected by signaling events. Most importantly, these chromosomal events are responsible for setting the stage for chromosome disjunction, and we will discuss the current view about how perturbation in meiotic prophase I events may contribute to chromosome nondisjunction.

5.2 Homologous Recombination

5.2.1 DSB Formation

One key feature of meiosis is that the cell forms programmed DSBs in order to initiate homologous recombination (Fig. 5.2). Recombination maintains genetic diversity among species by breaking linkage disequilibria and further facilitates the efficient removal of deleterious mutations (Hill and Robertson 2007). The process of meiotic recombination must lead to repair of the genome in an error-free pathway, as error-prone pathways risk introducing chromosomal aberrations. Meiosis harbors a complex mechanism of DNA repair through recombination that bears resemblance to somatic DNA damage repair of DSBs.

Chromosomal pairing and DSB repair are interdependent in some organisms, yet separable in others. In mice (*Mus musculus*), homolog pairing at chromosomal ends initiates before DSBs are formed (Boateng et al. 2013). While not initially thought to be common, this mechanism of initiating pairing independently of DSB formation is also found to hold true in *Caenorhabditis elegans* and *Drosophila melanogaster* (Dernburg et al. 1998; McKim and Hayashi-Hagihara 1998). In budding yeast, the majority of cells do not undergo pairing until DSBs are formed, and in the absence of DSBs, limited synaptonemal complex assembly occurs (Bhuiyan and Schmekel 2004). Even though some chromosomal pairing and synapsis may occur without meiotic DSBs, viable offspring are not produced due to random chromosome segregation in the absence of crossovers.

Homologous recombination (HR) begins when the topoisomerase-like protein, Spo11, is recruited to the DNA to initiate DSBs (Keeney et al. 1997). Spo11 is the catalytic subunit of a multi-subunit protein complex, as found in a number of studies in *Saccharomyces cerevisiae* (Table 5.1). Spo11 binds directly to the DNA, creates a DSB, and then remains covalently bound to the DSB until resection, the formation of ssDNA from dsDNA via nuclease activity, takes place (Neale et al. 2005). Spo11-induced DSBs are not random, as there are regions with little to no breaks (i.e., telomeres and centromeres in *S. cerevisiae*), while others are enriched with DSBs ("hot spots") (Gerton et al. 2000). DSB formation is influenced by epigenetic marks that can determine DSB levels and/or positioning.

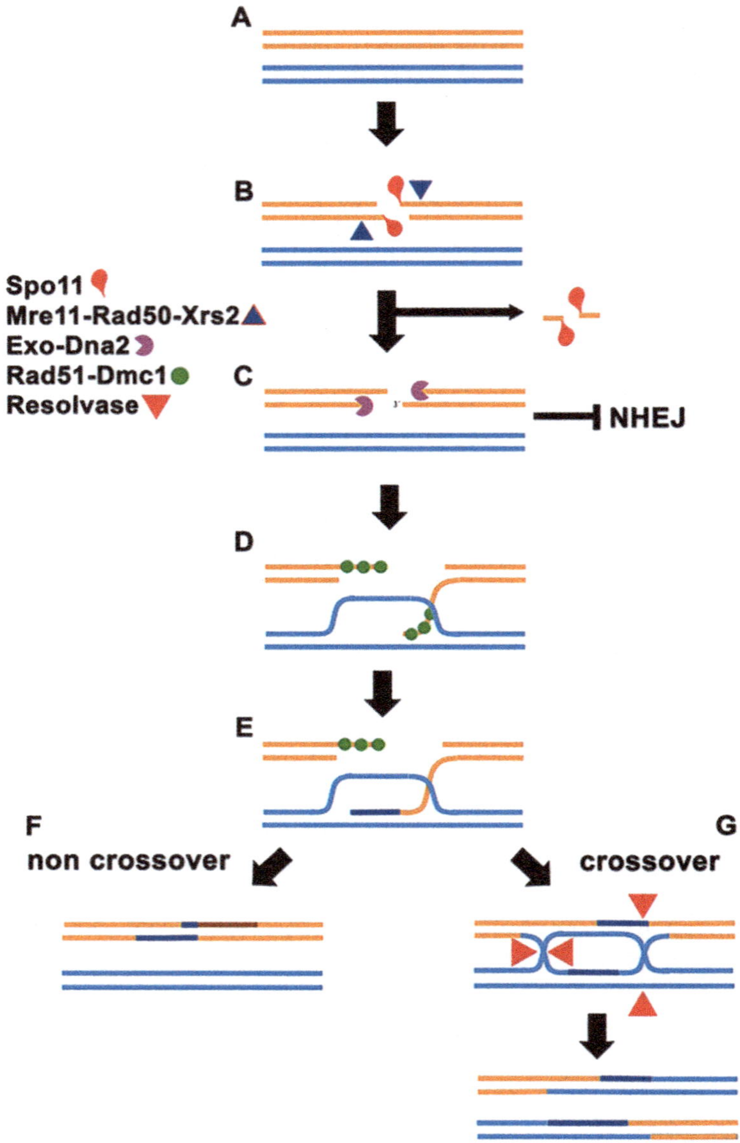

Fig. 5.2 Homologous recombination during meiotic prophase I. (A) Two homologous chromosomes have been replicated and are about to undergo meiosis (*orange* and *blue lines*). For simplicity, sister chromatids are not shown here, and two lines with similar colors represent the two strands of DNA. (B) The topoisomerase-like protein Spo11 (*red balloon*) and the MRX (*blue triangle*) complex localize to specific sites on the DNA, and Spo11 creates a double-strand break in the DNA. MRX then removes Spo11 by cleaving roughly 35 bp upstream of the DSB. This results in an oligonucleotide with Spo11 attached at one end being removed from the break site. (C) The nucleases Exo1 and Dna2 create longer ssDNA. This long-term resection may be executed by different protein complexes in different organisms; some may not require Exo–Dna2 for wild-type resection. Once ssDNA is formed, Ku proteins can no longer bind the cut site, thus inhibiting a mutagenic repair pathway known as NHEJ. (D) RPA, followed by Dmc1/Rad51 (here shown as a

PRDM9 was identified as a gene essential for meiosis that regulates recombination hot spots (Baudat et al. 2010; Hayashi et al. 2005; Parvanov et al. 2010). Germ cells that lack PRDM9 have an aberrant DSB pattern with a very small effect on DSB levels (inferred directly or by ssDNA-binding protein localization) (Brick et al. 2012; Sun et al. 2015). Most mammals studied express the PRDM9 protein that recognizes a specific DNA sequence via its Zn-finger motif, both DNA sequences and the Zn finger motif are rapidly evolving (Berg et al. 2010; Ponting 2011; Ségurel et al. 2011). In humans and mice, allelic variants of PRDM9 Zn-finger motifs are associated with hot spot usage (Baudat et al. 2010). Another protein domain found in PRDM9 is an H3K4 trimethyltransferase (me3) domain (Hayashi et al. 2005, ix; Koh-Stenta et al. 2014). Hot spots have an open chromatin structure (Getun et al. 2010). PRDM9 me3 of H3K4 causes chromatin restructuring, which leaves a nucleosome-free region (Baker et al. 2014). Altogether, these data suggest that PRDM9-dependent H3K4me3 may expose a DNA region to make them more amenable to Spo11-induced DSBs that will otherwise be induced elsewhere, thus determining hot spot positioning. Mammalian PRDM9 sites are frequently found outside promoters in intragenic regions (Brick et al. 2012). Conversely, nonmammalian eukaryotes from yeast to birds have DSB and/or recombination hot spots that are normally located in promoter regions of genes (Lam and Keeney 2015; Singhal et al. 2015) or within genes (Adrian and Comeron 2013). These locations are under much greater evolutionary constraint; therefore, these hot spots evolve at a much slower rate (Lam and Keeney 2015).

Studies done in the yeast *S. cerevisiae* and in *C. elegans* agree with studies done in mice, showing that epigenetic marks play an important role in hot spot definition. The chromatin's overall structure comprises a series of large loops attached at their base to the meiotic axis (Zickler and Kleckner 1999). DSBs are formed in the loop regions and are scarcely found in proximity to the axis (Blat et al. 2002). Based on the presence of "recombination nodules" (seen on EM images of meiotic chromosome spreads), it was suggested that the recombination process occurs in a location embedded in the synaptonemal complex (Carpenter 1975). These studies altogether suggest that DSBs formed in the loops are moved to the synaptonemal complex for repair. Spo11 in *S. cerevisiae* forms DSBs via interaction of proteins in the holocomplex with the axis as well as with Spp1, a member of the H3K4me3 Set1 complex (Acquaviva et al. 2013; Sommermeyer et al. 2013). Spp1 physically

Fig. 5.2 (continued) single green circle, though composing distinct complexes in vivo), coats the ssDNA and protects it from degradation. (E) Dmc1/Rad51 also conducts a homology search along the homolog. Once homology is found, polymerase extends the ssDNA to close the gap made by the initial exonucleases. (F) The crossover is resolved as either a "noncrossover" which results in gene conversion (*dark blue line* in the *yellow* homolog) when the newly synthesized DNA is removed from the invaded D-loop and is ligated to its original (*yellow*) homolog. Polymerase can then use the DNA of the invading strand (*yellow*) as a template for the top missing DNA (*purple*). (G) The "crossover" pathway. Proteins called resolvases cut the DNA (two pairs of *red arrows*), which, when religated, forms crossovers, DNA molecules that contain sequences originating from two different homologs

Table 5.1 Proteins involved in DSB formation

Protein name	Function	Species present	Species absent
Rec102	Binds chromatin, may form complex with Rec104, biochemical function unknown but important for Spo11 localization.	*S. cerevisiae*	Mice, human, *D. melanogaster*, *C. elegans*
Rec104	Binds chromatin, may form complex with Rec102, biochemical function unknown but important for Spo11 localization.	*S. cerevisiae*	Mice, human, *D. melanogaster*, *C. elegans*
Mei4	Biochemical function is unknown; however, it does form a complex with Rec114 and Mer2.	*S. cerevisiae*, mouse, human	*C. elegans*, *D. melanogaster*
Ski8	Scaffold protein that directly interacts with Spo11 in meiosis. Localizes directly to the chromatin. Has a separate function in mRNA metabolism not related to meiosis.	*S. cerevisiae*, human, mouse, *D. melanogaster*	*C. elegans*
Rec8	A meiosis-specific cohesin protein that holds chromatin together during DSB formation and HR.	*S. cerevisiae*, mouse, human, *C. elegans*, *D. melanogaster*	None
Rec114	Biochemical function is unknown; however, it does form a complex with Mei4 and Mer2.	*S. cerevisiae*	Mice, Human, *D. melanogaster*, *C. elegans*
Mer2/ Rec107	Localizes to chromatin prior to DSB formation. Is phosphorylated by Cdc28 to initiate recombination and forms a complex with Rec114 and Mei4.	*S. cerevisiae*	Mice, Human, *D. melanogaster*, *C. elegans*
Mre11	Has an endo- and exonuclease function to initiate DSBs and resect DNA. Initial resection removes Spo11 from broken ends. Forms a complex with Rad50 and Xrs2/Nbs1.	*S. cerevisiae*, mice, human, *D. melanogaster*, *C. elegans*	None
Rad50	Contains coiled-coiled domains to hold broken DNA ends together as well as an ATPase domain that acts as a motor to move the MRX/N complex along DNA during resection.	*S. cerevisiae*, mice, human, *D. melanogaster*, *C. elegans*	None
Xrs2/ Nbs1	Forms a complex with Mre11 and Rad50. Important for protein–protein interactions and has a role in localizing the complex to the DNA during DSB initiation.	*S. cerevisiae*, mice, human, *D. melanogaster*	*C. elegans*

interacts with the promoter-enriched H3K4me3, bringing the DSB to the synaptonemal complex (Acquaviva et al. 2013; Sommermeyer et al. 2013). H3K4me3 seems important for DSB activity since methyl transferase mutants that result in reduction of H3K4me3 also result in reduced hot spot activity (Sollier et al. 2004). In *C. elegans*, the H3K4me3 was not shown to play a role in DSB formation, but other marks do appear to play an important role, through the action of proteins such as HIM-17 and XND-1. HIM-17 is required for H3K9 methylation in early meiotic prophase (Reddy and Villeneuve 2004). In the absence of HIM-17, DSB formation is decreased but not eliminated, as indicated by reduction in the number of bivalents formed (Saito et al. 2012; Tsai et al. 2008). Global histone acetylation levels are associated with changes in the frequency of DSB formation in *C. elegans* (Gao et al. 2015). This pathway likely involves CRA-1 and XND-1 that are suggested to promote crossover formation on the autosomes and the X chromosomes, respectively (Gao et al. 2015, Wagner et al. 2010).

5.2.2 DSB Processing

Once DSBs are formed, nucleolytic activity removes the covalently bound Spo11 and processes the DNA ends in a process termed resection. This leads to the formation of long single-stranded DNA at meiotic DSBs (Sun et al. 1991). These single-stranded DNA overhangs are required for the strand invasion step that follows. A key protein complex acting in this process is the MRN/X protein complex composed of Mre11, Rad50, and NBS1/Xrs2 (Usui et al. 1998). Homologs of Mre11 and Rad50 exist between species, despite varying levels of sequence similarity (Badugu et al. 2015; Hopfner and Tainer 2003). The third member of the MRN complex is NBS1 in mammals or Xrs2 in *S. cerevisiae* (Carney et al. 1998; Ivanov et al. 1992). NBS1/Xrs2 is the least conserved member of the MRN complex; the conservation between *S. cerevisiae* Xrs2 and human NBS1 is low and restricted to the N-terminus of the protein (Carney et al. 1998). *C. elegans* on the other hand lacks an identifiable NBS1/Xrs2 homolog.

The MRN/X complex is thought to hold the broken ends of the chromosome together through the long coiled-coil domains of Rad50, while Mre11 resects the 5′ end of the broken DNA to promote the formation of 3′ single-stranded DNA overhangs (Paull and Gellert 1998). In *S. cerevisiae* and *C. elegans*, MRE11 is required both for DSB formation and resection, while in mice, it is known to act only in DSB processing (Ajimura et al. 1993; Cherry et al. 2007; Johzuka and Ogawa 1995). DSB formation and resection are two separate activities of MRE11. For example, in *C. elegans* and in *S. cerevisiae*, mutations in MRE11 have been isolated that are proficient in DSB initiation but are unable to resect broken DNA (Nairz and Klein 1997; Yin and Smolikove 2013). Mre11 has both endo- and exonuclease activities (Furuse et al. 1998; Moreau et al. 1999; Paull and Gellert 1998). This nuclease activity of Mre11 is suggested to be sufficient for Spo11 removal, which is required for resection of meiotic DSBs (Moreau et al. 1999; Neale et al. 2005). Mre11 has an additional meiotic function outside of its role in nucleolytic activity: it promotes the assembly of the MRX complex by recruitment of polySUMO chains (Chen et al. 2016). NBS1/Xrs2 is primarily involved in protein–protein interactions. In mitotic cells, it is posttranslationally modified through phosphorylation, ubiquitination, and SUMOylation (Cremona et al. 2012; Falck et al. 2012; Stiff et al. 2006; Wu et al. 2012).

Another key nuclease that participates in DSB processing in meiosis is Ctp1/Sae2/CtIP/COM-1. CtIP is an endonuclease required for Spo11 removal from the ends of broken DNA (Prinz et al. 1997; Rothenberg et al. 2009). Following Spo11 removal, CtIP is required for further end resection, a function that is separable from its role in Spo11 removal (Ma et al. 2015). CtIP was also suggested to act by facilitating the nuclease activity of Mre11 (Cannavo and Cejka 2014). It has been observed that CtIP is phosphorylated and that this posttranslational modification is necessary for its recruitment to DSBs through CtIP binding to Nbs1 (Dodson et al. 2010).

Once initial resection is underway, other proteins assist in further DNA processing to reform long-range resection. Compared to non-meiotic DNA damage resection, meiotic resection produces much smaller regions of single-stranded DNA. In *S. cerevisiae* mitosis, Dna2 and Exo1 contribute to resection (Mimitou and Symington 2008; Zhu et al. 2008). In *S. cerevisiae* meiosis, Dna2 is also capable of meiotic resection, but only in the absence of the single-stranded DNA-binding protein Dmc1 (Manfrini et al. 2010). DNA-2 likely does not play a role in meiotic resection in *C. elegans*. In *S. cerevisiae*, Exo1 is a 5′–3′ exonuclease that collaborates with Mre11 in bidirectional resection following nick formation by Mre11 (Zakharyevich et al. 2010). In *C. elegans*, Exo-1 functions in resection only when Mre11 or COM-1 resection is impaired and NHEJ is inactive (Lemmens et al. 2013; Yin and Smolikove 2013).

Processive resection may also be involved in the repression of error-prone DSB repair pathways, such as nonhomologous end joining (NHEJ). In *C. elegans*, inhibiting DSB resection by *com-1* deletion or an *mre-11* separation of function allele permits the utilization of NHEJ as a DSB repair pathway (Yin and Smolikove 2013). In other organisms, regulation of NHEJ in meiosis is independent of resection; in *S. cerevisiae*, NHEJ proteins are exported out of the nucleus upon meiotic entry, and in mammals, NHEJ proteins are degraded, which effectively eliminates the possibility of NHEJ as a mode of repair during meiosis (Goedecke et al. 1999; Valencia et al. 2001).

5.2.3 *Strand Invasion*

ssDNA must be protected from both nucleolytic digest that would shorten its length and extensive resection that will increase the length of ssDNA. The first may lead to a loss of genetic information, while the latter may increase the chance of creating regions of microhomology that could be utilized for error-prone repair pathways resulting in deletions and/or translocations. In meiosis, three repair proteins are used to protect these ssDNA ends before homology search begins: RPA, Rad51, and Dmc1 (Bishop et al. 1992; Sung and Robberson 1995). The first protein complex to recognize and load onto the ssDNA is RPA, a three subunit protein complex:

RPA70/RPA1, RPA32/RPA2, and RPA14/RPA3. Each RPA complex binds to approximately 20 or 30 nucleotides of ssDNA, and each ssDNA strand is coated by multiple RPA complexes forming a filament (Chen and Wold 2014). The RPA filament inhibits strand exchange, thus keeping premature recombination from taking place, but RPA is also required for recombination due to its role in preventing secondary structures of ssDNA (Gasior et al. 2001). Interestingly, RPA does not bind to DSBs when CtIP is depleted, indicating that RPA is in some way regulated by CtIP (Costelloe et al. 2012).

Once RPA is bound, Rad51 and Dmc1 are loaded onto the ssDNA (Nešić et al. 2004; Wang and Haber 2004). Rad51 is associated with both mitotic and meiotic HR, whereas Dmc1 is the meiosis-specific single-stranded binding protein and a member of the RecA/Rad51 gene family (Lin et al. 2006). Not all organisms have a Dmc1 homolog; both *C. elegans* and *Drosophila* are missing a homolog (Schurko and Logsdon 2008). The Rad51–Dmc1 filament nucleates from the center of the single-stranded DNA, and the two proteins elongate in opposing directions: Rad51 in the 3′–5′ direction and the Dmc1 filament in the 5′–3′ direction (Brown and Bishop 2015). Dmc1 has DNA strand exchange activity (Sehorn et al. 2004). The molecular mechanism behind how Dmc1 promotes strand exchange involves recognition of three bases at a time (Lee et al. 2015). This base triplet mechanism is highly similar to that of Rad51-based strand exchange (Lee et al. 2015; Qi et al. 2015). It has been shown that in meiosis, the filament formed can be composed of both Rad51 and Dmc1, when Dmc1 is found at a terminal position (Brown et al. 2015).

Dmc1 is required for proper completion of meiotic homologous recombination in most organisms. In *S. cerevisiae*, Dmc1 is essential for sporulation, and Rad51 plays a minor role in homologous recombination (Bishop et al. 1992; Cloud et al. 2012). In *S. pombe*, however, Rad51 rather than Dmc1 plays a more dominant role in homologous recombination (Fukushima et al. 2000). However, Rad51 can partially substitute for Dmc1; when *S. cerevisiae* Rad51 is overexpressed in cells that lack Dmc1, meiosis is partially restored (Tsubouchi and Roeder 2003). Crossover interference, the mechanism that prevents adjacent crossovers, is not restored by Rad51 overexpression in Dmc1 mutants, suggesting a role for Dmc1 in crossover interference (Tsubouchi and Roeder 2003). Interhomolog recombination is strongly favored during meiosis. This homolog bias is required to promote accurate chromosome segregation at MI via suppression of recombination between sister chromatids. Dmc1 and Rad51 are required for homolog bias; intersister recombination frequency is drastically increased in the absence of Dmc1 or Rad51 in *S. cerevisiae* (Cloud et al. 2012; Lao et al. 2013). Rad51 is inhibited in *S. cerevisiae*, inhibition of meiotic cells by Hed1, which restricts the access of Rad51 to Rad54 (Rad54 is a protein that facilitates D loop formation) (Busygina et al. 2008, 2012). Furthermore, Dmc1 is necessary for the proper localizations of the synaptonemal complex protein Red1 and the kinase Mek1 that may indirectly contribute to homolog bias in *S. cerevisiae* (Lao et al. 2013).

Mediator proteins are needed for Rad51/Dmc1 proteins to be loaded onto the DNA. The conserved Mei5-Sae3/Swi5 complex promotes Dmc1 loading in

S. cerevisiae and Rad51 loading in *S. pombe* (Hayase et al. 2004). In *S. cerevisiae*, Rad52 is a primary mediator protein that assists with Rad51/Dmc1 loading onto the DNA, as well as playing a role in displacing RPA (Gasior et al. 1998; Sugiyama and Kowalczykowski 2002). Rad52 is conserved throughout eukaryotes with two known exceptions: *C. elegans* and *Drosophila*. These two organisms lack any known Rad52 homolog, which is intriguing due to its integral role in strand exchange in all other organisms studied to date. Vertebrates, as well as *C. elegans*, contain an alternative Rad51 loader: Brca2 (Sharan and Bradley 1997). Brca2 possesses Rad51 binding and DNA-binding sites that may act in Rad51 loading (Sharan et al. 1997). Brca2 is required for meiosis in mouse (Sharan et al. 2004). In *C. elegans*, the Brc2 homolog, BRC-2, functions in Rad51 loading and RPA displacement on Spo-11 generated DSBs (Martin et al. 2005).

Throughout eukaryote and archaea, Rad51/RecA contains multiple paralogs, in addition to eukaryotic Dmc1 (Lin et al. 2006). The *C. elegans* RFS-1/RIP-1 complex promotes Rad51 loading, and the *rfs-1* mutant shows meiotic defects when combined with helicase *helq-1* mutants (Taylor et al. 2015). Rad51C is found in mouse meiotic prophase nuclei and is required for meiosis (Kuznetsov et al. 2007; Liu et al. 2007). In *Drosophila*, RAD51C/SPN-D is required for DSB repair in meiosis (Joyce et al. 2012). The Rad55–Rad57 complex, other RAD51 paralogs, in *S. cerevisiae* promotes Rad51 assembly by preventing Rad51 filament disruption by anti-recombinase activity (Liu et al. 2011). Both Rad55 and Rad57 are required for meiosis in *S. cerevisiae* (Gasior et al. 1998; Lovett and Mortimer 1987), while in *S. pombe*, Rad55 has a small effect on meiotic recombination (Lorenz et al. 2014).

5.2.4 Crossover Formation

ssDNA invasion into the homologous sequences is followed by DNA displacement, creating a D-loop. DNA synthesis "captures" the loop leading to the formation of a double Holliday Junction (dHJ). The ZMM (ZIP–MSH–MER) proteins, identified in *S. cerevisiae*, promote the resolution of this structure into a crossover. ZMM proteins include synaptonemal complex proteins (as Zip1/2/3/4) and proteins directly involved in crossover formation (as Msh4/5) (Lynn et al. 2007). Msh4/5 creates a sliding clamp complex that utilizes ATP hydrolysis for movement (Snowden et al. 2004). In *C. elegans*, MSH-5 can be found uniformly throughout the chromatin, but once crossovers are formed in early to mid pachytene, MSH-5 begins to form distinct foci at the single interfering crossover on each chromosome (Yokoo et al. 2012). These two proteins are thought to be involved in physically holding the crossover intermediates until recombination is complete, and a loss of this function would affect the outcome of recombination. In *S. cerevisiae*, Msh4/5 mutants that are unable to interact with DNA have a decrease in the number of crossovers, whereas ATP hydrolysis defective Msh4 mutants have reduced crossover interference as well as crossover numbers (Krishnaprasad et al. 2015; Nishant et al. 2010; Novak et al. 2001; Rakshambikai et al. 2013). This indicates that the Msh4/5 complex in yeast may not only play a stabilizing role during recombination but also affect crossover numbers and interference via a distinct mechanism.

COSA-1, a protein discovered in *C. elegans*, co-localizes to crossover sites during meiosis and acts in the same pathway as the MSH-4/5 complex (Yokoo et al. 2012). COSA-1 bears a close resemblance to the human Cyclin B1 protein based on structural modeling. COSA-1 homologs are widely conserved throughout eukaryotes, except in the *Drosophila* lineages in which Msh4, Msh5, and COSA-1 are all absent (Schurko and Logsdon 2008). The precise mechanism of COSA-1 function is still unknown; however, a model has been proposed consisting of a two-stage licensing process: starting with MSH-4/5 and COSA-1 localization, followed by COSA-1 partnering or interacting with an unknown CDK to phosphorylate MSH-4/5 which promotes recombination progression (Yokoo et al. 2012). The mammalian COSA-1 homolog, CNTD1, exhibits similar phenotypes as those of the *C. elegans* homolog, which includes normal progression of early meiosis but a failure of late meiotic recombination markers to appear or be removed from the chromatin, thus indicating defects in crossover formation. These findings lead to the model that a CDK partner may work with CNTD1 in mammals to ensure normal meiosis progression, likely via phosphorylation of specific target proteins (Holloway et al. 2014).

Crossover formation is also dependent upon proteins involved in posttranslational modifications. RNF212 in mammals, Zip3 in *S. cerevisiae*, and ZHP-3 in *C. elegans* are all homologous E3 ligase proteins. These proteins have been implicated in SUMOylation in yeast and possibly ubiquitination in other organisms (Reynolds et al. 2013). RNF212 is required for crossover formation in mammals; in the absence of RNF212, synapsis can still occur relatively normally; however, localization of crossover promoting protein such as MLH1 and MSH4 is reduced or eliminated, suggesting a role for RNF212 in stabilizing crossover precursors (Reynolds et al. 2013). In mammals, RNF212 localizes as puncta in early pachytene, which progressively become more selective and co-localize with the MSH4/5 complex (Reynolds et al. 2013). The ubiquitin ligase HEI10 has been implicated in this process as an antagonistic factor to RNF212 (Qiao et al. 2014; Ward et al. 2007). In HEI10-deficient mice, RNF212 localizes to crossovers but fails to be removed, preventing crossover formation and arresting meiosis (Qiao et al. 2014). HEI10 also localizes to crossovers in wild-type cells, suggesting a direct role in this pathway. These studies point to an interplay between SUMO and ubiquitin in order to promote the correct number of meiotic crossovers and non-crossovers. The RNF212 homolog in yeast, Zip3, has been implicated in both the SUMOylation and ubiquitination pathways (Cheng et al. 2006; Perry et al. 2005). Consistent with the findings from mice, the *C. elegans* RNF212 homolog, ZHP-3, is also important for crossover formation (Jantsch et al. 2004). Despite the fact that Zip3 was first implicated in assembly of the synaptonemal complex, a role in synaptonemal complex assembly was not identified for RNF212 (mice) or ZHP-3 (*C. elegans*) (Agarwal and Roeder 2000). Thus, although the role of RNF212/Zip3/

ZHP-3 proteins in promoting crossover formation is conserved, their function in synaptonemal complex assembly is restricted to yeast.

A model based on studies in *S. cerevisiae* meiosis indicates that crossovers can arise as an outcome of two pathways: class I crossovers/interfering crossovers and class II crossovers/non-interfering crossovers. Class I crossovers make up the majority of meiotic crossovers and are promoted by the ZMM proteins. Class I and class II crossovers originate by the action of resolvases, structure-specific nucleases that can act on HJ. In *S. cerevisiae*, Mlh1 and Mlh3 form a heterodimer that acts to resolve dHJs in a pathway that leads exclusively to crossover formation (Hunter and Borts 1997; Wang et al. 1999). This pathway also involves the nuclease Exo1 (Zakharyevich et al. 2012). Similar observations were made in mouse: most meiotic crossovers arise by the action of the Mlh1–Mlh3 complex (Baker et al. 1996; Lipkin et al. 2002).

Class II crossovers originate from the combined activities of several endonucleases. These nucleases can resolve dHJs forming either crossovers or noncrossovers, although some resolvases can act on the D-loop intermediate as well and promote only crossovers (Osman et al. 2003; Svendsen and Harper 2010). Nuclease complexes that are involved in generation of class II crossovers include Mus81–Eme1/Mms4, Slx-1–Slx-4, Xpf1–ERCC1, and Gen1/Yen1 (Bailly et al. 2010; Boddy et al. 2001; Saito et al. 2012). These nuclease complexes exhibit partially redundant functions in HJ resolution, and the part each nuclease plays in HJ resolution varies from organism to organism. In *S. cerevisiae*, *mms4* as well as *mus81* mutants have a significant effect on meiosis, while *yen1* (*gen1* in mammals), *slx1*, and *slx4* have no effect as shown in single mutant analysis (Mullen et al. 2001). The redundancy between these nucleases is revealed in double mutant analysis; combining *mus81* with *yen1*, *slx1*, or *slx4* deletion mutants exacerbates the *mus81* mutant phenotype (Matos et al. 2011). A major role for these nucleases is observed in resolving unregulated joint molecules (e.g., HJ between sister or multiple partners), as found in *sgs1* mutants (Oh et al. 2008; Zakharyevich et al. 2012). *mus81* mutants produce no viable spores, suggesting that MUS81 has a key role in meiosis in *S. pombe* (Boddy et al. 2000). In mice, MUS81 appears to play a minor role; both BTBD12/*slx4* and *mus81* mutants have reduced fertility and meiotic progression defects, but cells that progress to diakinesis show no crossover defects (Holloway et al. 2011). The effect of *mlh3* null mutant on crossover formation is enhanced (but not eliminated) by deleting *mus81* in this background, an indication that Mus81 is required for crossover formation in the absence of Class I resolvases (Holloway et al. 2008). Moreover, MLH1 foci numbers are increased in *mus81* mutants, likely indicating a cross talk between the class I and II crossover pathways (Holloway et al. 2008). Despite the fact that each of the three complexes (Mus81–Eme1, Slx-1–Slx-4, Xpf1–ERCC1) possesses a separable nuclease activity and frequently functions redundantly, evidence from mitotic cells suggests that in fact these complexes associate to form one multi-protein complex. In *Drosophila*, xpf1/Mei9-Ercc1 physically interacts with *mus312* and *slx4*, and this interaction is mediated by Mei9 (Radford et al. 2005; Yildiz et al. 2002). In *C. elegans*, HIM18/Slx-4 interact with XPF-1, SLX-1, and MUS-81, while EME-1 and

ERCC1 are brought to the complex by interaction with MUS-81 and XPF-1, respectively (Saito et al. 2013), and in human cells, a Mus81–Eme1–Slx-1–Slx-4 complex was identified to function in mitotic resolution (Wyatt et al. 2013).

Investigation of the role that structure-specific nucleases play in meiosis in *C. elegans* has added complexity to the picture. In *C. elegans* meiosis, all crossovers are MSH-4/5-dependent crossovers, suggesting that they are interfering (Kelly et al. 2000). Interestingly, however, the nucleases involved in each of these events utilize a class of nucleases named Class II in *S. cerevisiae*, challenging the classification of resolvases (Saito et al. 2013). In *C. elegans*, *him-18/slx-4* mutants have reduced crossover numbers, bivalent stability is compromised, and bivalent differentiation is delayed (Saito et al. 2009). As found in *S. cerevisiae*, worm *gen1* (yeast *yen1*) and *slx-1* have mild effects on meiosis as single mutants (Bailly et al. 2010; Saito et al. 2012). However, double mutant analysis is consistent with a role in crossover resolution for all of these nucleases (Saito et al. 2013). These studies are consistent with a model for two main resolution pathways in *C. elegans*: one for MUS-81 and SLX-1 and one for XPF-1, with a very minor role for GEN-1. Also similarly to *S. cerevisiae*, *C. elegans mus-81* and *slx-1* mutant phenotypes are exacerbated in the absence of Sgs1/HIM-6 (Saito et al. 2013).

5.3 Chromosome Structure

5.3.1 Sister Chromatid Cohesion

As chromosomes enter meiosis, sister chromatids are connected to each other and held together by the sister chromatid cohesion complex (Fig. 5.3A). The composition of this complex resembles, but is not identical, to that of the mitotic sister chromatid cohesion complex. It is therefore not surprising that recruitment of mitotic cohesion to meiotic chromosomes is unable to compensate for the lack of meiotic cohesion (Yokobayashi et al. 2003). The cohesion complex is composed of four protein subunits: Smc1, Smc3, Scc3/SA/STAG, and Scc1/Rad21 (Klesin). In meiosis, Scc1 is replaced by meiotic specific subunit(s), and in some organisms other subunits are replaced as well. For example, Scc1 is replaced by Rec8, ubiquitously among organisms, while in mammals, Smc1 is replaced by Smc1β and Smc3/STAG1 is replaced by STAG3 (Pezzi et al. 2000; Revenkova et al. 2001; Watanabe and Nurse 1999). Cohesion protein components are evolutionarily conserved (Harvey et al. 2002; Koshland and Strunnikov 1996) and are related to other complexes that act in maintaining the function of chromosomes: the condensin complex (chromosome condensation), the Smc4/5 complex (DNA damage repair), and in *Drosophila*, the synaptonemal complex protein C(2)M shares sequence similarity to Scc1 (Heidmann et al. 2004).

The cohesion subunits assemble to form a ring structure (Anderson et al. 2002), most of which is accounted for by the Smc subunits, with each Smc subunit

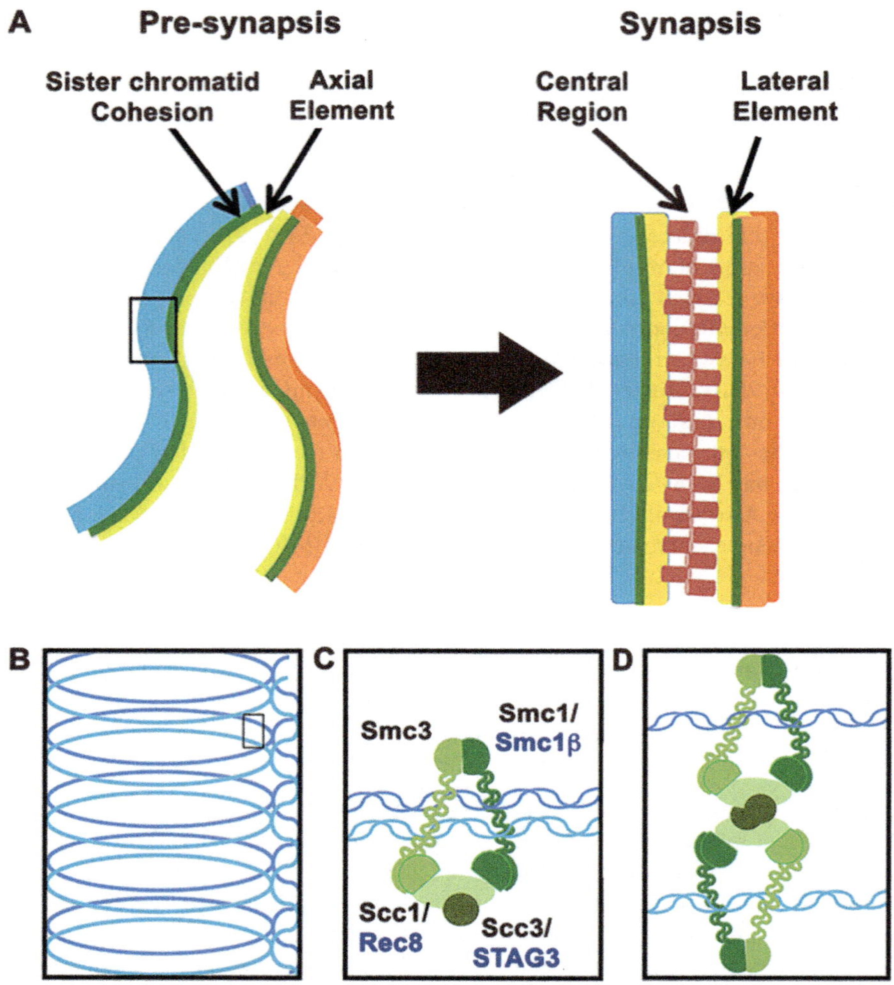

Fig. 5.3 The synaptonemal complex and sister chromatid cohesion. (A) Two sister chromatids (*blue pair* and *orange pair*) are held together by sister chromatid cohesion complexes. During pairing, axial elements (*yellow*) are first to bind and then central region proteins assemble to form the synaptonemal complex (*red*). In this context, axial elements are called lateral elements. (B) Close-up on the blue homolog region *boxed* in (A) of chromatin arranged in large loops. (C) Model one of the cohesion complex working during meiosis: a single cohesion complex consisting of Smc1, Smc3/Smc3β, Scc3/SA/STAG3, and Scc1/Rec8 forms a loop-like structure. Names of mitotic–meiotic subunits are in *black* and meiosis-specific subunits in *blue*. This single loop can fit two DNA strands within it, holding sisters together. (D) Model two: two complete cohesion complexes each have one strand of DNA within their loop and the two complexes interact with each other to hold the two DNA strands together. Both C and D are close-ups of *black rectangle* in B

containing two long coiled-coil domains. A wide hole enclosed by the cohesion ring structure is compatible with different models explaining how sister chromatids are connected (Huang et al. 2005). Model 1 (Fig. 5.3C) suggests that the ~40 nM ring structure can fit two DNA molecules (10 nM fiber that contain nucleosomes) (Haering et al. 2002). This led to the proposal that the cohesion complex can connect two sister chromatids by passing the two DNA strands through a single ring in multiple positions along the chromosomes (Haering et al. 2002). An alternative model (model 2, Fig. 5.3D) suggests that each chromatid is captured by its own ring and two adjacent rings interact to establish cohesion (Zhang et al. 2008b). The establishment of sister chromatid cohesion typically occurs during S phase (Uhlmann and Nasmyth 1998). However, most of the cohesion complex is loaded onto the chromosomes prior to replication (Michaelis et al. 1997), which suggests that modification of the ring structure (opening/closing) is an important part of establishment of sister chromatid cohesion.

The variety in cohesion subunits suggests that multiple cohesion complexes may exist with specific functions. This is supported by a specific localization pattern temporally and spatially for some of the cohesion complexes. In the mouse, the Scc1/Rad21 meiotic subunit can be replaced not only by Rec8 but also Rad21L (Herrán et al. 2011; Ishiguro et al. 2011; Lee and Hirano 2011). In early prophase, Rad21L's presence is predominant, compared to that of Rec8, and these two proteins occupy nonoverlapping positions on the chromosomes (Ishiguro et al. 2011; Lee and Hirano 2011). In *C. elegans*, five Scc1 subunits exist: COH-1, COH-2/SCC-1, COH-3, COH-4, and REC-8; all five subunits have suggested meiotic functions based on localization and/or mutant phenotypes (Severson and Meyer 2014). However, REC-8 and COH-3/4 seem to play the major role in meiotic cohesion (Pasierbek et al. 2001; Severson et al. 2009). The REC-8 and the COH-3/4 cohesin complexes play redundant roles, as breakdown of prophase cohesion requires depletion of both complexes (Severson et al. 2009). These functions are not identical as (1) REC-8 and COH-3/4 localize to different chromosomal domains on pre-metaphase chromosomes, (2) the COH-3/4 complex is able to bi-orient sister chromatids in the absence of REC-8, while REC-8 cannot bi-orient sisters in the absence of COH-3/4 (Severson and Meyer 2014; Severson et al. 2009). The picture became even more complex with the discovery that the mitotic cohesion complex may play a role in meiosis as well [Scc1/Rad21: Prieto et al. (2002), Smcα: Gutiérrez-Caballero et al. (2011), COH-1 and COH-2/SCC-1: Severson and Meyer (2014), STAG2: Prieto et al. (2002)]. However, in some cases, their roles in meiosis are debated. For example, despite the presence of Scc1/Rad21 on meiotic chromosomes, it is not required for meiotic division (Tachibana-Konwalski et al. 2010). Therefore, some cohesion proteins may "wait" on meiotic chromosomes to perform their role later on in mitotic divisions within the zygote.

Cohesin complexes are essential for chromosome function in meiosis. In early stages of meiosis, cohesion is essential for the formation and maintenance of the synaptonemal complex at the level of axis formation (the substructure of the synaptonemal complex, Fig. 5.3A, first to assemble). Yeast mutants that lack Rec8 do not form an axis (Klein et al. 1999). In organisms in which several cohesin complexes exist, partially overlapping function of cohesin complexes in synaptonemal complex assembly is revealed. In the absence of mouse Rec8 or Rad21L, synapsis is severely impaired but axes still form (Herrán et al. 2011; Xu et al. 2005). Loss of axis is observed only when both mouse Rec8 and Rad21L are depleted (Llano et al. 2012). Similarly to what is found in mice, in *C. elegans* mutants in which all meiotic cohesion is removed (*rec-8*, *coh-3/4*) the homologs are unable to form an axis. However, removing only one of these complexes restricts the effects only in to the central region of the synaptonemal complex (Severson et al. 2009). Effects on synaptonemal complex maintenance, as opposed to establishment, were also observed in Rad21L mutants in *Drosophila* and Rec8 mutants in *Sordaria* (Urban et al. 2014, Storlazzi et al. 2008). The role of cohesion in synaptonemal complex assembly may be indirect due to cohesion's function in modifying chromosome structure or direct by interacting with synaptonemal complex proteins. The latter was demonstrated in *Drosophila* and mice: in *Drosophila*, mitotic cohesion protein Rad21 physically interacts with C(2)M, and in mouse, SMC1 and SMC3 interact with synaptonemal complex proteins SYC3 and SYC2 (Urban et al. 2014; Vallaster et al. 2011). In addition to its importance for synaptonemal complex assembly, cohesion plays a role in shaping the length of the meiotic axis. Increasing cohesion on chromosomes by inhibiting mechanisms for its removal shortens axis length in *C. elegans* and yeast (Challa et al. 2016; Crawley et al. 2016).

The key role of cohesion in synaptonemal complex assembly is likely to enable repair of meiotic DSBs, since the synaptonemal complex is required for and promotes interhomolog repair. However, cohesion may also control the repair of DSBs independently of its role in synaptonemal complex assembly. For example, cohesion is required for the repair of DSBs using sister chromatid recombination in the absence of the synaptonemal complex in *C. elegans* (Crawley et al. 2016; Smolikov et al. 2007). In *C. elegans*, defects in cohesion loading impair early steps in DSB processing (Lightfoot et al. 2011), whereas an increase in cohesion loading leads to a delay in DSB repair (Crawley et al. 2016). In yeast, increased cohesion also leads to DSB repair defects (Challa et al. 2016), and sister chromatid cohesion plays a role in implementing homolog bias in meiosis (Hong et al. 2013). DSB formation is also dependent on cohesion, as complete removal of the meiotic cohesion complexes prevents axis assembly which is required for DSB formation (Severson et al. 2009).

The most notable function of cohesion is in holding sister chromatids together, thus allowing proper chromosome segregation in mitosis and meiosis. Cohesion counteracts the action of forces generated by the spindle through the kinetochore that try to pull the chromatids apart. By this action, cohesion promotes proper alignment of chromosomes on the metaphase plate. Therefore, release of cohesion requires a regulated process: in mitosis, cohesion is released first from the chromosome arms (prophase to pre-metaphase, the "prophase pathway") followed by release from centromeres (in anaphase) (Waizenegger et al. 2000). Release of meiotic cohesion occurs in a similar order, but the release of cohesion from chromosome arms occurs at MI releasing homologous chromosomes, while release of centromeric cohesion occurs at MII, releasing sister chromatids. Release of meiotic cohesion in anaphase of both meiotic divisions occurs by proteolytic cleavage of the cohesion ring by similar mechanisms to what is found in mitotic anaphase (Buonomo et al. 2000; Kudo et al. 2006; Salah and Nasmyth 2000). Protection of centromeric cohesion by Shugoshin and PP2A/Air2 prevents this mechanism from releasing centromeric cohesion in MI (Kitajima et al. 2004; Lee et al. 2008). Premature release of cohesion results in chromosome missegregation (as will be discussed in Sect. 5.3).

Many of the proteins that assist cohesion loading and establishment in mitosis have conserved functions in meiosis. These functions include the opening and closing of the cohesion ring without the need for proteolytic cleavage via mechanisms that are currently poorly understood. The NIPBL/Scc2-Mau2/Scc4 cohesion loader complex was suggested to act by a mechanism involving the ATP-dependent opening of the cohesin ring structure (Arumugam et al. 2003; Hu et al. 2011; Murayama and Uhlmann 2014; Unal et al. 2008). NIPBL/Nipped-B localizes to the synaptonemal complex axis in mouse, *C. elegans*, and *Drosophila* (Gause et al. 2008; Kuleszewicz et al. 2013; Lightfoot et al. 2011; Visnes et al. 2014), and Mau2/Scc4 additionally localizes to the synaptonemal complex in mouse (Visnes et al. 2014). Similar to its role in mitosis (Ciosk et al. 2000), NIPBL/Scc2 promotes sister chromatid cohesion in meiosis (Lin et al. 2011). However, in yeast, it was shown that at least part of the role that Scc2 plays in formation of meiotic cohesion is due to increasing transcription of the Rec8 gene (Lin et al. 2011). Nipped-B mutants in *Drosophila* have defects in the maintenance of cohesion (and the synaptonemal complex), while *scc-2* mutants in *C. elegans* do not load cohesion and phenotypically resemble cohesion null mutants (Gause et al. 2008; Lightfoot et al. 2011). The protein Ctf7/Eco1 promotes cohesion establishment by acetylating Smc3 (Ivanov et al. 2002; Skibbens et al. 1999; Toth et al. 1999). RNAi for Arabidopsis Ctf7 in meiosis prevents Scc3 localization to the chromosome axis and impairs sister chromatid cohesion in meiosis (Singh et al. 2013). Pds5 is a cohesion-associated protein that promotes Eco1-mediated acetylation and thus is suggested to play a role in cohesion maintenance (Vaur et al. 2012). Pds5 is associated with the meiotic axis (Fukuda and Hoog 2010; Storlazzi et al. 2008; Zhang et al. 2005). Studies in several model organisms support an important, but not essential, role for pds5 in meiotic cohesion. The *C. elegans* pds5/evl-14 point mutant has a small effect on meiotic cohesion (Wang et al. 2003). In *S. cerevisiae*, Pds5 mutants show impaired loading of Rec8, but these defects are sufficient to prevent synaptonemal complex assembly (Jin et al. 2009; Zhang et al. 2005). In *S. pombe* and *Sordaria*, Rec8 loading is reduced but not eliminated in Pds5 mutants (Ding et al. 2006; Storlazzi et al. 2008). While in most organisms, the effects of cohesion loading on recombination can be explained as stemming from a function in synaptonemal complex assembly, studies in *Arabidopsis* suggest a direct role in recombination, as axis formation is not affected in Pds5 mutants but recombination is affected (Pradillo et al. 2015). Pds5 also affects chromosome structures in fission yeast, and Pds5 mutants in budding yeast lead to chromosomal hyper-condensation (Ding et al. 2016; Jin et al. 2009). One protein was shown to counteract cohesion maintenance in meiosis: WAPL is necessary for unloading the cohesins in mitosis. The activity of WAPL is regulated by Eco1-mediated acetylation of Smc3 (Rolef Ben-Shahar et al. 2008; Sutani et al. 2009; Unal et al. 2008). It is likely that WAPL plays a similar role in meiotic prophase I, although the biochemical mechanism is not yet resolved. WAPL localizes to the synaptonemal complex in mouse (Zhang et al. 2008a). In *C. elegans*, WAPL-1 regulates cohesion removal of the COH-3/4 complexes but not the REC-8 cohesion complex (Crawley et al. 2016). *Drosophila wapl* mutants lead to nondisjunction of achaismata chromosomes, likely via heterochromatin pairing (Vernì et al. 2000).

5.3.2 The Synaptonemal Complex: An Introduction

The synaptonemal complex, a multimeric protein complex, mediates the association of homologous chromosomes in meiotic prophase I (Fig. 5.3; Fawcett 1956; Moses 1956, 1958). The proteinaceous synaptonemal complex is composed of lateral elements which bind to the axis of each homologous chromosome pair and central region proteins that connect to the lateral elements. Together they form a tripartite, ladder-like structure which holds homologous chromosomes together (Zickler and Kleckner 1999). Components of the synaptonemal complex contain coiled-coil domains which are conserved across all Eukarya synaptonemal complex proteins, but the amino acid sequence of the proteins is highly variable across species (Fig. 5.4; Table 5.2; Fraune et al. 2012; Page and Hawley 2004). Complete formation of the synaptonemal complex occurs when central region proteins have loaded and form a bridge between the lateral element proteins; this is known as synapsis of homologous chromosomes. Full synapsis is important for stable pairing, which allows for the repair of DSBs and the formation of crossovers, one of the outcomes of this repair. Crossovers create a physical tether between homologous chromosomes. This tether is important for holding the homologous chromosomes together until segregation; therefore, complete synapsis is necessary for proper chromosome segregation during meiosis. Loss of the synaptonemal complex leads to unstable pairing and deficient homologous recombination (e.g., de Vries et al. 2005; MacQueen et al. 2002; Sym et al. 1993). Homologous recombination is crucial for proper segregation of homologous chromosomes, and improper segregation leads to the formation of aneuploid gametes.

5.3.2.1 Synaptonemal Complex Assembly Dynamics, Pairing, and Signaling

Prior to synaptonemal complex assembly and complete synapsis along homologous chromosomes, the homologs need to pair. In general, the method by which the homologs find their partners is conserved across metazoans: chromosome movement prevents nonhomologous chromosome pairing while promoting homologs to

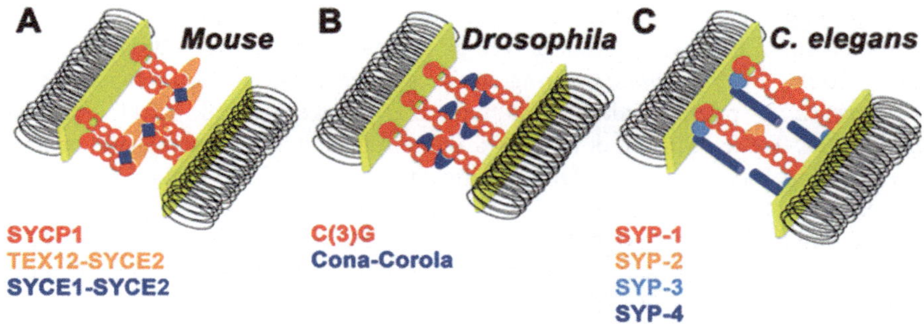

Fig. 5.4 Synaptonemal complex proteins vary among organisms. (A) SC in mouse consists of lateral proteins (*yellow*) and central region proteins, SYCP1 (*red*), TEX12 (*orange*), SYCE2 (*orange/blue*), and SYCE1 (*blue*). SYCP1 is the only central region protein to interact with the lateral elements, whereas TEX12, SYCE2, and SYCE1 interact only with each other and the opposite end of SYCP1. (B) SC in *D. melanogaster* consists of three proteins, C(3)G (*red*), Cona (*blue*), and Corola (*blue*). Cona and Corola form a complex in the middle of the central region and do not contact the lateral elements. C(3)G is the only protein to interact with lateral elements similar to mammalian SYCP1. (C) *C. elegans* SC is composed of four central region proteins: SYP-1 (*red*), SYP-2 (*orange*), SYP-3 (*light blue*), and SYP-4 (*blue*). Unlike other organisms, multiple *C. elegans* proteins interact with lateral elements, including SYP-1, SYP-3, and SYP-4

Table 5.2 Synaptonemal complex proteins

Species	Lateral elements	Central region proteins
S. cerevisiae	Red1 Hop1 Mek1	Zip1 Zip2 Zip3 Zip4 (Emc11-Gmc2)
C. elegans	HIM-3 HTP-1/2 HTP-3	SYP-1 SYP-2 SYP-3 SYP-4
D. melanogaster	C(2)M	C(3)G COROLLA CONA
M. musculus	SYCP3 SYCP2 HORMAD1/2	SYCP1 SYCE1/2 TEX12 SYCE3

pair, which acts through chromosome connections to the nuclear envelope and the cytoskeleton. Chromosomes are attached to the nuclear envelope through one or more ends in a configuration typically defined as a "bouquet" (Dresser and Giroux 1988; Harper 2004). *Drosophila* and *C. elegans* are exceptions in that they do not form a typical bouquet (Harper 2004). In *C. elegans*, initial pairing of chromosomes occurs at pairing centers, repetitive DNA sequences that are found on one end of each chromosome (Phillips et al. 2009). These sequences are recognized and bound by zinc-finger proteins, which are suggested to link the pairing centers of the chromosomes to the nuclear envelope (MacQueen et al. 2005; Phillips et al. 2005). In *Drosophila,* pairing initiates at centromeric locations, while in *S. cerevisiae*, pairing initiates at telomeres (Takeo et al. 2011; Tanneti et al. 2011; Trelles-Sticken et al. 2000). In the mouse, both telomeres and centromeres (telocentric in mouse) show significant levels of pairing in the initiation of meiosis (Boateng et al. 2013). SUN-KASH domain proteins were shown to be involved in the anchoring of chromosomes to the nuclear envelope in *S. cerevisiae*, *S. pombe*, *C. elegans*, and mouse model systems (Boateng et al. 2013; Conrad et al. 2008; Ding et al. 2007; Morimoto et al. 2012; Niwa et al. 2000). To disturb nonhomologous chromosome interactions, a whipping movement is created allowing for chromosomes to continue the search for homologs (Chikashige et al. 1994; Parvinen and Söderström 1976). In *Drosophila* and mice, a rotational movement is used to promote proper homologous chromosome pairing (Christophorou et al. 2015; Scherthan et al. 1996). In *S. pombe*, *C. elegans*, and *Drosophila*, the mechanism for homology search requires dynein-mediated movement of microtubules, while in *S. cerevisiae*, actin plays a role in the chromosomal movements in meiosis (Christophorou et al. 2015; Horn et al. 2013; Yamamoto et al. 1999).

Synaptonemal complex assembly initiates when the lateral elements (referred to as axial elements prior to synaptonemal complex assembly) bind to the chromosomes (Zickler and Kleckner 1999). The recruitment of axis proteins to chromosomes may be mediated by a direct interaction with the cohesion complex, providing a scaffold for synaptonemal complex assembly (Urban et al. 2014; Vallaster et al. 2011). Once chromosomes have found their homologs, central region protein nucleation occurs at the site of homolog pairing. In *S. cerevisiae* and *Drosophila*, the first synapsis initiation events were shown to coincide with centromeres, while in *C. elegans*, synapsis most likely initiates at pairing centers (Rog and Dernburg 2015). In mouse, however, centromeres are not the sites of synapsis initiation (Qiao et al. 2012). Once the central region proteins fully align along the chromosomes, the synaptonemal complex is considered to be fully synapsed. It is not clear yet if central region proteins interact directly with lateral elements or if the lateral elements create the right chromosomal environment for their assembly. The type and number of proteins that make up the two different parts of the synaptonemal complex (lateral and central elements) vary between organisms (see Table 5.2). Synaptonemal complex disassembly occurs at the end of prophase after DSB formation, repair, and the presence of crossovers that has now created a physical tether between homologs. Synaptonemal complex disassembly occurs in a stepwise manner, typically starting with the removal of central elements of the synaptonemal complex, followed by some elements of the axis (Eijpe et al. 2003; Nabeshima et al. 2005). Some synaptonemal complex proteins are retained on chromosomes in restricted locations and disassemble only during meiosis I, when homologous chromosomes segregate (Bisig et al. 2012; de Carvalho et al. 2008).

Just as the proteins of the synaptonemal complex vary from organism to organism, so does the interplay between recombination and the synaptonemal complex. In all cases, the synaptonemal complex is crucial for crossover formation between homologs. However, while in *Drosophila* and *C. elegans*, the synaptonemal complex is absolutely required for crossover formation, in mouse and in *S. cerevisiae*, the requirement is only partial (e.g., de Vries et al. 2005; MacQueen et al. 2002; Page and Hawley 2001; Sym et al. 1993). Another difference between organisms is in how much the synaptonemal complex requires DSBs for its formation. In *C. elegans* and *Drosophila*, the loss of DSB formation does not affect synaptonemal complex assembly, disassembly, or its structure (Colaiácovo et al. 2003; McKim et al. 1998). In mouse and in *S. cerevisiae*, in the absence of DSBs, the synaptonemal complex fails to properly form and very limited synapsis is observed (Bhuiyan and Schmekel 2004; Romanienko and Camerini-Otero 2000).

5.3.2.2 The Contribution of the Synaptonemal Complex to Recombination

The main role of the synaptonemal complex is considered to be to hold the homologous chromosomes together to stabilize the pairing interactions between them and to ensure crossover formation. Crossovers create a physical tether that holds together homologs at points where the synaptonemal complex disassembles. If the synaptonemal complex assembles improperly, such as partial assembly or aggregate formation, the ability to form crossovers is impaired and can lead to aneuploidy or cell death. The axial elements of the synaptonemal complex were shown to be required for DSB formation. In *S. cerevisiae*, both axis proteins Red1 and Hop1 are required for DSB formation (Mao-Draayer et al. 1996; Woltering et al. 2000; Xu et al. 1997), while in *C. elegans*, HTP-3 is required for DSB formation (Goodyer et al. 2008).

Other studies have suggested further roles for the synaptonemal complex in the regulation of the recombination process. The synaptonemal complex may promote the crossover/noncrossover decision by its specific effect on supporting crossovers. Mutants in the ZMM class of proteins in *S. cerevisiae*, which includes SC components, are required for crossover formation, but have no role in noncrossover formation (Börner et al. 2004). Axial element proteins of the synaptonemal complex act in ensuring homolog bias. Hop1 and Red1 from *S. cerevisiae* and SYCP3 from mouse impose homolog bias by serving as a barrier for sister chromatid recombination (Li et al. 2011), while in *S. pombe*, the axis imposes interhomolog recombination by promoting recombination with the homolog (Latypov et al. 2010). In *C. elegans*, reducing the level of synaptonemal complex central region proteins impairs interference (Libuda et al. 2013). However, it is not clear if the synaptonemal complex plays a role in crossover interference in other organisms since interference acts prior in *S. cerevisiae* (Bishop and Zickler 2004) or concurrently in *Sordaria macrospora* (Zhang et al. 2014) to synaptonemal complex assembly.

5.3.2.3 Posttranslational Modification

The proteins that make up the synaptonemal complex have been well studied, but how these proteins are regulated by posttranslational modifications is still not fully understood. One function of posttranslational modification is to regulate the assembly of the synaptonemal complex. *S. cerevisiae* provides the most complete understanding of synaptonemal complex assembly, a process which is regulated by SUMOylation in this organism. Zip3, one of the ZMM proteins, is an E3 SUMO ligase (Cheng et al. 2006). SUMO localizes to Zip1, and the absence of SUMO causes defects in synaptonemal complex assembly (Cheng et al. 2006; Hooker and Roeder 2006; Voelkel-Meiman et al. 2013). An additional component in this pathway includes UBC9, an E2 ligase that forms SUMO chains that are required for synaptonemal complex assembly (Klug et al. 2013). The C-terminal domains of the SC proteins Zip1 and Red1 contain SUMO-interacting domains (SIMs) (Cheng et al. 2006; Lin et al. 2010). Red1 is SUMOylated, in a Zip1-dependent manner, and SUMOylation facilitates the interaction between the two, promoting SC assembly (Eichinger and Jentsch 2010). A second step in SC assembly involves Emc11 SUMOylation (Humphryes et al. 2013; Leung et al. 2015; Voelkel-Meiman et al. 2013). SUMOylation of Emc11 promotes Zip1 assembly, via the central region of the SC, which is located at the Zip1 N-terminus (Leung et al. 2015). In *C. elegans*, synaptonemal complex aggregation is prevented by the activity of a neddylation regulated ubiquitin ligase, but it is still not clear if this effect on synaptonemal complex assembly is direct (Brockway et al. 2014). Posttranslational modification of synaptonemal complex proteins is also important for their disassembly. In *M. musculus*, PLK-1 localizes to and is required for the phosphorylation of SYCP1, TEX12, and SYCE1 in spermatocytes and is important for synaptonemal complex disassembly (Jordan et al. 2012). Phosphorylation of synaptonemal complex central region proteins also can initiate disassembly, as was shown for *C. elegans* SYP-2 (Nadarajan et al. 2016).

Another function of posttranslational modification of the synaptonemal complex is in regulation of recombination in meiosis. In *S. cerevisiae*, phosphorylation of the axial element, Hop1, is required for interhomolog recombination bias (Carballo et al. 2008), while the phosphorylation of Zip1, central region protein, promotes the formation of interfering crossovers as well as synaptonemal complex assembly (Chen et al. 2015). Axial element proteins of the synaptonemal complex are phosphorylated in a temporally regulated manner in mice, but the function of this phosphorylation is currently unknown (Fukuda et al. 2012). Since posttranslational modifications play an important role in the regulation of protein complexes, more in-depth understanding of how posttranslational protein modifications regulate the synaptonemal complex is required.

5.4 Signaling in Meiotic Prophase I

The accurate execution of meiotic events requires signaling pathways to communicate between different steps of the meiotic program. This communication is thought to occur between many different components of the meiotic program within each cell, although intercellular communication cannot be excluded. In terms of signaling related to recombination events, many of the pathway components are borrowed from DNA damage repair signaling. This is not surprising due to the similarity between meiotic and mitotic recombination. Pathways that regulate homologous chromosome pairing and synapsis are intertwined with recombination-related signaling, as expected from the interdependency between recombination and synapsis found in many organisms. Overall, meiotic signaling involves a combination of feedback loops acting within a pathway (e.g., DSB repair signals to DSB formation), with cross talk between pathways (e.g., crossover formation signals to chromosome pairing), and altogether regulating the complex molecular events that take place in meiosis.

DSB formation and repair are linked by signaling mechanisms that ensure when recombination proceeds normally, DSB formation is halted. ATM/Tel1 and ATR/Mec1 are two related protein kinases that play an important role in mitotic DSB repair (Guleria and Chandna 2016). In *Drosophila* and mouse meiosis, DSB formation is downregulated by ATM, while in *S. cerevisiae*, ATM and ATR are both involved in downregulating DSB formation (Joyce et al. 2011; Lange et al. 2011; Zhang et al. 2011). ATM and ATR's regulation of DSB formation may involve phosphorylation of members of the Spo11 complex, as was shown for Rec114 in *S. cerevisiae* (Carballo et al. 2013). This mechanism of downregulating DSB formation may be controlling interference at the level of DSB formation, as in *S. cerevisiae* ATM preferentially downregulates proximal DSBs, preventing the positioning of adjacent DSBs (Garcia et al. 2015). ATM and ATR are also involved in regulating DSB repair. In *S. cerevisiae*, activation of ATM/Tel1 depends on Mre11, a key player in DSB resection (Usui et al. 2001). However, ATM is not only activated by Mre11 but also positively regulates resection with ATR (Cartagena-Lirola et al. 2006; Terasawa et al. 2008; Usui et al. 2001). ATM and ATR promote differential timing of resection on a different subset of DSBs; ATM regulation is earlier and acts upon a few DSBs, whereas ATR regulates the late and abundant DSBs (Joshi et al. 2015). The synaptonemal complex axis is also a target of ATM and ATR. In mouse, axis and cohesion proteins are phosphorylated, in an ATM- and ATR-dependent manner, in chromosomal regions not yet synapsed, suggesting a role in silencing of unsynapsed chromatin (Fukuda et al. 2012; Royo et al. 2013). In *S. cerevisiae*, ATM/ATR phosphorylates Hop1 to promote interhomolog recombination (Carballo et al. 2008; Penedos et al. 2015). Mutating a predicted ATM/ATR phosphorylation site on Zip3 reduces crossover levels in *S. cerevisiae* (Serrentino et al. 2013). These effects on crossover levels are in agreement with the role for ATM and (more so) for ATR in imposing homolog bias (Joshi et al. 2015). ATR1/Mec1 also mediates Zip1 phosphorylation, but this event is specifically required for the early meiotic role of Zip1 in binding centromeres (centromere coupling) (Falk et al. 2010). In *C. elegans*, the early meiotic role for ATM/ATR has not been described as these proteins function mainly in the context of externally induced DNA damage and the mitotic germ line (Garcia-Muse and Boulton 2005; Stergiou et al. 2007). In *C. elegans* meiosis, ATM is involved in the restoration of chromosomal structure after DNA damage is induced in meiotic nuclei (Couteau and Zetka 2011).

Similarly to ATM and ATR, CHK2/Mek1 acts primarily in the DNA damage response in mitotic cells (Zannini et al. 2014). In mice ovaries, Chk2 is regulated by ATM and in *S. pombe*, CHK2 is a target of ATR, thus linking the two DNA damage signaling pathways (Miles et al. 2010; Tougan et al. 2010). The activation of CHK2 in mouse ovaries is essential for DNA damage-induced apoptosis of oocytes (Bolcun-Filas et al. 2014). Work done in *S. cerevisiae* and *S. pombe* is consistent with a role for Chk2 in the DNA damage repair pathway, but in a different step: promoting meiotic DSB repair (Niu et al. 2007; Tougan et al. 2010; Xu et al. 1997). In *S. cerevisiae*, Chk2 was shown to be involved in inhibiting sister chromatid recombination (Niu et al. 2007). In *C. elegans*, CHK-2 acts both in promoting DSB formation and homologous chromosome pairing. These pathways add to the growing list of noncanonical functions of CHK2 beyond DNA damage signaling (Zannini et al. 2014). CHK-2 promotes the formation of meiotic DSBs by controlling the recruitment of proteins required for DSB formation (DSB-1 and DSB-2) to chromosomes (Rosu et al. 2013; Stamper et al. 2013). In its second meiotic function, CHK-2 is required for localization of most pairing center proteins, promoting both chromosome pairing and synapsis (MacQueen and Villeneuve 2001; Phillips and Dernburg 2006). These mechanisms involve the phosphorylation of SUN-1 (part of the protein complex tethering chromosomes to the nuclear envelope) by CHK-2 (Penkner et al. 2009). CHK-2 also acts as a "sensor" protein. Defects in chromosome synapsis or recombination are relayed to CHK-2 via proteins forming the synaptonemal complex axis; this process promotes prolongation of CK activities (Kim et al. 2015). Chk2 may have a meiotic function outside DSB repair in mouse as well, as it is required in spermatogenesis for a specific H3 phosphorylation mark (Govin et al. 2010).

Polo-like kinases (PLKs) are involved in the regulation of cell division in both mitosis and meiosis that includes a wide variety of cellular functions (Archambault et al. 2015). In *C. elegans*, two PLKs act in meiosis: the meiotic functioning PLK-2 and the mainly mitotic functioning PLK-1 which can partially substitute for PLK-2 (Chase et al. 2000; Harper et al. 2011). In *C. elegans*, SUN-1 is targeted by PLK-2 (and CHK-2, see above). PLK-2 localizes to SUN-1 and phosphorylates it, promoting pairing and inhibiting nonhomologous synapsis (Harper et al. 2011; Labella et al. 2011). PLK-2's localization to SUN-1 patches is partially dependent upon the phosphorylation of SUN-1, suggesting a positive feedback loop between the two proteins (Woglar et al. 2013). In *S. cerevisiae*, the polo-like kinase homolog Cdc5 appears to perform a later meiotic function. Cdc5 is required for pachytene exit and synaptonemal complex disassembly and is the target of the transcriptional regulator Ndt80 (Sourirajan and Lichten 2008). Ndt80 was also suggested to be part of the signal transduction pathway signaling between the recruitment of the Spo11 complex to the MI division (Malone et al. 2004).

The MAP kinase (MAPK) pathway plays an important role in oocyte maturation, allowing for the transition from diplotene arrest to MI in vertebrates (Fan and Sun 2004; Fan et al. 2012; Sato 2015). In *C. elegans*, where cells do not normally arrest at diplotene, MAPK signaling is required for meiotic progression and the development of functional oocytes (Church et al. 1995; Lee et al. 2007). MAPK plays a role in meiotic DSB repair by serving as a signal for the transition from the meiotic repair mechanism to a mitotic-like repair pathway (Hayashi et al. 2007). The effect of the MAPK pathway on DSB repair may even take place prior to crossover formation, as defects in the timing of MPK activation are correlated with an alteration in the timing of key DSB repair events (Yin et al. 2016). Moreover, the MAPK pathway may regulate DSB repair directly, since proteins required for homologous recombination repair are among the suggested MAPK targets (Arur et al. 2009). MAPK inactivation is also required for the disassembly of synaptonemal complex central region proteins, thus acting to preserve synaptonemal complex structure until crossovers are formed (Nadarajan et al. 2016). Altogether, these studies show that the MAPK pathway may serve to connect the various molecular processes taking place in the pachytene stage and coordinate them with meiotic progression.

5.5 Defects Originating in Prophase that Lead to Aneuploidy

5.5.1 Meiosis and Aneuploidy

The inheritance of an abnormal number of chromosomes is termed aneuploidy. In humans, an aneuploid embryo with only one autosomal chromosome will die before a pregnancy is recognized, while a trisomic embryo will either end as a miscarriage or lead to the birth of a child with developmental disabilities (Herbert et al. 2015). Human oocytes show a high rate of aneuploidy [reviewed in Nagaoka et al. (2012)]: for example, 54 % of MII oocytes from young donors with no known fertility issues, and 62 % of oocytes from advanced maternal age in vitro fertilization donors show aneuploidy (Garcia-Cruz et al. 2010; Ottolini et al. 2015). These levels of aneuploidy are much higher than what is observed for oocytes in model organisms [despite varying levels of aneuploidy between studies in the same model organism, *Drosophila*: Traut (1980; Traut and Schröder (1978) and mice: (Hodges et al. (2005)]. Another feature of human meiosis is that aneuploidy rates increase as women age (Hassold and Chiu 1985; Pellestor et al. 2003). Oocytes of model organisms also exhibit a mild age-dependent increase in aneuploidy (Traut 1980; Traut and Schröder 1978). It was shown that this age effect is stronger for a long-lived mouse model (Lister et al. 2010). These findings suggest that the exceptional increase in aneuploidy in humans as they age is revealed as an outcome of the extension of reproductive age that follows the increase in life span of humans in the last century.

Missegregation of chromosomes during meiotic divisions is a leading cause for aneuploidy. It is conceivable that aneuploidy could originate from mitotic errors.

However, studies of all three meiotic products of human oocyte meiosis establish that aneuploidy is exclusively caused by meiotic errors (Ottolini et al. 2015). Aneuploidy can arise from chromosome missegregation (nondisjunction) at MI or MII. In the first meiotic division, homologous chromosomes segregate away from each other. Since meiotic prophase establishes the crossovers connecting homologous chromosomes, aneuploidy arising from prophase I defects is more likely to lead to chromosome missegregation in the first meiotic division. However, missegregation of a whole chromosome is less common than predicted, and defects leading to premature separation of sister chromatids are common (Angell 1991). These events include reverse segregation and precocious separation of sister chromatids (PSSCs) and mostly occur in MI (Ottolini et al. 2015). In advanced maternal age women, single chromatid nondisjunction, including MII errors, is more frequently found (Pellestor et al. 2003) (Fragouli et al. 2011, 2013). MI versus MII origins of aneuploidy are also dependent on the chromosome analyzed, for example, chromosome 16 aneuploidy involves mainly MI events, while chromosome 18 aneuploidy involves mainly MII errors (Bugge et al. 1998). Analysis of chromosomal rates of nondisjunction, prior to implantation, indicates that smaller chromosomes (15, 16, 19, 21, and 22) are more likely to missegregate (Fragouli et al. 2013). As found in human oocytes, smaller chromosomes of mice are more likely to lack chiasma (Hodges et al. 2005). Although aneuploidy is the most common form of meiotic chromosomal aberration, partial losses of chromosome segments (large deletions) were also observed in low frequency in human oocytes (Fragouli et al. 2011).

Unlike sperm cells, oocytes are arrested in meiotic prophase I during embryonic development, when resumption of meiosis occurs following ovulation (MI at ovulation and MII at fertilization). Thus, in humans, oocytes arrest at a stage following crossover formation and prior to MI division for decade(s). Autosomal aneuploidy is predominant in oogenesis, as compared to spermatogenesis (Martin et al. 1991; Pacchierotti et al. 2007), suggesting that the causes of infertility may be connected to the unique properties of female meiosis. The fact that aneuploidy is increased as women age may suggest that the process that is delayed in female meiosis is the one sensitive to perturbation. These observations suggest that an early meiotic event that is maintained up until fertilization (pre-diplotene) is "time sensitive" and perturbed as oocytes age. The fact that chromosomes that undergo nondisjunction have altered recombination rates and patterns (see below) led to the proposal that aneuploidy arises from a "two hit" model: one hit, originating at embryonic development and a second hit that is the "age-dependent" component (Lamb et al. 1996). There are likely many causes for aneuploidy, but here we will focus on the connection between the proper execution of prophase events and aneuploidy.

5.5.2 Crossover Formation and Aneuploidy

Studies in model organisms establish that a lack of crossovers leads to missegregation of chromosomes in MI. Some exceptions to this rule, however, exist. For example, (a) backup mechanisms of achiasmatic chromosome segregation and (b) meiotic programs that do not rely on crossovers (Wolf 1994). Current models hold that crossovers are a prerequisite for ensuring that homologous chromosomes will segregate away from each other to opposing poles. It was therefore suggested that oocyte nondisjunction might arise from an inability to form crossovers on a particular chromosome.

Both the reduction and absence of recombination is found in trisomic chromosomes. However, the magnitude of the effect varies between chromosomes and studies. Analysis of spontaneous abortions due to trisomy 16 or 21 showed 30 % and 16–40 % reduction in map length on respective chromosomes (Hassold et al. 1995; Lamb et al. 2005). Chromosomes 21 and 22, which are prone to nondisjunction, also show reduced recombination rate by cytological analysis of fetal oocytes [~5 % and 6 % of oocytes with no MLH1 foci, respectively (Cheng et al. 2009)]. 25–47 % of the trisomies involve achiasmate bivalents for the chromosome effected [13: Bugge et al. (2007), Hall et al. (2007a); 15: Robinson et al. (1998); 18: Bugge et al. (1998); 21: Lamb et al. (1997, 2005); 22: Hall et al. (2007b); X: Thomas et al. (2001)]. However, despite the fact that the frequency of trisomy 21 increases with maternal age (Penrose 2009), oocytes isolated from advanced maternal age individuals are as likely to have achiasmatic chromosomes as younger donors (Lamb et al. 1997). Moreover, trisomies for chromosome 15 are less likely to show noncrossovers when originating in older mothers (Robinson et al. 1998). These findings do not exclude the possibility that achiasmatic chromosomes may contribute to the high baseline level of aneuploidy found in humans.

Studies in oocytes from advanced maternal age donors indicate that recombination rate is a factor that accounts for 18 % of the variation in the incidence of aneuploidy (Ottolini et al. 2015). These aged oocytes have ~6 less crossovers overall compared to euploid oocytes (Ottolini et al. 2015). It is intriguing that the overall reduction of crossover rates over the whole genome, and not just on chromosomes 21, predisposes trisomy 21 (Brown et al. 2000). Oocytes that enter meiosis earlier in development will ovulate earlier as well (Polani and Crolla 1991), which leads to the hypothesis that global rates of recombination are lower in oocytes from late fetal stage, accounting for the age-dependent effect on aneuploidy. This "production line" model is likely incorrect, as oocytes from early and late fetal stage have similar recombination rates [MLH1 foci counts (Rowsey et al. 2014)].

A single crossover may not be sufficient to promote proper chromosome segregation. Analysis of fetal oocytes indicate that chromosomes 21 and 22 are less likely to contain crossover markers on both p and q arms of the chromosome [MLH1 foci (Cheng et al. 2009)], and chromosome 16 is less likely to have multiple crossover events (Garcia-Cruz et al. 2010). It is possible that a single crossover will be sufficient for proper segregation; however, analysis of all meiotic products has indicated that two crossovers involving all sister chromatids are required for proper segregation (Ottolini et al. 2015). According to this study, nonrecombinant chromatids are more likely to be subjected to precocious separation of sister chromatids (at MI) or missegregation at MII, even if they were originally part of a bivalent (Ottolini et al. 2015).

5.5.3 Crossover Distribution and Aneuploidy

Crossover positioning plays an important role in chromosome missegregation. Terminal crossovers on chromosome 21 are more likely to be found on chromosomes that have undergone MI nondisjunction, whereas pericentromeric crossovers are more likely to be found on MII nondisjunction chromosomes (Lamb et al. 1997, 2005). Analysis of spontaneous abortions due to trisomy 16 showed that most of the reduction in map length was within the pericentromeric regions (Hassold et al. 1995). Oocytes from advanced maternal age donors (obtained from in vitro fertilization) show variation in the positions of the crossover events, which in some oocytes were pericentromeric, predisposing these oocytes for segregation defects (Ottolini et al. 2015). Maternal age also affects the distribution of crossovers on chromosomes with nondisjunction events (Ghosh et al. 2009; Oliver et al. 2008, 2012). Analysis of recombination maps of chromosomes 21 from Down syndrome patients revealed an increase in telomeric-proximal recombination in all age groups (Oliver et al. 2008, 2012). These events are found in a higher proportion in children with chromosome 21 trisomies born to young mothers (Oliver et al. 2008). These studies suggest that the distal crossovers may contribute to the basal high levels of human nondisjunction, regardless of mother's age. However, the effect on MII was predominant in older mothers and mainly involve a centromeric exchange (Oliver et al. 2008), indicating that centromeric-proximal crossovers are subjected to the age-dependent effect on nondisjunction. Despite this correlation between altered recombination patterns and nondisjunction of autosomes, the X chromosome did not show this same association, which may indicate that the causes for nondisjunction vary between autosomes and the X chromosome (Thomas et al. 2001).

5.5.4 Cohesion Maintenance and Aneuploidy

Crossovers link homologous chromosomes only as long as sister chromatid cohesion is maintained. The age-related component of oocyte aneuploidy acts upon processes that take place after crossover formation, suggesting that compromised sister chromatid cohesion leads to a loss of chiasmata and the breakdown of the bivalent. Immunostainings for cohesins reveal a decreased association with chromosomes in aged mice oocytes (Liu and Keefe 2008; Tsutsumi et al. 2014). In humans, meiotic, but not mitotic, cohesion protein levels are reduced with age, as measured by immunofluorescence staining, despite having no effect on mRNA levels (Garcia-Cruz et al. 2010; Tsutsumi et al. 2014). Several studies, both in *Drosophila* and mouse, addressed this model by examining whether loss of cohesion predisposes chromosomes to nondisjunction in an age-dependent manner. *Drosophila* females with reduced levels of Ord (a protein required for meiotic cohesion) show an increase in nondisjunction which is notably augmented in aged oocytes (Jeffreys et al. 2003). Mice with a point mutation in the meiotic-specific cohesion complex subunit Smc1β show a severe age-dependent increase in the proportion of homologs with univalents, including a premature onset of these defects (Hodges et al. 2005). Altogether, these studies suggest that a reduction in cohesion levels contributes to the maternal-age effect on chromosome missegregation. This effect may be chromosomal region-dependent [centromeric-proximal but not telomeric-proximal crossovers are associated with age-dependent nondisjunction (Oliver et al. 2008)], suggesting that cohesion erosion in pericentromeric regions sensitizes the chromosome to age-dependent nondisjunction.

Cohesion proteins are loaded onto chromosomes prior and during meiotic prophase I; however, it is plausible that they will be replenished later, during the prolonged post-pachytene arrest. However, despite the expression of meiotic cohesion complex proteins in oocytes during their arrest, they do not turn over and their expression during the arrest is not required for maintenance of the chiasma (Tachibana-Konwalski et al. 2010). These studies suggest that the cohesion complexes loaded prior to oocyte arrest, during embryonic development, are the crucial proteins required for chiasma maintenance. How cohesion function is impaired is not yet clear. In addition to a direct effect on cohesion stability/structure, deregulation of pathways that regulate cohesion removal may come into play (e.g., the cohesion protector Sgo2).

5.5.5 *Models Suggesting How Prophase I Defects Lead to Aneuploidy*

As discussed above, the synaptonemal complex is required for the formation of all obligatory crossover events and may regulate their distribution. Therefore, it is conceivable that defects in the formation of the synaptonemal complex may lead to bivalents lacking a chiasma or having a non-favorable chiasma positioning. Defects in synapsis of chromosomes activate checkpoints that lead to cell death. If defects in synapsis are ignored, this may lead to the formation of oocytes predisposed to aneuploidy. Oocytes were shown to be less sensitive to the activation of these checkpoints compared to spermatocytes; similar synapsis defects will trigger apoptosis in all sperm cells, while a significant proportion of oocytes will survive (Morelli and Cohen 2005). Therefore, it is possible that a "weak checkpoint" may contribute to aneuploidy.

There are several possible explanations as to how crossover positioning may affect chromosome segregation. The crossover between homologous chromosomes in the bivalent structure is maintained by sister chromatid cohesion. Sister chromatid cohesion is established between the crossover site and each telomere, creating two cohesion domains in the presence of a single crossover. Therefore, cohesion loss on either of these domains is sufficient to break the bivalent structure into two univalents (separated homologous chromosomes; Fig. 5.5). In a bivalent with a terminal crossover, the segment between the crossover and the telomere region is shorter than a bivalent with a more central crossover. Assuming the amount of

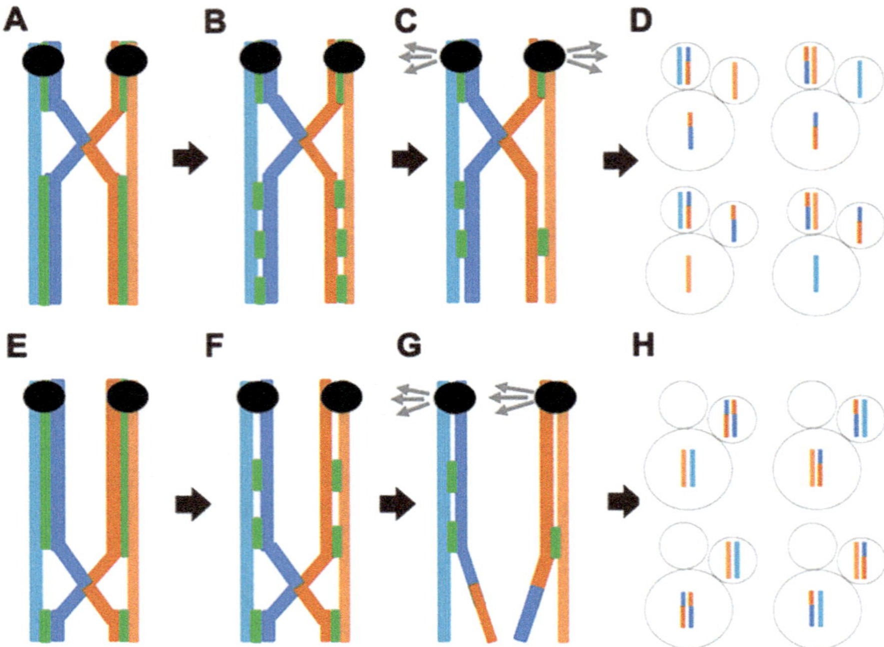

Fig. 5.5 Crossover position relative to the telomere and cohesion are both critical to proper meiotic segregation. A–C and E–G represent the bivalent prior to the meiotic divisions (chromosomes in *orange* and *blue*, cohesion in *green*, centromeres in *black*, and pulling forces in *gray*, *arrows*), while D and H are the outcomes of both meiotic divisions (chromosomes in oocyte and polar bodies). (A–C) In aged meiotic nuclei, sister chromatid synapsis can be lost prematurely before MI segregation (*green* region is depleted over time, A–C are sequentially older oocytes). If the crossover is far from the telomere region, even if cohesion deteriorates, some protein complexes are likely to be left on both sides of the crossover holding the sisters together. This reduction in cohesion will still permit proper segregation. (D) Four possible outcomes after MII: *big circle* represents the oocyte and the two *smaller circles* represent the two polar bodies. The *circle on the top* is polar body I (PBI) extruded after MI and the *small one on the side* is the second polar body (PBII) extruded after fertilization. (E–G) Aged meiotic nuclei that have crossovers closer to telomeres may lose all synapsis on one side of the crossover early, leading to missegregation (*gray arrows*—direction of pulling forces at MI). Resulting oocytes can become trisomic after fertilization. By chance, 50 % of segregation events of that chromosome may be "normal," leading to segregation patterns as in D

cohesion lost with age is similar, a bivalent with a terminal crossover will be more likely to lose all its cohesion between the crossover site and the telomere, increasing the likelihood of MI nondisjunction. Another possibility for terminal crossovers to cause MI nondisjunction is that the terminal crossover may compromise bi-orientation of kinetochores at MI. The contribution of pre-centromeric region crossovers to missegregation may also stem from interrupting centromere–spindle attachments and may interfere with centromere separation (Fig. 5.6). This can lead to sister chromatid nondisjunction at MI or MII. MI sister chromatid nondisjunction

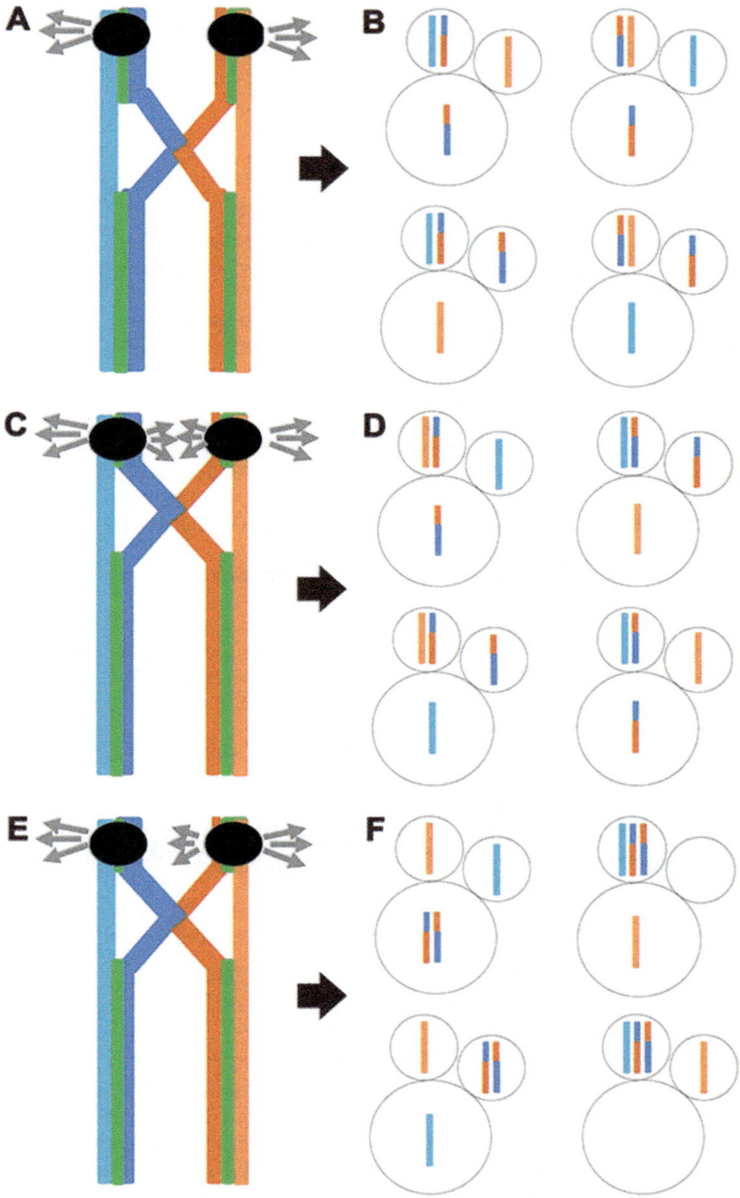

Fig. 5.6 Aneuploidy as a result of centromere–spindle attachment defects. A, C, and E represent the bivalent prior to the meiotic divisions (chromosomes in *orange* and *blue*, cohesion in *green*, centromeres in *black*, and pulling forces in *gray*, *arrows*), while B, D, and F are the outcomes of both meiotic divisions (chromosomes in oocyte and polar bodies). (A) Normal segregation is shown where centromere–spindle attachments correctly separate the homologs at the end of MI, and then sister chromatids separate after MII. *Gray arrows* point in the direction to which separating homologs are moving. (B) Four possible outcomes after MII: *big circle* represents the oocyte and the *two smaller circles* represent the two polar bodies. The *small circle on the top* is

may contribute to errors associated with reverse meiosis [separation of sisters at MI, homologs at MII, Fig. 5.6B (Ottolini et al. 2015)].

References

Acquaviva L, Székvölgyi L, Dichtl B, Dichtl BS, de La Roche Saint-André C, Nicolas A, Géli V (2013) The COMPASS subunit Spp1 links histone methylation to initiation of meiotic recombination. Science 339:215–218

Adrian AB, Comeron JM (2013) The Drosophila early ovarian transcriptome provides insight to the molecular causes of recombination rate variation across genomes. BMC Genomics 14:794

Agarwal S, Roeder GS (2000) Zip3 provides a link between recombination enzymes and synaptonemal complex proteins. Cell 102:245–255

Ajimura M, Leem SH, Ogawa H (1993) Identification of new genes required for meiotic recombination in *Saccharomyces cerevisiae*. Genetics 133:51–66

Anderson DE, Losada A, Erickson HP, Hirano T (2002) Condensin and cohesin display different arm conformations with characteristic hinge angles. J Cell Biol 156:419–424

Angell RR (1991) Predivision in human oocytes at meiosis I: a mechanism for trisomy formation in man. Hum Genet 86:383–387

Archambault V, Lépine G, Kachaner D (2015) Understanding the Polo Kinase machine. Oncogene 34:4799–4807

Arumugam P, Gruber S, Tanaka K, Haering CH, Mechtler K, Nasmyth K (2003) ATP hydrolysis is required for cohesin's association with chromosomes. Curr Biol 13:1941–1953

Arur S, Ohmachi M, Nayak S, Hayes M, Miranda A, Hay A, Golden A, Schedl T (2009) Multiple ERK substrates execute single biological processes in *Caenorhabditis elegans* germ-line development. Proc Natl Acad Sci USA 106:4776–4781

Badugu SB, Nabi SA, Vaidyam P, Laskar S, Bhattacharyya S, Bhattacharyya MK (2015) Identification of *Plasmodium falciparum* DNA repair protein Mre11 with an evolutionarily conserved nuclease function. PLoS One 10(5):e0125358. PMID: 25938776

Bailly AP, Freeman A, Hall J, Déclais A-C, Alpi A, Lilley DMJ, Ahmed S, Gartner A (2010) The *Caenorhabditis elegans* homolog of Gen1/Yen1 resolvases links DNA damage signaling to DNA double-strand break repair. PLoS Genet 6:e1001025

Baker SM, Plug AW, Prolla TA, Bronner CE, Harris AC, Yao X, Christie DM, Monell C, Arnheim N, Bradley A et al (1996) Involvement of mouse Mlh1 in DNA mismatch repair and meiotic crossing over. Nat Genet 13:336–342

Fig. 5.6 (continued) polar body I (PBI) extruded after MI and the *small circle on the top right* is the second polar body (PBII) extruded after fertilization. (C) If there are multiple spindles attaching to a single centromere, reverse meiosis can occur, in which sister chromatids separate at MI instead of MII. (D) Non-sister chromatids are extruded in PBI (*small circle* with two chromatids), as opposed to sister chromatids making up one homolog being in PBI normally. (E) If there is precocious separation of sister chromatids based on weak or unbalanced separation during MI, then aneuploidy can occur. (F) After MII, the oocyte can end up with either two, one, or no chromatids and polar bodies can contain up to three chromatids or as few as zero. This defect may be associated with centromeric-proximal crossovers (C and E). The schematics represent only one type of aberrant attachment, but others may be just as likely (e.g., recombined orange chromosomes pulling away from *blue* and non-recombined *orange* chromosome pulling in the other direction)

Baker CL, Walker M, Kajita S, Petkov PM, Paigen K (2014) PRDM9 binding organizes hotspot nucleosomes and limits Holliday junction migration. Genome Res 24:724–732

Baudat F, Buard J, Grey C, Fledel-Alon A, Ober C, Przeworski M, Coop G, De Massy B (2010) PRDM9 is a major determinant of meiotic recombination hotspots in humans and mice. Science 327:836–840

Berg IL, Neumann R, Lam K-WG, Sarbajna S, Odenthal-Hesse L, May CA, Jeffreys AJ (2010) PRDM9 variation strongly influences recombination hot-spot activity and meiotic instability in humans. Nat Genet 42:859–863

Bhuiyan H, Schmekel K (2004) Meiotic chromosome synapsis in yeast can occur without spo11-induced DNA double-strand breaks. Genetics 168:775–783

Bishop DK, Zickler D (2004) Early decision; meiotic crossover interference prior to stable strand exchange and synapsis. Cell 117:9–15

Bishop DK, Park D, Xu L, Kleckner N (1992) DMC1: a meiosis-specific yeast homolog of *E. coli* recA required for recombination, synaptonemal complex formation, and cell cycle progression. Cell 69:439–456

Bisig CG, Guiraldelli MF, Kouznetsova A, Scherthan H, Hoog C, Dawson DS, Pezza RJ (2012) Synaptonemal complex components persist at centromeres and are required for homologous centromere pairing in mouse spermatocytes. PLoS Genet 8:e1002701

Blat Y, Protacio RU, Hunter N, Kleckner N (2002) Physical and functional interactions among basic chromosome organizational features govern early steps of meiotic chiasma formation. Cell 111:791–802

Boateng KA, Bellani MA, Gregoretti IV, Pratto F, Camerini-Otero RD (2013) Homologous pairing preceding SPO11-mediated double-strand breaks in mice. Dev Cell 24:196–205

Boddy MN, Gaillard PH, McDonald WH, Shanahan P, Yates JR 3rd, Russell P (2000) Mus81-Eme1 are essential components of a Holliday junction resolvase. Cell 107(4):537–548

Boddy MN, Gaillard PH, McDonald WH, Shanahan P, Yates JR, Russell P (2001) Mus81-Eme1 are essential components of a Holliday junction resolvase. Cell 107:537–548

Bolcun-Filas E, Rinaldi VD, White ME, Schimenti JC (2014) Reversal of female infertility by Chk2 ablation reveals the oocyte DNA damage checkpoint pathway. Science 343:533–536

Börner GV, Kleckner N, Hunter N (2004) Crossover/noncrossover differentiation, synaptonemal complex formation, and regulatory surveillance at the leptotene/zygotene transition of meiosis. Cell 117:29–45

Brick K, Smagulova F, Khil P, Camerini-Otero RD, Petukhova GV (2012) Genetic recombination is directed away from functional genomic elements in mice. Nature 485:642–645

Brockway H, Balukoff N, Dean M, Alleva B, Smolikove S (2014) The CSN/COP9 signalosome regulates synaptonemal complex assembly during meiotic prophase I of *Caenorhabditis elegans*. PLoS Genet 10:e1004757

Brown MS, Bishop DK (2015) DNA strand exchange and RecA homologs in meiosis. Cold Spring Harb Perspect Biol 7:a016659

Brown AS, Feingold E, Broman KW, Sherman SL (2000) Genome-wide variation in recombination in female meiosis: a risk factor for non-disjunction of chromosome 21. Hum Mol Genet 9:515–523

Brown MS, Grubb J, Zhang A, Rust MJ, Bishop DK (2015) Small Rad51 and Dmc1 complexes often co-occupy both ends of a meiotic DNA double strand break. PLoS Genet 11:e1005653

Bugge M, Collins A, Petersen MB, Fisher J, Brandt C, Hertz JM, Tranebjaerg L, de Lozier-Blanchet C, Nicolaides P, Brøndum-Nielsen K et al (1998) Non-disjunction of chromosome 18. Hum Mol Genet 7:661–669

Bugge M, Collins A, Hertz JM, Eiberg H, Lundsteen C, Brandt CA, Bak M, Hansen C, Delozier CD, Lespinasse J et al (2007) Non-disjunction of chromosome 13. Hum Mol Genet 16:2004–2010

Buonomo SB, Clyne RK, Fuchs J, Loidl J, Uhlmann F, Nasmyth K (2000) Disjunction of homologous chromosomes in meiosis I depends on proteolytic cleavage of the meiotic cohesin Rec8 by separin. Cell 103:387–398

Busygina V, Sehorn MG, Shi IY, Tsubouchi H, Roeder GS, Sung P (2008) Hed1 regulates Rad51-mediated recombination via a novel mechanism. Genes Dev 22:786–795

Busygina V, Saro D, Williams G, Leung WK, Say AF, Sehorn MG, Sung P, Tsubouchi H (2012) Novel attributes of Hed1 affect dynamics and activity of the Rad51 presynaptic filament during meiotic recombination. J Biol Chem 287:1566–1575

Cannavo E, Cejka P (2014) Sae2 promotes dsDNA endonuclease activity within Mre11-Rad50-Xrs2 to resect DNA breaks. Nature 514:122–125

Carballo JA, Johnson AL, Sedgwick SG, Cha RS (2008) Phosphorylation of the axial element protein Hop1 by Mec1/Tel1 ensures meiotic interhomolog recombination. Cell 132:758–770

Carballo JA, Panizza S, Serrentino ME, Johnson AL, Geymonat M, Borde V, Klein F, Cha RS (2013) Budding yeast ATM/ATR control meiotic double-strand break (DSB) levels by down-regulating Rec114, an essential component of the DSB-machinery. PLoS Genet 9:e1003545

Carney JP, Maser RS, Olivares H, Davis EM, Le Beau M, Yates JR, Hays L, Morgan WF, Petrini JH (1998) The hMre11/hRad50 protein complex and Nijmegen breakage syndrome: linkage of double-strand break repair to the cellular DNA damage response. Cell 93:477–486

Carpenter AT (1975) Electron microscopy of meiosis in *Drosophila melanogaster* females. I. Structure, arrangement, and temporal change of the synaptonemal complex in wild-type. Chromosoma 51:157–182

Cartagena-Lirola H, Guerini I, Viscardi V, Lucchini G, Longhese MP (2006) Budding yeast Sae2 is an in vivo target of the Mec1 and Tel1 checkpoint kinases during meiosis. Cell Cycle 5:1549–1559

Challa K, Lee M-S, Shinohara M, Kim KP, Shinohara A (2016) Rad61/Wpl1 (Wapl), a cohesin regulator, controls chromosome compaction during meiosis. Nucleic Acids Res. doi:10.1093/nar/gkw034

Chase D, Serafinas C, Ashcroft N, Kosinski M, Longo D, Ferris DK, Golden A (2000) The polo-like kinase PLK-1 is required for nuclear envelope breakdown and the completion of meiosis in *Caenorhabditis elegans*. Genesis 26:26–41

Chen R, Wold MS (2014) Replication protein A: single-stranded DNA's first responder: dynamic DNA-interactions allow replication protein A to direct single-strand DNA intermediates into different pathways for synthesis or repair. Bioessays 36:1156–1161

Chen X, Suhandynata RT, Sandhu R, Rockmill B, Mohibullah N, Niu H, Liang J, Lo H-C, Miller DE, Zhou H et al (2015) Phosphorylation of the synaptonemal complex protein Zip1 regulates the crossover/noncrossover decision during yeast meiosis. Plos Biol 13:e1002329

Chen Y-J, Chuang Y-C, Chuang C-N, Cheng Y-H, Chang C-R, Leng C-H, Wang T-F (2016) *S. cerevisiae* Mre11 recruits conjugated SUMO moieties to facilitate the assembly and function of the Mre11-Rad50-Xrs2 complex. Nucleic Acids Res. doi:10.1093/nar/gkv1523

Cheng C-H, Lo Y-H, Liang S-S, Ti S-C, Lin F-M, Yeh C-H, Huang H-Y, Wang T-F (2006) SUMO modifications control assembly of synaptonemal complex and polycomplex in meiosis of *Saccharomyces cerevisiae*. Genes Dev 20:2067–2081

Cheng EY, Hunt PA, Naluai-Cecchini TA, Fligner CL, Fujimoto VY, Pasternack TL, Schwartz JM, Steinauer JE, Woodruff TJ, Cherry SM et al (2009) Meiotic recombination in human oocytes. PLoS Genet 5:e1000661

Cherry SM, Adelman CA, Theunissen JW, Hassold TJ, Hunt PA, Petrini JHJ (2007) The Mre11 complex influences DNA repair, synapsis, and crossing over in murine meiosis. Curr Biol 17:373–378

Chikashige Y, Ding DQ, Funabiki H, Haraguchi T, Mashiko S, Yanagida M, Hiraoka Y (1994) Telomere-led premeiotic chromosome movement in fission yeast. Science 264:270–273

Christophorou N, Rubin T, Bonnet I, Piolot T, Arnaud M, Huynh J-R (2015) Microtubule-driven nuclear rotations promote meiotic chromosome dynamics. Nat Cell Biol 17:1388–1400

Church DL, Guan KL, Lambie EJ (1995) Three genes of the MAP kinase cascade, mek-2, mpk-1/sur-1 and let-60 ras, are required for meiotic cell cycle progression in *Caenorhabditis elegans*. Development 121:2525–2535

Ciosk R, Shirayama M, Shevchenko A, Tanaka T, Toth A, Nasmyth K (2000) Cohesin's binding to chromosomes depends on a separate complex consisting of Scc2 and Scc4 proteins. Mol Cell 5:243–254

Cloud V, Chan Y-L, Grubb J, Budke B, Bishop DK (2012) Rad51 is an accessory factor for Dmc1-mediated joint molecule formation during meiosis. Science 337:1222–1225

Colaiácovo MP, MacQueen AJ, Martinez-Perez E, McDonald K, Adamo A, La Volpe A, Villeneuve AM (2003) Synaptonemal complex assembly in *C. elegans* is dispensable for loading strand-exchange proteins but critical for proper completion of recombination. Dev Cell 5:463–474

Conrad MN, Lee C-Y, Chao G, Shinohara M, Kosaka H, Shinohara A, Conchello J-A, Dresser ME (2008) Rapid telomere movement in meiotic prophase is promoted by NDJ1, MPS3, and CSM4 and is modulated by recombination. Cell 133:1175–1187

Costelloe T, Louge R, Tomimatsu N, Mukherjee B, Martini E, Khadaroo B, Dubois K, Wiegant WW, Thierry A, Burma S et al (2012) The yeast Fun30 and human SMARCAD1 chromatin remodellers promote DNA end resection. Nature 489:581–584

Couteau F, Zetka M (2011) DNA damage during meiosis induces chromatin remodeling and synaptonemal complex disassembly. Dev Cell 20:353–363

Crawley O, Barroso C, Testori S, Ferrandiz N, Silva N, Castellano-Pozo M, Jaso-Tamame AL, Martinez-Perez E (2016) Cohesin-interacting protein WAPL-1 regulates meiotic chromosome structure and cohesion by antagonizing specific cohesin complexes. eLife 5:563

Cremona CA, Sarangi P, Yang Y, Hang LE, Rahman S, Zhao X (2012) Extensive DNA damage-induced sumoylation contributes to replication and repair and acts in addition to the mec1 checkpoint. Mol Cell 45:422–432

de Carvalho CE, Zaaijer S, Smolikov S, Gu Y, Schumacher JM, Colaiácovo MP (2008) LAB-1 antagonizes the Aurora B kinase in *C. elegans*. Genes Dev 22:2869–2885

de Vries FAT, de Boer E, van den Bosch M, Baarends WM, Ooms M, Yuan L, Liu J-G, van Zeeland AA, Heyting C, Pastink A (2005) Mouse Sycp1 functions in synaptonemal complex assembly, meiotic recombination, and XY body formation. Genes Dev 19:1376–1389

Dernburg AF, McDonald K, Moulder G, Barstead R, Dresser M, Villeneuve AM (1998) Meiotic recombination in *C. elegans* initiates by a conserved mechanism and is dispensable for homologous chromosome synapsis. Cell 94:387–398

Ding D-Q, Sakurai N, Katou Y, Itoh T, Shirahige K, Haraguchi T, Hiraoka Y (2006) Meiotic cohesins modulate chromosome compaction during meiotic prophase in fission yeast. J Cell Biol 174:499–508

Ding X, Xu R, Yu J, Xu T, Zhuang Y, Han M (2007) SUN1 is required for telomere attachment to nuclear envelope and gametogenesis in mice. Dev Cell 12:863–872

Ding D-Q, Matsuda A, Okamasa K, Nagahama Y, Haraguchi T, Hiraoka Y (2016) Meiotic cohesin-based chromosome structure is essential for homologous chromosome pairing in *Schizosaccharomyces pombe*. Chromosoma 125(2):205–214

Dodson GE, Limbo O, Nieto D, Russell P (2010) Phosphorylation-regulated binding of Ctp1 to Nbs1 is critical for repair of DNA double-strand breaks. Cell Cycle 9:1516–1522

Dresser ME, Giroux CN (1988) Meiotic chromosome behavior in spread preparations of yeast. J Cell Biol 106:567–573

Eichinger CS, Jentsch S (2010) Synaptonemal complex formation and meiotic checkpoint signaling are linked to the lateral element protein Red1. Proc Natl Acad Sci USA 107:11370–11375

Eijpe M, Offenberg H, Jessberger R, Revenkova E, Heyting C (2003) Meiotic cohesin REC8 marks the axial elements of rat synaptonemal complexes before cohesins SMC1beta and SMC3. J Cell Biol 160:657–670

Falck J, Forment JV, Coates J, Mistrik M, Lukas J, Bartek J, Jackson SP (2012) CDK targeting of NBS1 promotes DNA-end resection, replication restart and homologous recombination. EMBO Rep 13:561–568

Falk JE, Chan AC-H, Hoffmann E, Hochwagen A (2010) A Mec1- and PP4-dependent checkpoint couples centromere pairing to meiotic recombination. Dev Cell 19:599–611

Fan H-Y, Sun Q-Y (2004) Involvement of mitogen-activated protein kinase cascade during oocyte maturation and fertilization in mammals. Biol Reprod 70:535–547

Fan H-Y, Liu Z, Mullany LK, Richards JS (2012) Consequences of RAS and MAPK activation in the ovary: the good, the bad and the ugly. Mol Cell Endocrinol 356:74–79

Fawcett DW (1956) The fine structure of chromosomes in the meiotic prophase of vertebrate spermatocytes. J Biophys Biochem Cytol 2:403–406

Fragouli E, Alfarawati S, Goodall N-N, Sánchez-García JF, Colls P, Wells D (2011) The cytogenetics of polar bodies: insights into female meiosis and the diagnosis of aneuploidy. Mol Hum Reprod 17:286–295

Fragouli E, Alfarawati S, Spath K, Jaroudi S, Sarasa J, Enciso M, Wells D (2013) The origin and impact of embryonic aneuploidy. Hum Genet 132:1001–1013

Fraune J, Schramm S, Alsheimer M, Benavente R (2012) The mammalian synaptonemal complex: protein components, assembly and role in meiotic recombination. Exp Cell Res 318:1340–1346

Fukuda T, Hoog C (2010) The mouse cohesin-associated protein PDS5B is expressed in testicular cells and is associated with the meiotic chromosome axes. Genes (Basel) 1:484–494

Fukuda T, Pratto F, Schimenti JC, Turner JMA, Camerini-Otero RD, Hoog C (2012) Phosphorylation of chromosome core components may serve as axis marks for the status of chromosomal events during mammalian meiosis. PLoS Genet 8:e1002485

Fukushima K, Tanaka Y, Nabeshima K, Yoneki T, Tougan T, Tanaka S, Nojima H (2000) Dmc1 of *Schizosaccharomyces pombe* plays a role in meiotic recombination. Nucleic Acids Res 28:2709–2716

Furuse M, Nagase Y, Tsubouchi H, Murakami-Murofushi K, Shibata T, Ohta K (1998) Distinct roles of two separable in vitro activities of yeast Mre11 in mitotic and meiotic recombination. EMBO J 17:6412–6425

Gao J, Kim H-M, Elia AE, Elledge SJ, Colaiácovo MP (2015) NatB domain-containing CRA-1 antagonizes hydrolase ACER-1 linking acetyl-CoA metabolism to the initiation of recombination during *C. elegans* meiosis. PLoS Genet 11:e1005029

Garcia V, Gray S, Allison RM, Cooper TJ, Neale MJ (2015) Tel1(ATM)-mediated interference suppresses clustered meiotic double-strand-break formation. Nature 520:114–118

Garcia-Cruz R, Casanovas A, Brieno-Enriquez M, Robles P, Roig I, Pujol A, Cabero L, Durban M, Garcia Caldes M (2010) Cytogenetic analyses of human oocytes provide new data on non-disjunction mechanisms and the origin of trisomy 16. Hum Reprod 25:179–191

Garcia-Muse T, Boulton SJ (2005) Distinct modes of ATR activation after replication stress and DNA double-strand breaks in *Caenorhabditis elegans*. EMBO J 24:4345–4355

Gasior SL, Wong AK, Kora Y, Shinohara A, Bishop DK (1998) Rad52 associates with RPA and functions with rad55 and rad57 to assemble meiotic recombination complexes. Genes Dev 12:2208–2221

Gasior SL, Olivares H, Ear U, Hari DM, Weichselbaum R, Bishop DK (2001) Assembly of RecA-like recombinases: distinct roles for mediator proteins in mitosis and meiosis. Proc Natl Acad Sci USA 98:8411–8418

Gause M, Webber HA, Misulovin Z, Haller G, Rollins RA, Eissenberg JC, Bickel SE, Dorsett D (2008) Functional links between Drosophila Nipped-B and cohesin in somatic and meiotic cells. Chromosoma 117:51–66

Gerton JL, DeRisi J, Shroff R, Lichten M, Brown PO, Petes TD (2000) Inaugural Article: Global mapping of meiotic recombination hotspots and coldspots in the yeast *Saccharomyces cerevisiae*. Proc Natl Acad Sci USA 97:11383–11390

Getun IV, Wu ZK, Khalil AM, Bois PRJ (2010) Nucleosome occupancy landscape and dynamics at mouse recombination hotspots. EMBO Rep 11:555–560

Ghosh S, Feingold E, Dey SK (2009) Etiology of Down syndrome: evidence for consistent association among altered meiotic recombination, nondisjunction, and maternal age across populations. Am J Med Genet 149A:1415–1420

Goedecke W, Eijpe M, Offenberg HH, van Aalderen M, Heyting C (1999) Mre11 and Ku70 interact in somatic cells, but are differentially expressed in early meiosis. Nat Genet 23:194–198

Goodyer W, Kaitna S, Couteau F, Ward JD, Boulton SJ, Zetka M (2008) HTP-3 links DSB formation with homolog pairing and crossing over during *C. elegans* meiosis. Dev Cell 14:263–274

Govin J, Dorsey J, Gaucher J, Rousseaux S, Khochbin S, Berger SL (2010) Systematic screen reveals new functional dynamics of histones H3 and H4 during gametogenesis. Genes Dev 24:1772–1786

Guleria A, Chandna S (2016) ATM kinase: much more than a DNA damage responsive protein. DNA Repair 39:1–20

Gutiérrez-Caballero C, Herrán Y, Sánchez-Martín M, Suja JA, Barbero JL, Llano E, Pendás AM (2011) Identification and molecular characterization of the mammalian α-kleisin RAD21L. Cell Cycle 10:1477–1487

Haering CH, Löwe J, Hochwagen A, Nasmyth K (2002) Molecular architecture of SMC proteins and the yeast cohesin complex. Mol Cell 9:773–788

Hall HE, Chan ER, Collins A, Judis L, Shirley S, Surti U, Hoffner L, Cockwell AE, Jacobs PA, Hassold TJ (2007a) The origin of trisomy 13. Am J Med Genet 143A:2242–2248

Hall HE, Surti U, Hoffner L, Shirley S, Feingold E, Hassold T (2007b) The origin of trisomy 22: evidence for acrocentric chromosome-specific patterns of nondisjunction. Am J Med Genet 143A:2249–2255

Harper L (2004) A bouquet of chromosomes. J Cell Sci 117:4025–4032

Harper NC, Rillo R, Jover-Gil S, Assaf ZJ, Bhalla N, Dernburg AF (2011) Pairing centers recruit a Polo-like kinase to orchestrate meiotic chromosome dynamics in *C. elegans*. Dev Cell 21:934–947

Harvey SH, Krien MJE, O'Connell MJ (2002) Structural maintenance of chromosomes (SMC) proteins, a family of conserved ATPases. Genome Biol 3:REVIEWS3003

Hassold T, Chiu D (1985) Maternal age-specific rates of numerical chromosome abnormalities with special reference to trisomy. Hum Genet 70:11–17

Hassold T, Merrill M, Adkins K, Freeman S, Sherman S (1995) Recombination and maternal age-dependent nondisjunction: molecular studies of trisomy 16. Am J Hum Genet 57:867–874

Hayase A, Takagi M, Miyazaki T, Oshiumi H, Shinohara M, Shinohara A (2004) A protein complex containing Mei5 and Sae3 promotes the assembly of the meiosis-specific RecA homolog Dmc1. Cell 119:927–940

Hayashi K, Yoshida K, Matsui Y (2005) A histone H3 methyltransferase controls epigenetic events required for meiotic prophase. Nature 438:374–378

Hayashi M, Chin GM, Villeneuve AM (2007) *C. elegans* germ cells switch between distinct modes of double-strand break repair during meiotic prophase progression. PLoS Genet 3:e191

Heidmann D, Horn S, Heidmann S, Schleiffer A, Nasmyth K, Lehner CF (2004) The Drosophila meiotic kleisin C(2)M functions before the meiotic divisions. Chromosoma 113:177–187

Herbert M, Kalleas D, Cooney D, Lamb M, Lister L (2015) Meiosis and maternal aging: insights from aneuploid oocytes and trisomy births. Cold Spring Harb Perspect Biol 7:a017970

Herrán Y, Gutiérrez-Caballero C, Sánchez-Martín M, Hernández T, Viera A, Barbero JL, de Álava E, de Rooij DG, Suja JA, Llano E et al (2011) The cohesin subunit RAD21L functions in meiotic synapsis and exhibits sexual dimorphism in fertility. EMBO J 30:3091–3105

Hill WG, Robertson A (2007) The effect of linkage on limits to artificial selection. Genet Res 89:311–336

Hodges CA, Revenkova E, Jessberger R, Hassold TJ, Hunt PA (2005) SMC1β-deficient female mice provide evidence that cohesins are a missing link in age-related nondisjunction. Nat Genet 37:1351–1355

Holloway JK, Booth J, Edelmann W, McGowan CH, Cohen PE (2008) MUS81 generates a subset of MLH1-MLH3-independent crossovers in mammalian meiosis. PLoS Genet 4:e1000186

Holloway JK, Mohan S, Balmus G, Sun X, Modzelewski A, Borst PL, Freire R, Weiss RS, Cohen PE (2011) Mammalian BTBD12 (SLX4) protects against genomic instability during mammalian spermatogenesis. PLoS Genet 7:e1002094

Holloway JK, Sun X, Yokoo R, Villeneuve AM, Cohen PE (2014) Mammalian CNTD1 is critical for meiotic crossover maturation and deselection of excess precrossover sites. J Cell Biol 205:633–641

Hong S, Sung Y, Yu M, Lee M, Kleckner N, Kim KP (2013) The logic and mechanism of homologous recombination partner choice. Mol Cell 51:440–453

Hooker GW, Roeder GS (2006) A role for SUMO in meiotic chromosome synapsis. Curr Biol 16:1238–1243

Hopfner K-P, Tainer JA (2003) Rad50/SMC proteins and ABC transporters: unifying concepts from high-resolution structures. Curr Opin Struct Biol 13:249–255

Horn HF, Kim DI, Wright GD, Wong ESM, Stewart CL, Burke B, Roux KJ (2013) A mammalian KASH domain protein coupling meiotic chromosomes to the cytoskeleton. J Cell Biol 202:1023–1039

Hu B, Itoh T, Mishra A, Katoh Y, Chan K-L, Upcher W, Godlee C, Roig MB, Shirahige K, Nasmyth K (2011) ATP hydrolysis is required for relocating cohesin from sites occupied by its Scc2/4 loading complex. Curr Biol 21:12–24

Huang CE, Milutinovich M, Koshland D (2005) Rings, bracelet or snaps: fashionable alternatives for Smc complexes. Philos Trans R Soc Lond B Biol Sci 360:537–542

Humphryes N, Leung W-K, Argunhan B, Terentyev Y, Dvorackova M, Tsubouchi H (2013) The Ecm11-Gmc2 complex promotes synaptonemal complex formation through assembly of transverse filaments in budding yeast. PLoS Genet 9:e1003194

Hunter N, Borts RH (1997) Mlh1 is unique among mismatch repair proteins in its ability to promote crossing-over during meiosis. Genes Dev 11:1573–1582

Ishiguro K-I, Kim J, Fujiyama-Nakamura S, Kato S, Watanabe Y (2011) A new meiosis-specific cohesin complex implicated in the cohesin code for homologous pairing. EMBO Rep 12:267–275

Ivanov EL, Korolev VG, Fabre F (1992) XRS2, a DNA repair gene of *Saccharomyces cerevisiae*, is needed for meiotic recombination. Genetics 132:651–664

Ivanov D, Schleiffer A, Eisenhaber F, Mechtler K, Haering CH, Nasmyth K (2002) Eco1 is a novel acetyltransferase that can acetylate proteins involved in cohesion. Curr Biol 12:323–328

Jantsch V, Pasierbek P, Mueller MM, Schweizer D, Jantsch M, Loidl J (2004) Targeted gene knockout reveals a role in meiotic recombination for ZHP-3, a Zip3-related protein in *Caenorhabditis elegans*. Mol Cell Biol 24:7998–8006

Jeffreys CA, Burrage PS, Bickel SE (2003) A model system for increased meiotic nondisjunction in older oocytes. Curr Biol 13:498–503

Jin H, Guacci V, Yu H-G (2009) Pds5 is required for homologue pairing and inhibits synapsis of sister chromatids during yeast meiosis. J Cell Biol 186:713–725

Johzuka K, Ogawa H (1995) Interaction of Mre11 and Rad50: two proteins required for DNA repair and meiosis-specific double-strand break formation in *Saccharomyces cerevisiae*. Genetics 139:1521–1532

Jordan PW, Karppinen J, Handel MA (2012) Polo-like kinase is required for synaptonemal complex disassembly and phosphorylation in mouse spermatocytes. J Cell Sci 125:5061–5072

Joshi N, Brown MS, Bishop DK, Börner GV (2015) Gradual implementation of the meiotic recombination program via checkpoint pathways controlled by global DSB levels. Mol Cell 57:797–811

Joyce EF, Pedersen M, Tiong S, White-Brown SK, Paul A, Campbell SD, McKim KS (2011) Drosophila ATM and ATR have distinct activities in the regulation of meiotic DNA damage and repair. J Cell Biol 195:359–367

Joyce EF, Paul A, Chen KE, Tanneti N, McKim KS (2012) Multiple barriers to nonhomologous DNA end joining during meiosis in Drosophila. Genetics 191:739–746

Keeney S, Giroux CN, Kleckner N (1997) Meiosis-specific DNA double-strand breaks are catalyzed by Spo11, a member of a widely conserved protein family. Cell 88:375–384

Kelly KO, Dernburg AF, Stanfield GM, Villeneuve AM (2000) *Caenorhabditis elegans* msh-5 is required for both normal and radiation-induced meiotic crossing over but not for completion of meiosis. Genetics 156:617–630

Kim Y, Kostow N, Dernburg AF (2015) The chromosome axis mediates feedback control of CHK-2 to ensure crossover formation in *C. elegans*. Dev Cell 35:247–261

Kitajima TS, Kawashima SA, Watanabe Y (2004) The conserved kinetochore protein shugoshin protects centromeric cohesion during meiosis. Nature 427:510–517

Klein F, Mahr P, Galova M, Buonomo SB, Michaelis C, Nairz K, Nasmyth K (1999) A central role for cohesins in sister chromatid cohesion, formation of axial elements, and recombination during yeast meiosis. Cell 98:91–103

Klug H, Xaver M, Chaugule VK, Koidl S, Mittler G, Klein F, Pichler A (2013) Ubc9 sumoylation controls SUMO chain formation and meiotic synapsis in *Saccharomyces cerevisiae*. Mol Cell 50:625–636

Koh-Stenta X, Joy J, Poulsen A, Li R, Tan Y, Shim Y, Min J-H, Wu L, Ngo A, Peng J et al (2014) Characterization of the histone methyltransferase PRDM9 using biochemical, biophysical and chemical biology techniques. Biochem J 461:323–334

Kolodkin AL, Klar AJ, Stahl FW (1986) Double-strand breaks can initiate meiotic recombination in *S. cerevisiae*. Cell 46:733–740

Koshland D, Strunnikov A (1996) Mitotic chromosome condensation. Annu Rev Cell Dev Biol 12:305–333

Krishnaprasad GN, Anand MT, Lin G, Tekkedil MM, Steinmetz LM, Nishant KT (2015) Variation in crossover frequencies perturb crossover assurance without affecting meiotic chromosome segregation in *Saccharomyces cerevisiae*. Genetics 199:399–412

Kudo NR, Wassmann K, Anger M, Schuh M, Wirth KG, Xu H, Helmhart W, Kudo H, Mckay M, Maro B et al (2006) Resolution of chiasmata in oocytes requires separase-mediated proteolysis. Cell 126:135–146

Kuleszewicz K, Fu X, Kudo NR (2013) Cohesin loading factor Nipbl localizes to chromosome axes during mammalian meiotic prophase. Cell Div 8:12

Kuznetsov S, Pellegrini M, Shuda K, Fernandez-Capetillo O, Liu Y, Martin BK, Burkett S, Southon E, Pati D, Tessarollo L et al (2007) RAD51C deficiency in mice results in early prophase I arrest in males and sister chromatid separation at metaphase II in females. J Cell Biol 176:581–592

Labella S, Woglar A, Jantsch V, Zetka M (2011) Polo kinases establish links between meiotic chromosomes and cytoskeletal forces essential for homolog pairing. Dev Cell 21:948–958

Lam I, Keeney S (2015) Nonparadoxical evolutionary stability of the recombination initiation landscape in yeast. Science 350:932–937

Lamb NE, Freeman SB, Savage-Austin A, Pettay D, Taft L, Hersey J, Gu Y, Shen J, Saker D, May KM et al (1996) Susceptible chiasmate configurations of chromosome 21 predispose to non-disjunction in both maternal meiosis I and meiosis II. Nat Genet 14:400–405

Lamb NE, Feingold E, Savage A, Avramopoulos D, Freeman S, Gu Y, Hallberg A, Hersey J, Karadima G, Pettay D et al (1997) Characterization of susceptible chiasma configurations that increase the risk for maternal nondisjunction of chromosome 21. Hum Mol Genet 6:1391–1399

Lamb NE, Yu K, Shaffer J, Feingold E, Sherman SL (2005) Association between maternal age and meiotic recombination for trisomy 21. Am J Hum Genet 76:91–99

Lange J, Pan J, Cole F, Thelen MP, Jasin M, Keeney S (2011) ATM controls meiotic double-strand-break formation. Nature 479:237–240

Lao JP, Cloud V, Huang C-C, Grubb J, Thacker D, Lee C-Y, Dresser ME, Hunter N, Bishop DK (2013) Meiotic crossover control by concerted action of Rad51-Dmc1 in homolog template bias and robust homeostatic regulation. PLoS Genet 9:e1003978

Latypov V, Rothenberg M, Lorenz A, Octobre G, Csutak O, Lehmann E, Loidl J, Kohli J (2010) Roles of Hop1 and Mek1 in meiotic chromosome pairing and recombination partner choice in *Schizosaccharomyces pombe*. Mol Cell Biol 30:1570–1581

Lee J, Hirano T (2011) RAD21L, a novel cohesin subunit implicated in linking homologous chromosomes in mammalian meiosis. J Cell Biol 192:263–276

Lee MH, Ohmachi M, Arur S, Nayak S, Francis R, Church D, Lambie E, Schedl T (2007) Multiple functions and dynamic activation of MPK-1 extracellular signal-regulated kinase signaling in *Caenorhabditis elegans* germline development. Genetics 177:2039–2062

Lee J, Kitajima TS, Tanno Y, Yoshida K, Morita T, Miyano T, Miyake M, Watanabe Y (2008) Unified mode of centromeric protection by shugoshin in mammalian oocytes and somatic cells. Nature 10:42–52

Lee JY, Terakawa T, Qi Z, Steinfeld JB, Redding S, Kwon Y, Gaines WA, Zhao W, Sung P, Greene EC (2015) DNA recombination. Base triplet stepping by the Rad51/RecA family of recombinases. Science 349:977–981

Lemmens BBLG, Johnson NM, Tijsterman M (2013) COM-1 promotes homologous recombination during *Caenorhabditis elegans* meiosis by antagonizing Ku-mediated non-homologous end joining. PLoS Genet 9:e1003276

Leung W-K, Humphryes N, Afshar N, Argunhan B, Terentyev Y, Tsubouchi T, Tsubouchi H (2015) The synaptonemal complex is assembled by a polySUMOylation-driven feedback mechanism in yeast. J Cell Biol 211:785–793

Li XC, Bolcun-Filas E, Schimenti JC (2011) Genetic evidence that synaptonemal complex axial elements govern recombination pathway choice in mice. Genetics 189:71–82

Libuda DE, Uzawa S, Meyer BJ, Villeneuve AM (2013) Meiotic chromosome structures constrain and respond to designation of crossover sites. Nature 502:703–706

Lightfoot J, Testori S, Barroso C, Martinez-Perez E (2011) Loading of meiotic cohesin by SCC-2 is required for early processing of DSBs and for the DNA damage checkpoint. Curr Biol 21:1421–1430

Lin Z, Kong H, Nei M, Ma H (2006) Origins and evolution of the recA/RAD51 gene family: evidence for ancient gene duplication and endosymbiotic gene transfer. Proc Natl Acad Sci USA 103:10328–10333

Lin F-M, Lai Y-J, Shen H-J, Cheng Y-H, Wang T-F (2010) Yeast axial-element protein, Red1, binds SUMO chains to promote meiotic interhomologue recombination and chromosome synapsis. EMBO J 29:586–596

Lin W, Jin H, Liu X, Hampton K, Yu H-G (2011) Scc2 regulates gene expression by recruiting cohesin to the chromosome as a transcriptional activator during yeast meiosis. Mol Biol Cell 22:1985–1996

Lipkin SM, Moens PB, Wang V, Lenzi M, Shanmugarajah D, Gilgeous A, Thomas J, Cheng J, Touchman JW, Green ED et al (2002) Meiotic arrest and aneuploidy in MLH3-deficient mice. Nat Genet 31:385–390

Lister LM, Kouznetsova A, Hyslop LA, Kalleas D, Pace SL, Barel JC, Nathan A, Floros V, Adelfalk C, Watanabe Y et al (2010) Age-related meiotic segregation errors in mammalian oocytes are preceded by depletion of cohesin and Sgo2. Curr Biol 20:1511–1521

Liu L, Keefe DL (2008) Defective cohesin is associated with age-dependent misaligned chromosomes in oocytes. Reprod Biomed Online 16:103–112

Liu Y, Tarsounas M, O'regan P, West SC (2007) Role of RAD51C and XRCC3 in genetic recombination and DNA repair. J Biol Chem 282:1973–1979

Liu J, Renault L, Veaute X, Fabre F, Stahlberg H, Heyer W-D (2011) Rad51 paralogues Rad55-Rad57 balance the antirecombinase Srs2 in Rad51 filament formation. Nature 479:245–248

Llano E, Herrán Y, García-Tuñón I, Gutiérrez-Caballero C, de Álava E, Barbero JL, Schimenti J, de Rooij DG, Sánchez-Martín M, Pendás AM (2012) Meiotic cohesin complexes are essential for the formation of the axial element in mice. J Cell Biol 197:877–885

Lorenz A, Mehats A, Osman F, Whitby MC (2014) Rad51/Dmc1 paralogs and mediators oppose DNA helicases to limit hybrid DNA formation and promote crossovers during meiotic recombination. Nucleic Acids Res 42:13723–13735

Lovett ST, Mortimer RK (1987) Characterization of null mutants of the RAD55 gene of *Saccharomyces cerevisiae*: effects of temperature, osmotic strength and mating type. Genetics 116:547–553

Lynn A, Soucek R, Börner GV (2007) ZMM proteins during meiosis: crossover artists at work. Chromosome Res 15:591–605

Ma L, Milman N, Nambiar M, Smith GR (2015) Two separable functions of Ctp1 in the early steps of meiotic DNA double-strand break repair. Nucleic Acids Res 43:7349–7359

MacQueen AJ, Villeneuve AM (2001) Nuclear reorganization and homologous chromosome pairing during meiotic prophase require *C. elegans* chk-2. Genes Dev 15:1674–1687

MacQueen AJ, Colaiácovo MP, McDonald K, Villeneuve AM (2002) Synapsis-dependent and -independent mechanisms stabilize homolog pairing during meiotic prophase in *C. elegans*. Genes Dev 16:2428–2442

MacQueen AJ, Phillips CM, Bhalla N, Weiser P, Villeneuve AM, Dernburg AF (2005) Chromosome sites play dual roles to establish homologous synapsis during meiosis in *C. elegans*. Cell 123:1037–1050

Malone RE, Haring SJ, Foreman KE, Pansegrau ML, Smith SM, Houdek DR, Carpp L, Shah B, Lee KE (2004) The signal from the initiation of meiotic recombination to the first division of meiosis. Eukaryot Cell 3:598–609

Manfrini N, Guerini I, Citterio A, Lucchini G, Longhese MP (2010) Processing of meiotic DNA double strand breaks requires cyclin-dependent kinase and multiple nucleases. J Biol Chem 285:11628–11637

Mao-Draayer Y, Galbraith AM, Pittman DL, Cool M, Malone RE (1996) Analysis of meiotic recombination pathways in the yeast *Saccharomyces cerevisiae*. Genetics 144:71–86

Martin RH, Ko E, Rademaker A (1991) Distribution of aneuploidy in human gametes: comparison between human sperm and oocytes. Am J Med Genet 39:321–331

Martin JS, Winkelmann N, Petalcorin MIR, McIlwraith MJ, Boulton SJ (2005) RAD-51-dependent and -independent roles of a *Caenorhabditis elegans* BRCA2-related protein during DNA double-strand break repair. Mol Cell Biol 25:3127–3139

Matos J, Blanco MG, Maslen S, Skehel JM, West SC (2011) Regulatory control of the resolution of DNA recombination intermediates during meiosis and mitosis. Cell 147:158–172

McKim KS, Hayashi-Hagihara A (1998) mei-W68 in *Drosophila melanogaster* encodes a Spo11 homolog: evidence that the mechanism for initiating meiotic recombination is conserved. Genes Dev 12:2932–2942

McKim KS, Green-Marroquin BL, Sekelsky JJ, Chin G, Steinberg C, Khodosh R, Hawley RS (1998) Meiotic synapsis in the absence of recombination. Science 279:876–878

Michaelis C, Ciosk R, Nasmyth K (1997) Cohesins: chromosomal proteins that prevent premature separation of sister chromatids. Cell 91:35–45

Miles DC, van den Bergen JA, Sinclair AH, Western PS (2010) Regulation of the female mouse germ cell cycle during entry into meiosis. Cell Cycle 9:408–418

Mimitou EP, Symington LS (2008) Sae2, Exo1 and Sgs1 collaborate in DNA double-strand break processing. Nature 455:770–774

Moreau S, Ferguson JR, Symington LS (1999) The nuclease activity of Mre11 is required for meiosis but not for mating type switching, end joining, or telomere maintenance. Mol Cell Biol 19:556–566

Morelli MA, Cohen PE (2005) Not all germ cells are created equal: aspects of sexual dimorphism in mammalian meiosis. Reproduction 130:761–781

Morimoto A, Shibuya H, Zhu X, Kim J, Ishiguro K-I, Han M, Watanabe Y (2012) A conserved KASH domain protein associates with telomeres, SUN1, and dynactin during mammalian meiosis. J Cell Biol 198:165–172

Moses MJ (1956) Chromosomal structures in crayfish spermatocytes. J Biophys Biochem Cytol 2:215–218

Moses MJ (1958) The relation between the axial complex of meiotic prophase chromosomes and chromosome pairing in a salamander (*Plethodon cinereus*). J Biophys Biochem Cytol 4:633–638

Mullen JR, Kaliraman V, Ibrahim SS, Brill SJ (2001) Requirement for three novel protein complexes in the absence of the Sgs1 DNA helicase in *Saccharomyces cerevisiae*. Genetics 157:103–118

Murayama Y, Uhlmann F (2014) Biochemical reconstitution of topological DNA binding by the cohesin ring. Nature 505:367–371

Nabeshima K, Villeneuve AM, Colaiácovo MP (2005) Crossing over is coupled to late meiotic prophase bivalent differentiation through asymmetric disassembly of the SC. J Cell Biol 168:683–689

Nadarajan S, Mohideen F, Tzur YB, Ferrandiz N, Crawley O, Montoya A, Faull P, Snijders AP, Cutillas PR, Jambhekar A et al (2016) The MAP kinase pathway coordinates crossover designation with disassembly of synaptonemal complex proteins during meiosis. eLife 5: e12039

Nagaoka SI, Hassold TJ, Hunt PA (2012) Human aneuploidy: mechanisms and new insights into an age-old problem. Nat Rev Genet 13:493–504

Nairz K, Klein F (1997) mre11S--a yeast mutation that blocks double-strand-break processing and permits nonhomologous synapsis in meiosis. Genes Dev 11:2272–2290

Neale MJ, Pan J, Keeney S (2005) Endonucleolytic processing of covalent protein-linked DNA double-strand breaks. Nature 436:1053–1057

Nešić D, Hsu Y, Stebbins CE (2004) Assembly and function of a bacterial genotoxin. Nature 429:429–433

Nishant KT, Chen C, Shinohara M, Shinohara A, Alani E (2010) Genetic analysis of baker's yeast Msh4-Msh5 reveals a threshold crossover level for meiotic viability. PLoS Genet 6:e1001083

Niu H, Li X, Job E, Park C, Moazed D, Gygi SP, Hollingsworth NM (2007) Mek1 kinase is regulated to suppress double-strand break repair between sister chromatids during budding yeast meiosis. Mol Cell Biol 27:5456–5467

Niwa O, Shimanuki M, Miki F (2000) Telomere-led bouquet formation facilitates homologous chromosome pairing and restricts ectopic interaction in fission yeast meiosis. EMBO J 19:3831–3840

Novak JE, Ross-Macdonald PB, Roeder GS (2001) The budding yeast Msh4 protein functions in chromosome synapsis and the regulation of crossover distribution. Genetics 158:1013–1025

Oh SD, Lao JP, Taylor AF, Smith GR, Hunter N (2008) RecQ helicase, Sgs1, and XPF family endonuclease, Mus81-Mms4, resolve aberrant joint molecules during meiotic recombination. Mol Cell 31:324–336

Oliver TR, Feingold E, Yu K, Cheung V, Tinker S, Yadav-Shah M, Masse N, Sherman SL (2008) New insights into human nondisjunction of chromosome 21 in oocytes. PLoS Genet 4: e1000033–e1000039

Oliver TR, Tinker SW, Allen EG, Hollis N, Locke AE, Bean LJH, Chowdhury R, Begum F, Marazita M, Cheung V et al (2012) Altered patterns of multiple recombinant events are associated with nondisjunction of chromosome 21. Hum Genet 131:1039–1046

Osman F, Dixon J, Doe CL, Whitby MC (2003) Generating crossovers by resolution of nicked Holliday junctions: a role for Mus81-Eme1 in meiosis. Mol Cell 12:761–774

Ottolini CS, Newnham LJ, Capalbo A, Natesan SA, Joshi HA, Cimadomo D, Griffin DK, Sage K, Summers MC, Thornhill AR et al (2015) Genome-wide maps of recombination and chromosome segregation in human oocytes and embryos show selection for maternal recombination rates. Nature 47:727–735

Pacchierotti F, Adler ID, Eichenlaub-Ritter U, Mailhes JB (2007) Gender effects on the incidence of aneuploidy in mammalian germ cells. Environ Res 104:46–69

Page SL, Hawley RS (2001) c(3)G encodes a Drosophila synaptonemal complex protein. Genes Dev 15:3130–3143

Page SL, Hawley RS (2004) The genetics and molecular biology of the synaptonemal complex. Annu Rev Cell Dev Biol 20:525–558

Parvanov ED, Petkov PM, Paigen K (2010) Prdm9 controls activation of mammalian recombination hotspots. Science 327:835

Parvinen M, Söderström KO (1976) Chromosome rotation and formation of synapsis. Nature 260:534–535

Pasierbek P, Jantsch M, Melcher M, Schleiffer A, Schweizer D, Loidl J (2001) A *Caenorhabditis elegans* cohesion protein with functions in meiotic chromosome pairing and disjunction. Genes Dev 15:1349–1360

Paull TT, Gellert M (1998) The 3″ to 5″ exonuclease activity of Mre 11 facilitates repair of DNA double-strand breaks. Mol Cell 1:969–979

Pellestor F, Andréo B, Arnal F, Humeau C, Demaille J (2003) Maternal aging and chromosomal abnormalities: new data drawn from in vitro unfertilized human oocytes. Hum Genet 112:195–203

Penedos A, Johnson AL, Strong E, Goldman AS, Carballo JA, Cha RS (2015) Essential and checkpoint functions of budding yeast ATM and ATR during meiotic prophase are facilitated by differential phosphorylation of a meiotic adaptor protein, Hop1. PLoS One 10:e0134297

Penkner AM, Fridkin A, Gloggnitzer J, Baudrimont A, Machacek T, Woglar A, Csaszar E, Pasierbek P, Ammerer G, Gruenbaum Y et al (2009) Meiotic chromosome homology search involves modifications of the nuclear envelope protein Matefin/SUN-1. Cell 139:920–933

Penrose LS (2009) The relative effects of paternal and maternal age in mongolism. 1933. J Genet 88:9–14

Perry J, Kleckner N, Börner GV (2005) Bioinformatic analyses implicate the collaborating meiotic crossover/chiasma proteins Zip2, Zip3, and Spo22/Zip4 in ubiquitin labeling. Proc Natl Acad Sci USA 102:17594–17599

Pezzi N, Prieto I, Kremer L, Pérez Jurado LA, Valero C, Del Mazo J, Martínez-A C, Barbero JL (2000) STAG3, a novel gene encoding a protein involved in meiotic chromosome pairing and location of STAG3-related genes flanking the Williams-Beuren syndrome deletion. FASEB J 14:581–592

Phillips CM, Dernburg AF (2006) A family of zinc-finger proteins is required for chromosome-specific pairing and synapsis during meiosis in *C. elegans*. Dev Cell 11:817–829

Phillips CM, Wong C, Bhalla N, Carlton PM, Weiser P, Meneely PM, Dernburg AF (2005) HIM-8 binds to the X chromosome pairing center and mediates chromosome-specific meiotic synapsis. Cell 123:1051–1063

Phillips CM, Meng X, Zhang L, Chretien JH, Urnov FD, Dernburg AF (2009) Identification of chromosome sequence motifs that mediate meiotic pairing and synapsis in *C. elegans*. Nat Cell Biol 11:934–942

Polani PE, Crolla JA (1991) A test of the production line hypothesis of mammalian oogenesis. Hum Genet 88:64–70

Ponting CP (2011) What are the genomic drivers of the rapid evolution of PRDM9? Trends Genet 27:165–171

Pradillo M, Knoll A, Oliver C, Varas J, Corredor E, Puchta H, Santos JL (2015) Involvement of the cohesin cofactor PDS5 (SPO76) during meiosis and DNA repair in *Arabidopsis thaliana*. Front Plant Sci 6:1034

Prieto I, Pezzi N, Buesa JM, Kremer L, Barthelemy I, Carreiro C, Roncal F, Martinez A, Gomez L, Fernandez R et al (2002) STAG2 and Rad21 mammalian mitotic cohesins are implicated in meiosis. EMBO Rep 3:543–550

Prinz S, Amon A, Klein F (1997) Isolation of COM1, a new gene required to complete meiotic double-strand break-induced recombination in *Saccharomyces cerevisiae*. Genetics 146:781–795

Qi Z, Redding S, Lee JY, Gibb B, Kwon Y, Niu H, Gaines WA, Sung P, Greene EC (2015) DNA sequence alignment by microhomology sampling during homologous recombination. Cell 160:856–869

Qiao H, Chen JK, Reynolds A, Hoog C, Paddy M, Hunter N (2012) Interplay between synaptonemal complex, homologous recombination, and centromeres during mammalian meiosis. PLoS Genet 8:e1002790

Qiao H, Prasada Rao HBD, Yang Y, Fong JH, Cloutier JM, Deacon DC, Nagel KE, Swartz RK, Strong E, Holloway JK et al (2014) Antagonistic roles of ubiquitin ligase HEI10 and SUMO ligase RNF212 regulate meiotic recombination. Nature 46:194–199

Radford SJ, Goley E, Baxter K, McMahan S, Sekelsky J (2005) Drosophila ERCC1 is required for a subset of MEI-9-dependent meiotic crossovers. Genetics 170:1737–1745

Rakshambikai R, Srinivasan N, Nishant KT (2013) Structural insights into *Saccharomyces cerevisiae* Msh4-Msh5 complex function using homology modeling. PLoS One 8:e78753

Reddy KC, Villeneuve AM (2004) *C. elegans* HIM-17 links chromatin modification and competence for initiation of meiotic recombination. Cell 118:439–452

Revenkova E, Eijpe M, Heyting C, Gross B, Jessberger R (2001) Novel meiosis-specific isoform of mammalian SMC1. Mol Cell Biol 21:6984–6998

Reynolds A, Qiao H, Yang Y, Chen JK, Jackson N, Biswas K, Holloway JK, Baudat F, de Massy B, Wang J et al (2013) RNF212 is a dosage-sensitive regulator of crossing-over during mammalian meiosis. Nat Genet 45:269–278

Robinson WP, Kuchinka BD, Bernasconi F, Petersen MB, Schulze A, Brøndum-Nielsen K, Christian SL, Ledbetter DH, Schinzel AA, Horsthemke B et al (1998) Maternal meiosis I non-disjunction of chromosome 15: dependence of the maternal age effect on level of recombination. Hum Mol Genet 7:1011–1019

Rodman TC (1971) Chromosomes of the first polar body in mammalian meiosis. Exp Cell Res 68:205–210

Rog O, Dernburg AF (2015) Direct visualization reveals kinetics of meiotic chromosome synapsis. Cell Rep 10:1639–1645

Rolef Ben-Shahar T, Heeger S, Lehane C, East P, Flynn H, Skehel M, Uhlmann F (2008) Eco1-dependent cohesin acetylation during establishment of sister chromatid cohesion. Science 321:563–566

Romanienko PJ, Camerini-Otero RD (2000) The mouse Spo11 gene is required for meiotic chromosome synapsis. Mol Cell 6:975–987

Rosu S, Zawadzki KA, Stamper EL, Libuda DE, Reese AL, Dernburg AF, Villeneuve AM (2013) The *C. elegans* DSB-2 protein reveals a regulatory network that controls competence for meiotic DSB formation and promotes crossover assurance. PLoS Genet 9:e1003674

Rothenberg M, Kohli J, Ludin K (2009) Ctp1 and the MRN-complex are required for endonucleolytic Rec12 removal with release of a single class of oligonucleotides in fission yeast. PLoS Genet 5:e1000722

Rowsey R, Gruhn J, Broman KW, Hunt PA, Hassold T (2014) Examining variation in recombination levels in the human female: a test of the production-line hypothesis. Am J Hum Genet 95:108–112

Royo H, Prosser H, Ruzankina Y, Mahadevaiah SK, Cloutier JM, Baumann M, Fukuda T, Hoog C, Tóth A, de Rooij DG et al (2013) ATR acts stage specifically to regulate multiple aspects of mammalian meiotic silencing. Genes Dev 27:1484–1494

Saito TT, Youds JL, Boulton SJ, Colaiácovo MP (2009) *Caenorhabditis elegans* HIM-18/SLX-4 interacts with SLX-1 and XPF-1 and maintains genomic integrity in the germline by processing recombination intermediates. PLoS Genet 5:e1000735

Saito TT, Mohideen F, Meyer K, Harper JW, Colaiácovo MP (2012) SLX-1 is required for maintaining genomic integrity and promoting meiotic noncrossovers in the *Caenorhabditis elegans* germline. PLoS Genet 8:e1002888

Saito TT, Lui DY, Kim H-M, Meyer K, Colaiácovo MP (2013) Interplay between structure-specific endonucleases for crossover control during *Caenorhabditis elegans* meiosis. PLoS Genet 9:e1003586

Salah SM, Nasmyth K (2000) Destruction of the securin Pds1p occurs at the onset of anaphase during both meiotic divisions in yeast. Chromosoma 109:27–34

Sato K-I (2015) Transmembrane signal transduction in oocyte maturation and fertilization: focusing on *Xenopus laevis* as a model animal. Int J Mol Sci 16:114–134

Scherthan H, Weich S, Schwegler H, Heyting C, Härle M, Cremer T (1996) Centromere and telomere movements during early meiotic prophase of mouse and man are associated with the onset of chromosome pairing. J Cell Biol 134:1109–1125

Schurko AM, Logsdon JM (2008) Using a meiosis detection toolkit to investigate ancient asexual "scandals" and the evolution of sex. Bioessays 30:579–589

Ségurel L, Leffler EM, Przeworski M (2011) The case of the fickle fingers: how the PRDM9 zinc finger protein specifies meiotic recombination hotspots in humans. Plos Biol 9:e1001211

Sehorn MG, Sigurdsson S, Bussen W, Unger VM, Sung P (2004) Human meiotic recombinase Dmc1 promotes ATP-dependent homologous DNA strand exchange. Nature 429:433–437

Serrentino ME, Chaplais E, Sommermeyer V, Borde V (2013) Differential association of the conserved SUMO ligase Zip3 with meiotic double-strand break sites reveals regional variations in the outcome of meiotic recombination. PLoS Genet 9:e1003416

Severson AF, Meyer BJ (2014) Divergent kleisin subunits of cohesin specify mechanisms to tether and release meiotic chromosomes. eLife 3:e03467

Severson AF, Ling L, van Zuylen V, Meyer BJ (2009) The axial element protein HTP-3 promotes cohesin loading and meiotic axis assembly in *C. elegans* to implement the meiotic program of chromosome segregation. Genes Dev 23:1763–1778

Sharan SK, Bradley A (1997) Murine Brca2: sequence, map position, and expression pattern. Genomics 40:234–241

Sharan SK, Morimatsu M, Albrecht U, Lim DS, Regel E, Dinh C, Sands A, Eichele G, Hasty P, Bradley A (1997) Embryonic lethality and radiation hypersensitivity mediated by Rad51 in mice lacking Brca2. Nature 386:804–810

Sharan SK, Pyle A, Coppola V, Babus J, Swaminathan S, Benedict J, Swing D, Martin BK, Tessarollo L, Evans JP et al (2004) BRCA2 deficiency in mice leads to meiotic impairment and infertility. Development 131:131–142

Singh DK, Andreuzza S, Panoli AP, Siddiqi I (2013) AtCTF7 is required for establishment of sister chromatid cohesion and association of cohesin with chromatin during meiosis in Arabidopsis. BMC Plant Biol 13:117

Singhal S, Leffler EM, Sannareddy K, Turner I, Venn O, Hooper DM, Strand AI, Li Q, Raney B, Balakrishnan CN et al (2015) Stable recombination hotspots in birds. Science 350:928–932

Skibbens RV, Corson LB, Koshland D, Hieter P (1999) Ctf7p is essential for sister chromatid cohesion and links mitotic chromosome structure to the DNA replication machinery. Genes Dev 13:307–319

Smolikov S, Eizinger A, Hurlburt A, Rogers E, Villeneuve AM, Colaiacovo MP (2007) Synapsis-defective mutants reveal a correlation between chromosome conformation and the mode of double-strand break repair during *Caenorhabditis elegans* meiosis. Genetics 176:2027–2033

Snowden T, Acharya S, Butz C, Berardini M, Fishel R (2004) hMSH4-hMSH5 recognizes Holliday Junctions and forms a meiosis-specific sliding clamp that embraces homologous chromosomes. Mol Cell 15:437–451

Sollier J, Lin W, Soustelle C, Suhre K, Nicolas A, Géli V, de La Roche Saint-André C (2004) Set1 is required for meiotic S-phase onset, double-strand break formation and middle gene expression. EMBO J 23:1957–1967

Sommermeyer V, Béneut C, Chaplais E, Serrentino ME, Borde V (2013) Spp1, a member of the Set1 complex, promotes meiotic DSB formation in promoters by tethering histone H3K4 methylation sites to chromosome axes. Mol Cell 49:43–54

Sourirajan A, Lichten M (2008) Polo-like kinase Cdc5 drives exit from pachytene during budding yeast meiosis. Genes Dev 22:2627–2632

Stamper EL, Rodenbusch SE, Rosu S, Ahringer J, Villeneuve AM, Dernburg AF (2013) Identification of DSB-1, a protein required for initiation of meiotic recombination in *Caenorhabditis elegans*, illuminates a crossover assurance checkpoint. PLoS Genet 9:e1003679

Stergiou L, Doukoumetzidis K, Sendoel A, Hengartner MO (2007) The nucleotide excision repair pathway is required for UV-C-induced apoptosis in *Caenorhabditis elegans*. Cell Death Differ 14:1129–1138

Stiff T, Walker SA, Cerosaletti K, Goodarzi AA, Petermann E, Concannon P, O'Driscoll M, Jeggo PA (2006) ATR-dependent phosphorylation and activation of ATM in response to UV treatment or replication fork stalling. EMBO J 25:5775–5782

Storlazzi A, Tesse S, Ruprich-Robert G, Gargano S, Pöggeler S, Kleckner N, Zickler D (2008) Coupling meiotic chromosome axis integrity to recombination. Genes Dev 22:796–809

Sugiyama T, Kowalczykowski SC (2002) Rad52 protein associates with replication protein A (RPA)-single-stranded DNA to accelerate Rad51-mediated displacement of RPA and presynaptic complex formation. J Biol Chem 277:31663–31672

Sun H, Treco D, Szostak JW (1991) Extensive 3′-overhanging, single-stranded DNA associated with the meiosis-specific double-strand breaks at the ARG4 recombination initiation site. Cell 64:1155–1161

Sun F, Fujiwara Y, Reinholdt LG, Hu J, Saxl RL, Baker CL, Petkov PM, Paigen K, Handel MA (2015) Nuclear localization of PRDM9 and its role in meiotic chromatin modifications and homologous synapsis. Chromosoma 124:397–415

Sung P, Robberson DL (1995) DNA strand exchange mediated by a RAD51-ssDNA nucleoprotein filament with polarity opposite to that of RecA. Cell 82:453–461

Sutani T, Kawaguchi T, Kanno R, Itoh T, Shirahige K (2009) Budding yeast Wpl1(Rad61)-Pds5 complex counteracts sister chromatid cohesion-establishing reaction. Curr Biol 19:492–497

Svendsen JM, Harper JW (2010) GEN1/Yen1 and the SLX4 complex: solutions to the problem of Holliday junction resolution. Genes Dev 24:521–536

Sym M, Engebrecht JA, Roeder GS (1993) ZIP1 is a synaptonemal complex protein required for meiotic chromosome synapsis. Cell 72:365–378

Tachibana-Konwalski K, Godwin J, van der Weyden L, Champion L, Kudo NR, Adams DJ, Nasmyth K (2010) Rec8-containing cohesin maintains bivalents without turnover during the growing phase of mouse oocytes. Genes Dev 24:2505–2516

Takeo S, Lake CM, Morais-de-Sá E, Sunkel CE, Hawley RS (2011) Synaptonemal complex-dependent centromeric clustering and the initiation of synapsis in Drosophila oocytes. Curr Biol 21:1845–1851

Tamaki H (1965) Chromosome behaviour at meiosis in *Saccharomyces cerevisiae*. J Gen Microbiol 41:93–98

Tanneti NS, Landy K, Joyce EF, McKim KS (2011) A pathway for synapsis initiation during zygotene in Drosophila oocytes. Curr Biol 21:1852–1857

Taylor MRG, Spirek M, Chaurasiya KR, Ward JD, Carzaniga R, Yu X, Egelman EH, Collinson LM, Rueda D, Krejci L et al (2015) Rad51 paralogs remodel pre-synaptic Rad51 filaments to stimulate homologous recombination. Cell 162:271–286

Terasawa M, Ogawa T, Tsukamoto Y, Ogawa H (2008) Sae2p phosphorylation is crucial for cooperation with Mre11p for resection of DNA double-strand break ends during meiotic recombination in *Saccharomyces cerevisiae*. Genes Genet Syst 83:209–217

Thomas NS, Ennis S, Sharp AJ, Durkie M, Hassold TJ, Collins AR, Jacobs PA (2001) Maternal sex chromosome non-disjunction: evidence for X chromosome-specific risk factors. Hum Mol Genet 10:243–250

Toth A, Ciosk R, Uhlmann F, Galova M, Schleiffer A, Nasmyth K (1999) Yeast cohesin complex requires a conserved protein, Eco1p(Ctf7), to establish cohesion between sister chromatids during DNA replication. Genes Dev 13:320–333

Tougan T, Kasama T, Ohtaka A, Okuzaki D, Saito TT, Russell P, Nojima H (2010) The Mek1 phosphorylation cascade plays a role in meiotic recombination of *Schizosaccharomyces pombe*. Cell Cycle 9:4688–4702

Traut H (1980) X-chromosomal nondisjunction induced by aging oocytes of *Drosophila melanogaster*: the special susceptibility of mature eggs. Can J Genet Cytol 22:433–437

Traut H, Schröder FJ (1978) The increase in the frequency of X-chromosomal aneuploidy in *Drosophila melanogaster* as a consequence of suppressed oviposition. Mutat Res 49:225–232

Trelles-Sticken E, Dresser ME, Scherthan H (2000) Meiotic telomere protein Ndj1p is required for meiosis-specific telomere distribution, bouquet formation and efficient homologue pairing. J Cell Biol 151:95–106

Tsai CJ, Mets DG, Albrecht MR, Nix P, Chan A, Meyer BJ (2008) Meiotic crossover number and distribution are regulated by a dosage compensation protein that resembles a condensin subunit. Genes Dev 22:194–211

Tsubouchi H, Roeder GS (2003) The importance of genetic recombination for fidelity of chromosome pairing in meiosis. Dev Cell 5:915–925

Tsutsumi M, Fujiwara R, Nishizawa H, Ito M, Kogo H, Inagaki H, Ohye T, Kato T, Fujii T, Kurahashi H (2014) Age-related decrease of meiotic cohesins in human oocytes. PLoS One 9: e96710

Uhlmann F, Nasmyth K (1998) Cohesion between sister chromatids must be established during DNA replication. Curr Biol 8:1095–1101

Unal E, Heidinger-Pauli JM, Kim W, Guacci V, Onn I, Gygi SP, Koshland DE (2008) A molecular determinant for the establishment of sister chromatid cohesion. Science 321:566–569

Urban E, Nagarkar-Jaiswal S, Lehner CF, Heidmann SK (2014) The cohesin subunit Rad21 is required for synaptonemal complex maintenance, but not sister chromatid cohesion, during Drosophila female meiosis. PLoS Genet 10:e1004540

Usui T, Ohta T, Oshiumi H, Tomizawa J, Ogawa H, Ogawa T (1998) Complex formation and functional versatility of Mre11 of budding yeast in recombination. Cell 95:705–716

Usui T, Ogawa H, Petrini JH (2001) A DNA damage response pathway controlled by Tel1 and the Mre11 complex. Mol Cell 7:1255–1266

Valencia M, Bentele M, Vaze MB, Herrmann G, Kraus E, Lee SE, Schär P, Haber JE (2001) NEJ1 controls non-homologous end joining in *Saccharomyces cerevisiae*. Nature 414:666–669

Vallaster M, Vallaster CD, Wu SM (2011) Epigenetic mechanisms in cardiac development and disease. Acta Biochim Biophys Sin 44:92–102

Vaur S, Feytout A, Vazquez S, Javerzat JP (2012) Pds5 promotes cohesin acetylation and stable cohesin-chromosome interaction. EMBO Rep 13:645–652

Vernì F, Gandhi R, Goldberg ML, Gatti M (2000) Genetic and molecular analysis of wings apart-like (wapl), a gene controlling heterochromatin organization in *Drosophila melanogaster*. Genetics 154:1693–1710

Visnes T, Giordano F, Kuznetsova A, Suja JA, Lander AD, Calof AL, Ström L (2014) Localisation of the SMC loading complex Nipbl/Mau2 during mammalian meiotic prophase I. Chromosoma 123:239–252

Voelkel-Meiman K, Taylor LF, Mukherjee P, Humphryes N, Tsubouchi H, MacQueen AJ (2013) SUMO localizes to the central element of synaptonemal complex and is required for the full synapsis of meiotic chromosomes in budding yeast. PLoS Genet 9:e1003837

Waizenegger IC, Hauf S, Meinke A, Peters JM (2000) Two distinct pathways remove mammalian cohesin from chromosome arms in prophase and from centromeres in anaphase. Cell 103:399–410

Wang X, Haber JE (2004) Role of Saccharomyces single-stranded DNA-binding protein RPA in the strand invasion step of double-strand break repair. Plos Biol 2:E21

Wang TF, Kleckner N, Hunter N (1999) Functional specificity of MutL homologs in yeast: evidence for three Mlh1-based heterocomplexes with distinct roles during meiosis in recombination and mismatch correction. Proc Natl Acad Sci USA 96:13914–13919

Wang F, Yoder J, Antoshechkin I, Han M (2003) *Caenorhabditis elegans* EVL-14/PDS-5 and SCC-3 are essential for sister chromatid cohesion in meiosis and mitosis. Mol Cell Biol 23:7698–7707

Wagner CR, Kuervers L, Baillie DL, Yanowitz JL (2010) xnd-1 regulates the global recombination landscape in *Caenorhabditis elegans*. Nature 467(7317):839–843. PMID: 20944745

Ward JO, Reinholdt LG, Motley WW, Niswander LM, Deacon DC, Griffin LB, Langlais KK, Backus VL, Schimenti KJ, O'Brien MJ et al (2007) Mutation in mouse hei10, an e3 ubiquitin ligase, disrupts meiotic crossing over. PLoS Genet 3:e139

Watanabe Y, Nurse P (1999) Cohesin Rec8 is required for reductional chromosome segregation at meiosis. Nature 400:461–464

Woglar A, Daryabeigi A, Adamo A, Habacher C, Machacek T, La Volpe A, Jantsch V (2013) Matefin/SUN-1 phosphorylation is part of a surveillance mechanism to coordinate chromosome synapsis and recombination with meiotic progression and chromosome movement. PLoS Genet 9:e1003335

Wolf KW (1994) How meiotic cells deal with non-exchange chromosomes. Bioessays 16:107–114

Woltering D, Baumgartner B, Bagchi S, Larkin B, Loidl J, de los Santos T, Hollingsworth NM (2000) Meiotic segregation, synapsis, and recombination checkpoint functions require physical interaction between the chromosomal proteins Red1p and Hop1p. Mol Cell Biol 20:6646–6658

Wu J, Zhang X, Zhang L, Wu C-Y, Rezaeian AH, Chan C-H, Li J-M, Wang J, Gao Y, Han F et al (2012) Skp2 E3 ligase integrates ATM activation and homologous recombination repair by ubiquitinating NBS1. Mol Cell 46:351–361

Wyatt HDM, Sarbajna S, Matos J, West SC (2013) Coordinated actions of SLX1-SLX4 and MUS81-EME1 for Holliday junction resolution in human cells. Mol Cell 52:234–247

Xu L, Weiner BM, Kleckner N (1997) Meiotic cells monitor the status of the interhomolog recombination complex. Genes Dev 11:106–118

Xu H, Beasley MD, Warren WD, van der Horst GTJ, McKay MJ (2005) Absence of mouse REC8 cohesin promotes synapsis of sister chromatids in meiosis. Dev Cell 8:949–961

Yamamoto A, West RR, McIntosh JR, Hiraoka Y (1999) A cytoplasmic dynein heavy chain is required for oscillatory nuclear movement of meiotic prophase and efficient meiotic recombination in fission yeast. J Cell Biol 145:1233–1249

Yildiz O, Majumder S, Kramer B, Sekelsky JJ (2002) Drosophila MUS312 interacts with the nucleotide excision repair endonuclease MEI-9 to generate meiotic crossovers. Mol Cell 10:1503–1509

Yin Y, Smolikove S (2013) Impaired resection of meiotic double-strand breaks channels repair to nonhomologous end joining in *Caenorhabditis elegans*. Mol Cell Biol 33:2732–2747

Yin Y, Donlevy S, Smolikove S (2016) Coordination of recombination with meiotic progression in the *Caenorhabditis elegans* germline by KIN-18, a TAO kinase that regulates the timing of MPK-1 signaling. Genetics 202:45–59

Yokobayashi S, Yamamoto M, Watanabe Y (2003) Cohesins determine the attachment manner of kinetochores to spindle microtubules at meiosis I in fission yeast. Mol Cell Biol 23:3965–3973

Yokoo R, Zawadzki KA, Nabeshima K, Drake M, Arur S, Villeneuve AM (2012) COSA-1 reveals robust homeostasis and separable licensing and reinforcement steps governing meiotic crossovers. Cell 149:75–87

Zakharyevich K, Ma Y, Tang S, Hwang PY-H, Boiteux S, Hunter N (2010) Temporally and biochemically distinct activities of Exo1 during meiosis: double-strand break resection and resolution of double Holliday junctions. Mol Cell 40:1001–1015

Zakharyevich K, Tang S, Ma Y, Hunter N (2012) Delineation of joint molecule resolution pathways in meiosis identifies a crossover-specific resolvase. Cell 149:334–347

Zannini L, Delia D, Buscemi G (2014) CHK2 kinase in the DNA damage response and beyond. J Mol Cell Biol 6:442–457

Zhang Z, Ren Q, Yang H, Conrad MN, Guacci V, Kateneva A, Dresser ME (2005) Budding yeast PDS5 plays an important role in meiosis and is required for sister chromatid cohesion. Mol Microbiol 56:670–680

Zhang J, Håkansson H, Kuroda M, Yuan L (2008a) Wapl localization on the synaptonemal complex, a meiosis-specific proteinaceous structure that binds homologous chromosomes, in the female mouse. Reprod Domest Anim 43:124–126

Zhang N, Kuznetsov SG, Sharan SK, Li K, Rao PH, Pati D (2008b) A handcuff model for the cohesin complex. J Cell Biol 183:1019–1031

Zhang L, Kim KP, Kleckner NE, Storlazzi A (2011) Meiotic double-strand breaks occur once per pair of (sister) chromatids, and via Mec1/ATR and Tel1/ATM, once per quartet of chromatids. Proc Natl Acad Sci USA 108:20036–20041

Zhang L, Espagne E, De Muyt A, Zickler D, Kleckner NE (2014) Interference-mediated synaptonemal complex formation with embedded crossover designation. Proc Natl Acad Sci USA 111:E5059–E5068

Zhu Z, Chung W-H, Shim EY, Lee SE, Ira G (2008) Sgs1 helicase and two nucleases Dna2 and Exo1 resect DNA double-strand break ends. Cell 134:981–994

Zickler D, Kleckner N (1999) Meiotic chromosomes: integrating structure and function. Annu Rev Genet 33:603–754

Chapter 6
Translational Control of Germ Cell Decisions

Kumari Pushpa, Ganga Anil Kumar, and Kuppuswamy Subramaniam

Abstract Germline poses unique challenges to gene expression control at the transcriptional level. While the embryonic germline maintains a global hold on new mRNA transcription, the female adult germline produces transcripts that are not translated into proteins until embryogenesis of subsequent generation. As a consequence, translational control plays a central role in governing various germ cell decisions including the formation of primordial germ cells, self-renewal/differentiation decisions in the adult germline, onset of gametogenesis and oocyte maturation. Mechanistically, several common themes such as asymmetric localization of mRNAs, conserved RNA-binding proteins that control translation by 3′ UTR binding, translational activation by the cytoplasmic elongation of the polyA tail and the assembly of mRNA-protein complexes called mRNPs have emerged from the studies on *Caenorhabditis elegans*, *Xenopus* and *Drosophila*. How mRNPs assemble, what influences their dynamics, and how a particular 3′ UTR-binding protein turns on the translation of certain mRNAs while turning off other mRNAs at the same time and space are key challenges for future work.

6.1 Introduction

Differential gene expression is at the core of development. Although most cells in our body possess the same genome, gene expression differs among the different cell types, which provides each cell type its unique identity. Gene expression can be controlled at any of the different stages starting from chromatin status to the

K. Pushpa
Regional Centre for Biotechnology, Faridabad, Haryana, India

G.A. Kumar
Indian Institute of Technology-Kanpur, Kanpur, India

Indian Institute of Technology-Madras, Chennai, India

K. Subramaniam (✉)
Indian Institute of Technology-Madras, Chennai, India
e-mail: subbu@iitm.ac.in

S. Arur (ed.), *Signaling-Mediated Control of Cell Division*, Results and Problems in Cell Differentiation 59, DOI 10.1007/978-3-319-44820-6_6

posttranslational modification of proteins. Regulation at the transcriptional stage is energy efficient: when a gene product is not required, stopping transcription eliminates the need to degrade or store the gene product(s) in an inactive form. Consistent with this, most somatic development programs are controlled at the level of transcription. However, the transcriptional control as a sole mechanism of gene regulation is inadequate in certain important developmental and physiological situations. For example, the highly condensed metaphase chromatin is not readily accessible for transcription factors, yet specific proteins are required only at specific steps during the cell division cycle. Similarly, neurons need to rapidly produce new proteins at the synaptic ends of axons, which are often at considerable distance from the nucleus, where transcription occurs. Posttranscriptional mechanisms, such as translational control, play a key role in such circumstances. Several key decisions during germ cell development demand gene expression controls beyond the transcriptional step, and translational control plays a particularly crucial role at these decision points.

In the first part of this chapter, we discuss the germline-specific developmental contexts in which translational control of gene expression is preferred over the transcriptional control. The second part reviews our current understanding of the roles played by translational control during different stages of germ cell development, namely, specification, self-renewal/differentiation decision, meiotic progression, and oocyte maturation.

6.2 The Germline Challenges to Transcriptional Control of Gene Expression

6.2.1 Transcriptional Quiescence During Germline Specification

Transcriptional quiescence is a hallmark of the early embryonic germline in many organisms. In the early embryo, unlike the somatic blastomeres, which activate new mRNA transcription as soon as they are born, their germline counterparts do not initiate new mRNA transcription. These cells lack the actively transcribing form of RNA polymerase II, which indicates a genome-wide shutdown of mRNA transcription (Seydoux and Dunn 1997). Such a global suppression of transcription most likely helps the embryonic germline maintain totipotency by evading the influence of maternally inherited transcription factors, which would otherwise activate the various somatic differentiation programs. This is particularly well supported by the functioning of SKN-1 and PAL-1 transcription factors in the *C. elegans* embryo (see below). While the maternal SKN-1 protein is distributed equally to the somatic blastomere EMS and its germline sibling P_2, it activates the transcription of genes that specify somatic fates in EMS, but not in P_2 in which the PIE-1 protein maintains global transcriptional quiescence (Bowerman et al. 1993; Maduro et al.

2001; Seydoux et al. 1996). Similarly, PAL-1 activates muscle fate in the D blastomere but not in its germline sibling P_4, where PIE-1 is present (Hunter and Kenyon 1996).

6.2.2 *Transcriptional Quiescence During Meiosis and Gametogenesis*

The highly condensed state of the meiotic chromatin is not readily accessible to transcription factors. In *Drosophila* male germ cells completely shut down their transcription as they enter meiosis (Olivieri and Olivieri 1965; Schafer et al. 1995). In other organisms spermatocytes are transcriptionally silent during meiotic entry (leptotene and zygotene) and again during spermatid elongation that involves chromatin compaction along with histone exchange (Sassone-Corsi 2002). Similarly, during oogenesis, de novo mRNA synthesis does not occur during early stages of meiosis. It is briefly activated during pachytene and diplotene before being globally silenced again through the final stages of maturation (Walker et al. 2007). This transcriptional quiescence of germ cells during gametogenesis and of the mature gametes necessitates that the transcripts be premade, stored in a translationally dormant state until their protein products are needed post fertilization in the embryo.

6.2.3 *Making mRNAs for the Next Generation*

Early embryonic development proceeds in the absence of new mRNA transcription. As a consequence, genes that control early embryogenesis are transcribed in the maternal germline, and the mRNAs are deposited in the oocyte. Premature translation of these maternal mRNAs—many of their protein products direct somatic differentiation—will be detrimental to the mother's germline [e.g., see Ciosk et al. (2006)]. Therefore, translational control of these mRNAs is vital for gametogenesis.

6.2.4 *Germ Cells Share a Common Cytoplasm*

In many species, germ cells are connected by cytoplasmic bridges, which allow the sharing of cytoplasmic contents among germ cells during different developmental stages until they form mature gametes. This concept is stretched in certain species like *C. elegans* to an extent where all germ nuclei share a common cytoplasm—not an ideal condition for transcriptional control to be effective. Not surprisingly, translational control is the predominant mode of gene regulation in the *C. elegans* germline [(Merritt et al. 2008); see below].

6.3 Translational Control During Germ Cell Development

From the fate specification to gamete maturation, several key developmental events occur during the process of germ cell development. Germ cell fate determination is the first event in this long journey. In most species, germ cells are specified at a different location from the future gonad; as a consequence the primordial germ cells (PGCs) must migrate to, and get incorporated into, the somatic gonad. Since only a small number of PGCs are born in the early embryo, they enter a proliferative phase to establish a population of self-renewing germline stem cells (GSCs) in a micro-environment of the gonad called the GSC niche. Some of these GSCs then switch from the mitotic to the meiotic mode and finally differentiate into gametes. Translation regulation plays a central role in all these developmental stages.

6.3.1 Translational Control During Germ Cell Specification and PGC Development

There are two modes of germ cell specification in metazoans. In invertebrates and anuran amphibians, maternally inherited components specify germ cell fate. By contrast, no specific cell in the early embryo possesses germ cell fate in urodele amphibians and vertebrates; instead, inductive signals from neighboring cells induce it in certain cells (Extavour and Akam 2003; McLaren 1999). In organisms that follow the inheritance mode, maternally inherited germline components—principally mRNAs and proteins—reside in the same cytoplasm as the somatic factors at the zygotic stage. As a consequence, their asymmetric localization and translational activation of the localized transcripts are particularly crucial to germ cell fate specification in these organisms. The role of maternal mRNAs and their translational control during germ cell formation have been extensively studied in *Drosophila* and *C. elegans*.

6.3.1.1 Translational Activation of Localized mRNAs Determines Germ Cell Fate in *Drosophila*

Germ cell formation in *Drosophila* begins in the oocyte with the assembly of a special cytoplasm, called germplasm, at the posterior of the oocyte. Germplasm contains the polar granules consisting of proteins including Oskar, Tudor, Vasa, Valois, and Staufen and many mRNAs including *oskar*, which is required for PGC specification, and *nanos*, which is essential for PGC development (Forbes and Lehmann 1998; Kim-Ha et al. 1991; Markussen et al. 1995; Rongo et al. 1995). Although the mechanism of how the proteins get to the germplasm is not known, many of the mRNAs are transported by passive diffusion and localized by entrapment. One mechanism of entrapment involves partial base pairing between the mRNAs and piRNAs bound to the Piwi protein Aubergine (Forrest and Gavis 2003; Kugler and Lasko 2009; Vourekas et al. 2016). Oskar initiates germplasm formation; its expression alone is sufficient to induce germplasm formation at ectopic locations and thus serves as a central component of germ cell fate determination (Mahowald 2001). The *oskar* mRNA is transported in a translationally repressed state via microtubules to the oocyte posterior. This localization of *oskar* mRNA at the posterior end is essential for translational activation of Oskar (Jambor et al. 2014). Bruno, an RBP, inhibits *oskar* mRNA via both a cap-dependent and a cap-independent mechanism. For the cap-dependent function, Bruno binds to the 3′ UTR of *oskar* mRNA, resulting in the recruitment of Cup to the 5′ cap, which binds to eIF4E and prevents it from recruiting eIF4G, causing repression of *oskar* mRNA (Nakamura et al. 2004). Bruno also inhibits *oskar* translation in a cap-independent mechanism by packaging *oskar* mRNA into large particles that are inaccessible to the translational machinery (Chekulaeva et al. 2006; Nakamura et al. 2004). Further, Bicaudal-C, which generally regulates the polyA tail length of target transcripts, also contributes to the inhibition of *oskar* translation, possibly by inhibiting the association of cytoplasmic polyadenylation element binding protein (CPEB) Orb with *oskar* mRNA (Castagnetti and Ephrussi 2003; Chicoine et al. 2007; Saffman et al. 1998). This repression is alleviated at the posterior pole by a specific derepressing element in the 5′ UTR of *oskar* mRNA (Gunkel et al. 1998). Although the mechanism of derepression is not known, Vasa, Aubergine, Orb, and Staufen are required for the *oskar* mRNA's efficient translation (Castagnetti and Ephrussi 2003; Chang et al. 1999; Harris and Macdonald 2001; Markussen et al. 1997; Micklem et al. 2000; Wilson et al. 1996).

Oskar recruits other germplasm components such as Vasa and Tudor and about 200 maternal mRNAs to the germ granules (Frise et al. 2010; Trcek et al. 2015). Based on a few well-studied examples, such as *nanos* (*nos*), *germ cell-less* (*gcl*), *polar granule component* (*pgc*), and *cyclin B*, it is assumed that the translation of these mRNAs is specifically activated by the germ granule components and that their protein products direct PGC development. While the Gcl and Pgc proteins repress transcription in the *Drosophila* PGCs, Nos and its RNA-binding partner Pumilio prevent premature proliferation of PGCs by repressing the translation of cyclin B mRNA (Hanyu-Nakamura et al. 2008; Kadyrova et al. 2007; Leatherman et al. 2002). In addition, Nos and Pumilio are essential for PGC development and later GSC maintenance; they contribute to these processes possibly by suppressing other mRNAs (Forbes and Lehmann 1998; Harris et al. 2011). Like *osk* mRNA, *nos* mRNA is transported to the germplasm in a translationally silent state. The *nos* 3′ UTR contains several stem-loop structures which bind different RBPs at different stages of development. The RBPs Glorund and Smaug (Smg) suppress *nos* translation in the late-stage oocyte and embryo, respectively (Kalifa et al. 2006; Smibert

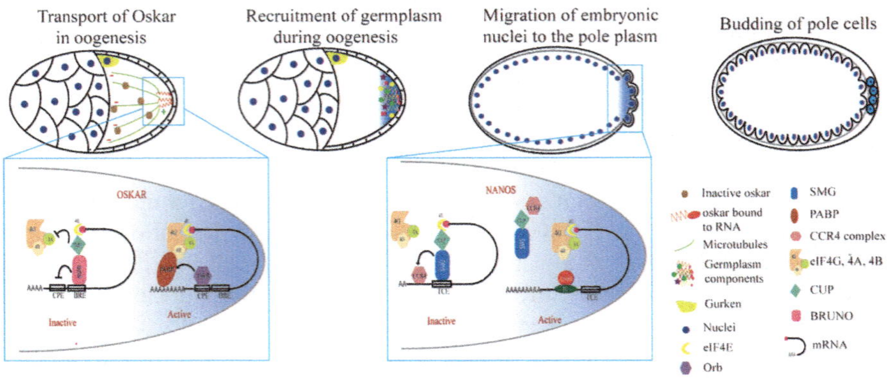

Fig. 6.1 Translational control during germ cell specification in *Drosophila*. Germplasm is assembled during oogenesis by Oskar. *oskar* mRNA is transported in an inactive form and deposited at the posterior pole where its translation is activated. *oskar* translation is suppressed by Bruno with the help of Cup. At the posterior, *oskar* translation is activated by ORB; ORB recruits PABP and facilitates translational activation. Oskar further recruits other germplasm components during late stages of oogenesis. The early *Drosophila* embryo is a syncytium; at the start of cellularization, some of the posterior nuclei and the surrounding germplasm form the pole cells (PGCs) by budding. *nanos* mRNA is translationally suppressed in the anterior region by Smaug (Smg) which recruits Cup and prevents translation initiation. At the posterior end, *nos* mRNA is bound by Oskar, which prevents Smg from binding to *nos-2* 3′ UTR, which derepresses *nos* mRNA. Smaller cells diagrammed in the *top left* two cartoons represent the nurse cells

et al. 1996). Smg recruits the deadenylase complex and targets *nos* mRNA for degradation in the bulk cytoplasm (Semotok et al. 2005). In addition, Smg recruits Cup and inhibits *nos* translation (Nelson et al. 2004). In the germplasm, Osk activates *nos* translation by dislodging Smg from its 3′ UTR (Dahanukar et al. 1999; Zaessinger et al. 2006). In summary, a complex interplay between various cis elements and trans-activating factors ensures localization-dependent translation of mRNAs that determine PGC formation and development in *Drosophila* in the transcriptionally quiescent early embryo (Fig. 6.1).

6.3.1.2 Asymmetric Segregation, Transcriptional Quiescence, and Sequential Translational Activation Lead to PGC Formation in *C. elegans*

In *C. elegans*, germ granules are present in the syncytial maternal germline even before oogenesis begins and are distributed initially throughout the zygotic cytoplasm (Strome and Wood 1982). Not surprisingly, an Oskar-like single factor that initiates germplasm assembly in the oocyte has not been identified in *C. elegans*. Although they are segregated to the germline during asymmetric cleavages (see below) and some of their protein and mRNA components are essential for PGC formation or development, the germ granules per se are not essential for the germ

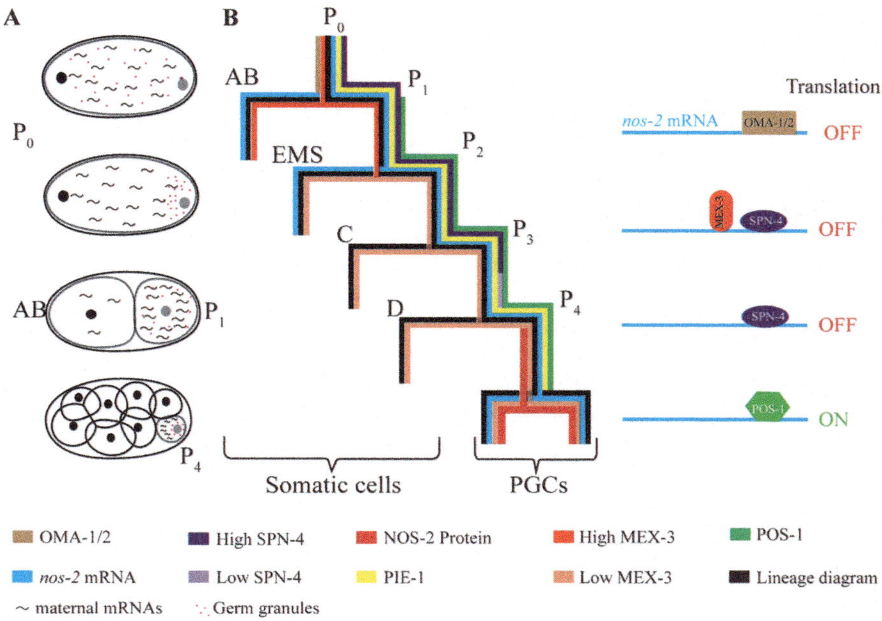

Fig. 6.2 Translational control during germ cell specification in *C. elegans*. (**a**) Schematic representation of asymmetric cleavage and asymmetric distribution of maternal components. *Pink dots* and *tilde-like structures* represent the germ granules and the maternal mRNAs, respectively. (**b**) Distribution patterns of RBPs and *nos-2* mRNA during early embryonic cleavages; *colors* representing the different components are indicated at the *bottom*. Translation of *nos-2* mRNA is suppressed sequentially by OMA-1 and OMA-2 in oocytes (not shown here), by MEX-3 in the AB blastomere, and by SPN-4 in the P lineage until P_3. The rapid decrease in the SPN-4 to POS-1 ratio in P_4 enables POS-1 to compete out SPN-4 for binding to the *nos-2* 3′ UTR, which depresses *nos-2* translation in P_4

cell fate (Gallo et al. 2010). In sharp contrast to the syncytial beginning of the *Drosophila* embryo, the *C. elegans* embryo undergoes complete cytokinesis during embryonic cell divisions. The zygote undergoes an asymmetric cleavage generating a large anterior cell called AB and a small posterior cell called P_1. While the *Drosophila* pole cells are committed to the germline as soon as they are formed, the P_1 blastomere is not fully committed to the germline. Instead P_1 undergoes three more rounds of asymmetric division generating one daughter cell at each of these divisions that acquires somatic fate (Fig. 6.2a), while the other daughter retains the germline (P) fate (Sulston et al. 1983). This process of asymmetry between the generation of the germline and somatic lineage necessitates (a) asymmetric segregation of maternally produced germline components at each of these divisions to the P lineage, (b) protection of germline components from degradation in the germline lineage, and (c) resistance of acquisition of somatic fate by the germline blastomeres P_1, P_2, P_3, and P_4 (Fig. 6.2b). These remarkable feats are accomplished by a combination of transcriptional quiescence and sequential translational activation of maternal mRNAs.

Transcriptional quiescence in the P lineage is maintained by the maternal protein PIE-1, which is segregated to, and maintained in, the P lineage (Mello et al. 1996; Seydoux et al. 1996). PIE-1 suppresses transcription by preventing phosphorylation of the carboxy-terminal domain (CTD) of RNA polymerase II, analogous to *Drosophila* Pgc (Batchelder et al. 1999; Seydoux and Dunn 1997; Zhang et al. 2003). Meanwhile, sequential translational activation of maternal mRNAs such as *pos-1*, *apx-1*, and *pal-1* helps the somatic blastomeres arising from the P lineage to acquire their respective somatic fates; however, due to the presence of PIE-1, the corresponding germline siblings are protected from somatic differentiation (Hunter and Kenyon 1996; Mello et al. 1992; Tabara et al. 1999). The role of PIE-1 in the protection of germline identity is further underscored by the acquisition of somatic fate by P_4 in the *pos-1* mutant embryos. *pos-1* mutant embryos exhibit a reduction in PIE-1 accumulation in the P_4 blastomere simultaneous to PAL-1 inducing the muscle fate (Tabara et al. 1999).

Although the mechanisms remain unclear, the persistence of maternal mRNAs in the P lineage appears to be a consequence of transcriptional quiescence. The *pie-1* mutant P blastomeres activate mRNA transcription but fail to maintain the maternal mRNAs. However, the degradation of maternal mRNAs can be suppressed by blocking RNA polymerase II activity (Tenenhaus et al. 2001). RNA-binding proteins produced from maternal mRNAs such as OMA-1, OMA-2, MEX-3, SPN-4, and POS-1 control the translation of other maternal mRNAs (Guven-Ozkan et al. 2010; Jadhav et al. 2008; Kaymak and Ryder 2013; Ogura et al. 2003; Pagano et al. 2009; Spike et al. 2014b; Tabara et al. 1999). A well-studied example is the translational control of *nanos-2* (*nos-2*) mRNA, which encodes a *C. elegans* ortholog of the *Drosophila* Nanos (Subramaniam and Seydoux 1999). Translation of *nos-2* mRNA is restricted to the P_4 blastomere by the sequential actions of the repressors OMA-1/2, MEX-3, SPN-4, and the derepressor POS-1 (Fig. 6.2b) (Jadhav et al. 2008). Each of these proteins mediate their effect by binding to the respective recognition sequences in the *nos-2* 3′ UTR. In P_4 the SPN-4/POS-1 ratio decreases enabling POS-1 to bind to *nos-2* 3′ UTR and derepressing *nos-2* translation. The translationally competent *nos-2* mRNA produces NOS-2 which promotes PGC development.

6.3.1.3 Role of Translational Control During PGC Development in Vertebrates

Akin to *Drosophila oskar*, the zebrafish *buckyball* (*buc*), a gene conserved across vertebrates, acts as the primary germplasm organizer in zebrafish oocytes. The *buc* mRNA is distributed throughout all the four blastomeres at the 4-cell stage, but the protein localizes to the germplasm at the cleavage furrows. However, it is not clear if translational regulation is responsible for this difference between the *buc* mRNA and protein distribution (Bontems et al. 2009). Similar to *C. elegans nanos-2* mRNA, zebrafish *nanos* mRNA is translationally repressed via its 3′ UTR in early embryos and expressed only in PGCs (Koprunner et al. 2001). The zebrafish *nanos* mRNA is repressed in the somatic cells of early embryo by the mir-430 miRNA, which binds to the 3′ UTR and promotes deadenylation (Mishima et al. 2006). This repression is relieved in PGCs by a vertebrate-specific RBP called Dead end, which inhibits the miRNA binding to the *nanos* 3′ UTR (Kedde et al. 2007). Dead end is conserved in all vertebrates and has been shown to be a component of germplasm and required for germ cell development (Weidinger et al. 2003; Youngren et al. 2005).

Germ cell specification in mouse is an example for the induction mode. Although this mode of germ cell specification is primarily regulated at the transcriptional level, recently it has been shown that Blimp1, a transcriptional repressor crucial for PGC development, is negatively regulated by *let-7* miRNA. The *let-7*-mediated suppression is eventually relieved by its negative regulator *let-28* during PGC specification (West et al. 2009). Separate studies have shown that the *Blimp1* 3′ UTR contains target sites for *let-7* binding and that the *let-7* miRNA can suppress its translation (Nie et al. 2008).

6.3.2 Translational Control During Mitosis–Meiosis Decision

Genes that drive differentiation are often transcribed in the progenitor or stem cells. Once transcribed these mRNAs are stored in the stem cells until the onset of differentiation in a translationally silent state. The preeminence of translational control in mitosis–meiosis decision in *C. elegans* and *Drosophila* has been recognized for quite some time; recent studies support a similar role for translational regulation in the mouse testis as well (Zhou et al. 2015). Furthermore, translational control plays vital roles even in somatic adult stem cell maintenance (Crist et al. 2012). In both invertebrates and vertebrates, niche-dependent transcriptional regulation and germ cell-intrinsic translational regulatory networks control the mitosis–meiosis decision (Kimble 2011; Lehmann 2012).

6.3.2.1 A Complex RNA Regulatory Network Guides the Mitosis–Meiosis Decision in *C. elegans*

The *C. elegans* gonad exhibits distal-proximal polarity with GSCs at the distal end and mature gametes in the proximal region. The somatic cell called distal tip cell (DTC) signals GSCs via the LAG-2 Delta ligand localized on its cell membrane and the GLP-1 Notch receptor expressed in GSCs (Austin and Kimble 1987; Crittenden et al. 1994; Henderson et al. 1994; Yochem and Greenwald 1989). GLP-1, together with an “RNA regulatory loop” that is initiated by GLP-1 signaling, promotes self-renewal of the GSCs by inhibiting premature meiotic differentiation (Fig. 6.3). The RNA regulatory loop consists of the nearly identical PUF paralogs FBF-1 and FBF-2 (collectively called FBF), the KH domain RBP GLD-1, and the *glp-1* mRNA. In the distal region, GLP-1 activates the transcription of FBF-2 (Lamont et al. 2004), which, along with FBF-1, represses the translation of *gld-1* mRNA (Crittenden et al. 2002). As a result, GLD-1 expression is restricted to more proximal cells where it activates meiotic differentiation. In addition, GLD-1 represses *glp-1* translation—thus completing the regulatory loop—and thereby reinforcing the fate switch (Marin and Evans 2003). Besides GLD-1, FBF inhibits differentiation by repressing the translation of meiotic entry promoters GLD-2 and GLD-3 and some of the meiotic machinery components (Eckmann et al. 2002; Merritt and Seydoux 2010; Millonigg et al. 2014). Further, FBF represses the translation of CKI-2, an inhibitor of CYE-1 cyclin E; CYE-1 and its kinase partner CDK-2, in turn, phosphorylate and inactivate GLD-1 (Jeong et al. 2011; Kalchhauser et al. 2011).

The size of the proliferation zone is controlled by both positive and negative regulation of *glp-1* mRNA translation. While GLD-1 represses *glp-1* translation (Marin and Evans 2003), the cytoplasmic polyA polymerase GLD-4 promotes it, in part, by extending the polyA tail length (Millonigg et al. 2014).

Preventing differentiation of the GSCs, however, is not sufficient for their self-renewal. PUF-8, another PUF family member, and the KH domain protein MEX-3 promote intrinsic GSC proliferation directly (Ariz et al. 2009). Germ cells lacking PUF-8 and MEX-3, although capable of meiotic entry in the absence of GLP-1 activity, do not enter meiosis; interestingly, they do not proliferate either. These observations suggest that the mRNAs regulated by PUF-8 and MEX-3 in the distal germline are likely to control mitotic proliferation independent of their effect on differentiation. The identities of these mRNAs are currently unknown. Interestingly, PUF-8 also represses the proliferative fate in the transition zone, where it functions redundantly with the GTPase-activating protein GAP-3 to inhibit the LET-60 RAS activity. While PUF-8 represses the translation of *let-60* mRNA, GAP-3 inactivates LET-60 protein by promoting GTP hydrolysis. In the *puf-8*, *gap-3* double-mutant adult germline, the active form of LET-60 RAS, increases, due to the loss of both mRNA- and protein-level controls, and activates the downstream MPK-1 ERK signaling. Consequently, cells continue to proliferate without entering differentiation (Vaid et al. 2013). However, the reasons underlying the yin-yang function of PUF-8 in promoting both proliferation and differentiation are currently an active area of investigation. Perhaps PUF-8 accomplishes this feat by interacting with different factor(s) in the different regions of the germline, which probably allows it to switch targets, analogous to its ortholog in *Drosophila* (see below).

At least three pathways dependent on translational control act redundantly to promote meiotic entry in *C. elegans* (Fig. 6.3). (1) The GLD-1 pathway, containing of the gene products GLD-1 and Nanos-3 (NOS-3) (Hansen et al. 2004; Kadyk and Kimble 1998). Here, NOS-3 promotes GLD-1 expression by an unknown mechanism. GLD-1 is a translational repressor and represses the translation of several mRNAs including *cye-1*, *pal-1*, *mex-3*, *cep-1*, *tra-2*, and *rme-2* mRNAs; however,

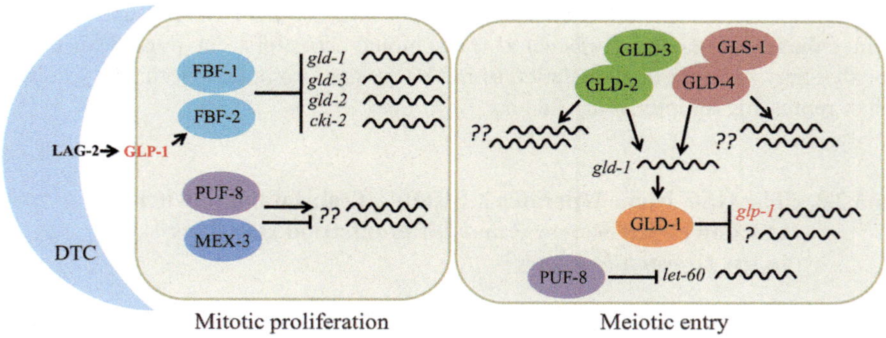

Fig. 6.3 Translational control of the mitosis–meiosis decision in *C. elegans*. The LAG-2 ligand produced by the somatic cell called the distal tip cell (DTC) activates the GLP-1 receptor present on germ cells. This results in the transcriptional activation of the RBP FBF-2, which along with FBF-1 inhibits meiotic entry by suppressing the translation of *gld-1*, *gld-3*, *gld-2*, and *cki-2* mRNAs. The RBPs PUF-8 and MEX-3 promote proliferation, possibly by regulating the translation of unknown mRNAs. Although a single proliferating cell is shown in this cartoon, the mitotic region extends to about 20-cell diameters from DTC. The entire proliferative zone comprises of a total of ~200 cells. The schematic on the right represents a cell from the transition zone. In the transition zone, GLP-1 activity and the levels of FBFs decrease, resulting in the expression of FBF targets such as GLD-1 and GLD-2. GLD-1 represses the translation of *glp-1* and unknown mRNAs to promote meiotic entry. GLD-2 and GLD-4 (PAPs) promote GLD-1 expression. In addition, these two PAPs promote meiotic entry independently of GLD-1 by regulating the translation of unknown mRNAs. Furthermore, PUF-8 facilitates meiotic entry by repressing the translation of *let-60*, which encodes RAS, a well-known proliferation-promoting factor, in this zone

the targets that may be directly relevant for meiotic entry have not been identified (Biedermann et al. 2009; Jan et al. 1999; Lee and Schedl 2001; Mootz et al. 2004; Schumacher et al. 2005). Since GLD-1 is as a translational repressor, the most consistent hypothesis is that GLD-1 probably promotes meiotic entry by repressing mRNA(s) that inhibit meiotic entry. (2) The GLD-1 pathway is composed of the polyA polymerase (PAP) GLD-2 and its RNA-binding partner GLD-3 function (Eckmann et al. 2004; Kadyk and Kimble 1998; Wang et al. 2002). Unlike the canonical PAPs, GLD-2 lacks RNA-binding domain. Instead, it depends on other RBPs such as GLD-3 and RNP-8 to polyadenylate its target mRNAs. GLD-2 can promote meiotic entry independently of GLD-1; however, it has also been known to partner with GLD-3 and promote *gld-1* mRNA translation, suggesting a complex interplay of these factors in maintaining the tight balance between mitosis and meiosis (Suh et al. 2006). In addition, another noncanonical PAP called GLD-4 and its RNA-binding partner GLS-1 also promote *gld-1* mRNA translation and regulate GLD-1 protein levels. Thus, GLD-2 and GLD-4 act redundantly to promote the translation of *gld-1* mRNA (Schmid et al. 2009). (3) The GLD-4–GLS-1 pathway promotes meiotic entry independent of GLD-1 and GLD-2, constituting the third meiotic entry pathway (Millonigg et al. 2014). Besides the *gld-1* mRNA, GLD-2 promotes the stability and translation of many other mRNAs, including *oma-2*, *egg-1*, *pup-2*, and *tra-2* mRNAs, in meiotic germ cells (Kim et al. 2010). However, the targets relevant to meiotic entry remain unknown for both GLD-2 and GLD-4. Since these cytoplasmic PAPs act as translational activators, one hypothesis could be that they promote the translation of mRNA(s) that promote meiotic entry, rather than repressing an interfering factor.

6.3.2.2 The GSC Niche Operates a Bistable Translational Switch to Control Self-Renewal and Differentiation Decisions in the *Drosophila* Ovary

Signaling from the GSC niche and an internal translational network is crucial to GSC maintenance in *Drosophila* ovary. Interestingly, different signaling cascades operate in the female and male germlines: BMP pathway in the ovary and JAK-STAT pathway in the testis [for a review see Lehmann (2012)]. Both promote the GSC fate by preventing precocious differentiation. In this chapter we will limit our discussion to the intrinsic factors acting downstream of the niche signaling in female GSCs as they are better understood. In the ovary, the BMP-type ligands Decapentaplegic (Dpp) and Glass bottom boat (Gbb) secreted by the niche repress the transcription of bag of marbles (bam), which is a differentiation-promoting factor, in GSCs (Chen and McKearin 2003; Song et al. 2004; Wharton et al. 1991; Xie and Spradling 1998). Separately, Nos and the PUF protein Pumilio (Pum) repress the translation of *brain tumor* (*brat*), *mei-P26*, and possibly a few other differentiation-promoting factors (Harris et al. 2011; Joly et al. 2013). The requirement of Nos and Pum is crucial for the maintenance of GSCs: in the absence of Nos or Pum, all GSCs differentiate, which depletes the ovary of GSCs (Forbes and Lehmann 1998; Gilboa and Lehmann 2004).

In the *Drosophila* ovary, GSCs divide such that one daughter cell (GSC) stays closer, and attached, to the niche, while the second daughter cell (cystoblast) forms away from the niche and receives less Dpp signal as a consequence, repression of *bam* is relieved in the cystoblast. Bam forms a complex with Sex-lethal, MeiP26, and BGCN and represses the translation of *nos* mRNA (Li et al. 2009, 2013). Reduction in Nos levels depresses *brat* translation, resulting in the accumulation of Brat in the cystoblast. Brat partners with Pum and represses the translation of Dpp signal transducers dMyc, Mad, Medea, and Schnurri, which reinforces the self-renewal-to-differentiation fate switch in the cystoblast (Harris et al. 2011; Newton et al. 2015). Thus, by changing its partner, from Nos in GSCs to Brat in cystoblasts, Pum is able to change its translational targets, from *mei-P26* and *brat* mRNAs in the GSC to *dMyc* and *mad* mRNAs in the cystoblast, and enable these two cell types to maintain their respective fates (Fig. 6.4). However, identities of the other translational targets of Nos-Pum and Brat-Pum, which are likely to be directly involved in the execution of self-renewal or differentiation programs, are currently unknown. In addition, miRNAs and the general transcriptional machinery have also been suggested to play a role in GSC maintenance in the *Drosophila* ovary [for a review, see Slaidina and Lehmann (2014)]. The function of Brat-Pum pair appears to be restricted to enacting the self-renewal-to-differentiation fate switch; the continued

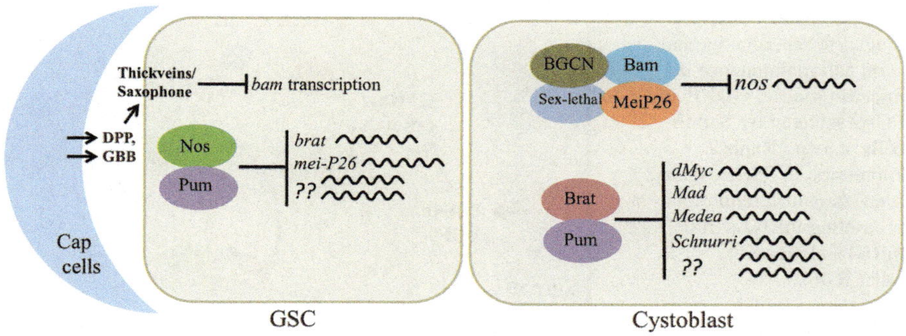

Fig. 6.4 Translational control of the self-renewal/differentiation decision in the *Drosophila* ovary. In the *Drosophila* ovary, signaling from the niche (cap cells) suppresses *bam* transcription in the GSC. Apart from this, Nos-Pum pair represses the translation of *brat*, *mei-P26*, and other unknown mRNAs to prevent premature differentiation. The GSC divides such that one daughter cell is oriented away from the niche and does not receive sufficient niche signals to suppress *bam*. Bam forms a multi-protein complex with Mei-P26, Sex-lethal, and BGCN and suppresses *nos* translation. Absence of Nos derepresses the *brat* mRNA leading to Brat expression, which partners with Pum and inhibits the translation of dMyc, Mad, Medea, and Schnurri to promote differentiation

presence of Pum is detrimental to the differentiation process, and its translation is repressed at later stages by Rbfox1 (Carreira-Rosario et al. 2016).

6.3.2.3 Assembly of Differentiation-Promoting mRNAs into Translationally Dormant mRNPs Maintains GSCs in the Mouse Testis

In mammalian females, germ cells enter meiosis soon after arriving at the gonad thus much of our current knowledge on mammalian GSC maintenance comes from studies on males. Spermatogonial stem cells (SSCs)—the GSCs in mouse testis—self-renew as well as differentiate amidst somatic cells called Sertoli cells. Fibroblast growth factor 9 (FGF9) and glial cell line-derived neurotrophic factor (GDNF) secreted by Sertoli cells are essential for the self-renewal of SSCs (Barrios et al. 2010; Bowles et al. 2010; Meng et al. 2000). Although these two signals induce the expression of a battery of transcription factors, one of the crucial downstream targets is Nanos2 (Fig. 6.5). Unlike in the *Drosophila* ovary, the mouse Nanos2 does not seem to associate with a PUF ortholog. Instead, Nanos2 interacts with another RBP, Dead end1 (DND1), to recruit differentiation-promoting mRNAs, such as Sohlh2, Dmrt1, Dazl, and Taf7l, into mRNP complexes, where it represses translation by interacting with the CCR4-NOT deadenylase complex and promoting the deadenylation of the target mRNAs (Suzuki et al. 2012, 2016; Zhou et al. 2015). Interestingly, not all mRNAs associated with the Nanos2 mRNPs are repressed at the posttranscriptional level. A few pre-mRNAs, such as Sohlh1 and Taf7l, increase in the absence of Nanos2, indicating a potential transcriptional control as well.

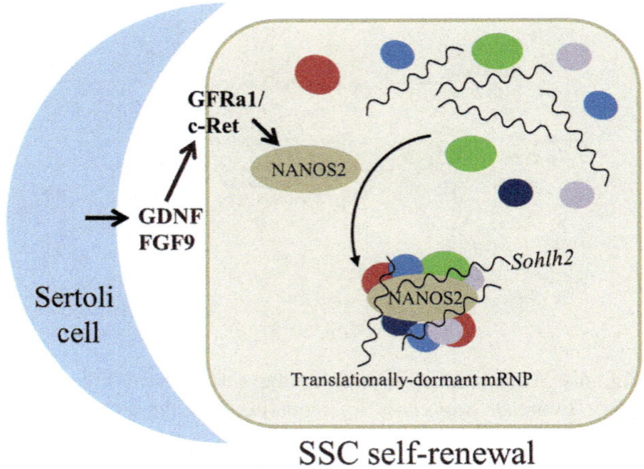

Fig. 6.5 Nanos mRNP is crucial for spermatogonial stem cell maintenance in mice. In mouse, GDNF and FGF9 secreted by Sertoli cells activate Nanos2 expression in SSCs. Nanos2 sequesters differentiation-promoting mRNAs into mRNPs along with other RBPs

Furthermore, mere recruitment by Nanos2 alone is not sufficient; proper assembly of the mRNP, which requires the core mRNP component Rck, is also essential for translational suppression (Zhou et al. 2015). Apart from mRNAs, Nanos2 sequesters the mTOR protein as well to mRNPs and inhibits mTORC1 signaling pathway, which is known to negatively regulate SSC self-renewal (Zhou et al. 2015). In summary, maintenance of differentiation-promoting mRNAs in a translationally dormant state by assembling them into mRNPs appears to be the principal mechanism by which the SSCs are maintained in mice.

6.3.3 Translational Control of Meiotic Progression and Oocyte Maturation

Germ cells that enter differentiation execute two different developmental programs simultaneously: one, progression through the meiotic divisions and the other, gametogenesis. Exquisite control of translation, with great spatial and temporal precision, governs meiotic progression and oocyte maturation. In fact, much of our current knowledge of translational control mechanisms comes from studies on oocyte maturation. In most organisms, germ cells progress through meiotic prophase I and arrest at some point—the exact stage varies among species—prior to metaphase I (M-I). During this arrest period, oocytes grow enormously in size accumulating mRNAs and proteins required for the rest of oogenesis and early embryogenesis and the various factors required for translation. However, only a small fraction of the stored mRNAs is translated; others are stored in a translationally dormant state waiting for appropriate stimuli. Upon induction by hormonal (*Xenopus* and mammals) or sperm-derived (*C. elegans*) signals, the oocyte completes meiosis I and arrests at metaphase II (M-II) till fertilization. The progression through M-I to M-II, termed as oocyte maturation, is regulated by the translational activation of some of the stored mRNAs. Premature translation of mRNAs is detrimental to the development of fertilization-competent oocyte; hence strict translational control is crucial during both meiotic progression and oocyte maturation. In organisms ranging from *C. elegans* to mammals, regulation of the polyA tail by cytoplasmic polyadenylation is the principal mechanism of translation regulation during oogenesis. Nevertheless, other mechanisms are likely employed as well. The role of translational control during meiotic progression has been more thoroughly studied and better understood in the case of oogenesis in *Xenopus* and *C. elegans*; therefore, we will restrict ourselves to these two examples here.

6.3.3.1 Multiple RNA-Binding Proteins Act to Maintain Meiotic Commitment in *C. elegans*

Intriguingly, germ cells maintain mitotic potential even after initiating the meiotic program. In female germ cells, GLD-1 is the key factor that prevents mitotic reentry (Francis et al. 1995). Mutations in other genes, such as *gld-2* and *gld-4*, also cause dedifferentiation (Kadyk and Kimble 1998; Schmid et al. 2009), but since these genes are known to promote *gld-1* mRNA translation, the actual cause is possibly the reduction in GLD-1 expression (Fig. 6.6). GLD-1 inhibits the reentry, at least in

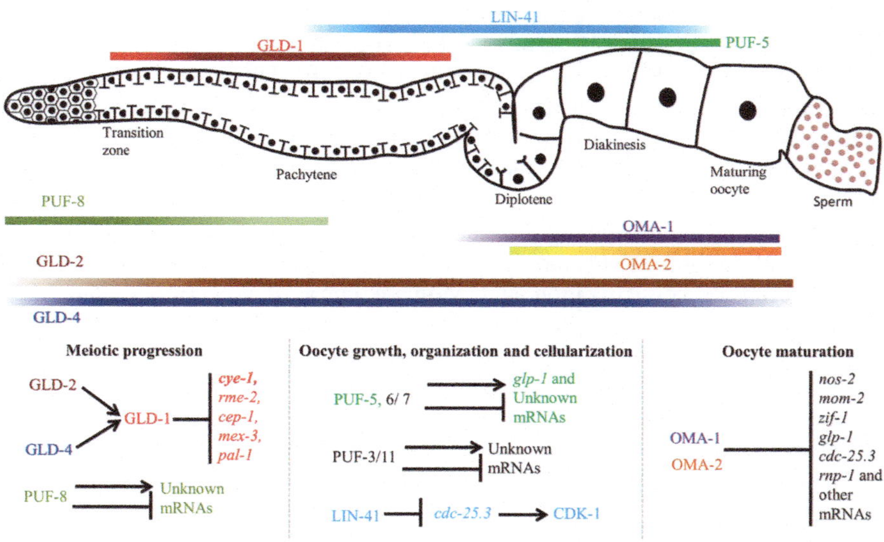

Fig. 6.6 Translational regulation during meiotic progression, oocyte growth, and maturation in *C. elegans*. A cartoon of the adult hermaphrodite gonad is shown at the *top*. Orientation: *distal to the left and proximal to the right*. *C. elegans* gonad with different regions marked. *Horizontal bars* indicate expression patterns of the RBPs that regulate meiotic progression and/or oocyte maturation. Intensity variations of the color reflect the concentrations of the corresponding proteins. The RBPs and their corresponding target mRNAs in the different stages of meiotic development are shown at the bottom. See text for a more detailed description

part, by repressing the translation of *cye-1* mRNA: while germ cells return to the mitotic mode after meiotic entry when GLD-1 is absent, they do not reenter mitosis if both GLD-1 and CYE-1 are absent (Biedermann et al. 2009). The translation of *cye-1* mRNA is inhibited by at least one other factor, because the depletion of CYE-1 increases the number of meiotic cells in *gld-1 gld-2* double-mutant germlines (Fox et al. 2011). Besides *cye-1*, GLD-1 represses a number of mRNAs whose protein products function at a later stage either during oocyte growth (RME-2) or embryonic patterning (MEX-3 and PAL-1). Premature translation of these mRNAs severely affects meiotic progression; for instance, when both GLD-1 and MEX-3 are absent, meiotic cells transdifferentiate into somatic lineages due to the ectopic translation of mRNAs encoding transcription factors such as PAL-1 (Ciosk et al. 2006). GLD-1 functions in spermatocytes as well, where it acts redundantly with PUF-8 to arrest dedifferentiation, but the downstream targets in this case are not known (Priti and Subramaniam 2015).

6.3.3.2 mRNPs Control the Translation of Maternal mRNAs During Oocyte Maturation in *C. elegans*

Meiosis in developing *C. elegans* oocytes arrests at diakinesis of meiosis I. Oocytes grow rapidly in size by taking up cytoplasmic contents from the syncytial germline before closing off the cytoplasmic bridge. During this phase, while GLD-1 expression ceases, the expression of other RBPs such as LIN-41, MEX-3, OMA-1, OMA-2, PUF-5, and SPN-4 begins (Detwiler et al. 2001; Draper et al. 1996; Lublin and Evans 2007; Ogura et al. 2003; Spike et al. 2014a). These proteins together form mRNPs and potentially control the translation of a large number of stored mRNAs (Spike et al. 2014b). Specifically, LIN-41 controls oocyte growth by inhibiting premature cellularization and M-phase entry; it accomplishes this, at least in part, by repressing the translational of the CDK-1 activator CDC-25.3 (Spike et al. 2014a, b). In contrast, the nearly identical paralogs OMA-1 and OMA-2 promote cellularization and M-phase entry in the arrested diakinetic oocytes. In the *oma-1 oma-2* double mutants, nuclear events such as the chromatin localization of the Aurora B kinase AIR-2 and nuclear envelope breakdown fail to occur, and the cytoplasmic streaming from the syncytial germline, which is required for oocyte growth, continues for longer than the wild type resulting in the formation of large oocytes (Detwiler et al. 2001; Govindan et al. 2009; Harris et al. 2006). The OMA proteins repress the translation of several mRNAs, including *nos-2*, *glp-1*, *mom-2*, *zif-1*, *cdc-25.3*, *rnf-5*, and *rnp-1* mRNAs, in the oocyte (Guven-Ozkan et al. 2010; Jadhav et al. 2008; Kaymak and Ryder 2013; Oldenbroek et al. 2013; Spike et al. 2014b). How the OMA-mediated translational repression of these mRNAs contributes to regulating oocyte growth and maturation is yet to be explored. The PUF family proteins also contribute to oocyte development. PUF-5, PUF-6, and PUF-7 function redundantly to promote oocyte formation. Another set of PUF proteins, PUF-3 and PUF-11, limit oocyte growth. Although these PUF proteins control the translation of several mRNAs, the ones involved in oocyte formation and growth are unknown (Hubstenberger et al. 2012; Lublin and Evans 2007).

6.3.3.3 Precise Temporal Sequence of Translational Activation Directs Oogenesis in *Xenopus*

Since *Xenopus* produces large number of relatively big oocytes and oogenesis proceeds in the absence of transcription, the *Xenopus* oocyte has been an excellent model to investigate the mechanisms of translational control. During the arrest at diplotene, oocytes produce and accumulate large amounts of mRNAs and other translational accessories. Most of these mRNAs are stored in a translationally silent state until the oocyte resumes meiosis. Translational quiescence is maintained by the binding of an RBP called the cytoplasmic polyadenylation element binding protein (CPEB), which binds to a specific sequence element in the 3′ UTR termed the cytoplasmic polyadenylation element (CPE) (Lin et al. 2010). CPEB interacts with other proteins such as Maskin, which interacts with the initiation factor eIF4E and prevents translational initiation (Barnard et al. 2005). In addition, CPEB interacts with the PAP Gld2, the *Xenopus* ortholog of GLD-2 discussed above, and the polyA ribonuclease PARN. While Gld2 extends the polyA tail, PARN promptly shortens it through its function as a deadenylase (Barnard et al. 2004; Copeland and Wormington 2001; Kim and Richter 2006; Korner et al. 1998).

When the oocyte resumes meiosis, translation of the stored mRNAs is activated in a precise temporal sequence, which is crucial for the meiotic progression and oocyte maturation (Fig. 6.7). Based on the timing of activation, the mRNAs are classified as "early," which are translated prior to the nuclear envelope breakdown [or the germinal vesicle breakdown (GBVD)] in *Xenopus*, and "late," which are translated after GBVD. Unlike most mRNAs that are controlled by CPEB, translation of Ringo mRNA, a cyclin B-type protein, is under the control of the PUF protein Pum2 (Cao et al. 2010; Padmanabhan and Richter 2006). Upon stimulation by progesterone secreted by the follicle cells, Pum2 dissociates from the Ringo mRNA, resulting in its translation. Ringo then activates Cdc2/Cdk1, which phosphorylate several proteins including CPEB and Musashi, an RBP that binds to another 3′ UTR element called MPE (Fig. 6.7b) (Arumugam et al. 2012b; Kim and Richter 2007; Mendez et al. 2002). Activated by phosphorylation, Musashi recruits Gld2, which extends the polyA tail of the "early" mRNAs, such as the ones encoding Mos and cyclin B5, leading to their translational activation (Cragle and MacNicol 2014). Expression of Mos and cyclin B5 is thus crucial for the translational activation of the "late" mRNAs. Mos activates the mitogen-activated protein kinase (MAPK) pathway, which phosphorylates CPEB at sites distinct from the ones phosphorylated by Ringo-Cdc2 (Keady et al. 2007; Posada et al. 1993; Sagata et al. 1988). Thus, CPEB is phosphorylated by both the MAPK pathway and Ringo/Cdc2. Depending on the CPEB-containing repression complex assembled on a particular mRNA, any of several possible mechanisms respond to the CPEB phosphorylation and activate translation [for a review see Charlesworth et al.

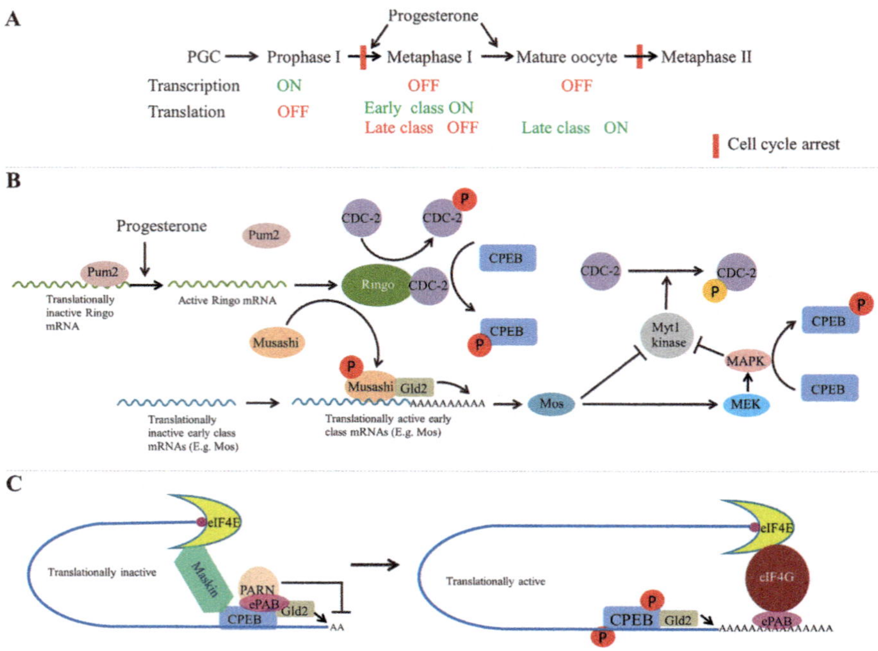

Fig. 6.7 Sequential activation of polyadenylation during oocyte maturation in *Xenopus*. (**a**) Outline of the transcriptional and translational status at the key stages of oocyte maturation. (**b**) Summary of the signaling cascade activated by progesterone during oocyte maturation. Progesterone stimulation releases Ringo mRNA from Pum2-mediated repression. Ringo associates with Cdc-2, phosphorylates CPEB, Cdc-2, and Musashi. Phosphorylated Musashi recruits the PAP Gld2 and activates translation of early class mRNAs. One of them is Mos mRNA. Mos activates the MAPK pathway by phosphorylating MEK. MAPK in turn phosphorylates CPEB at a site distinct from the one phosphorylated by Ringo/cyclin B-Cdc-2. (**c**) Phosphorylation of CPEB activates translation of masked mRNAs. Translationally inactive (masked) mRNAs are bound by CPEB, ePAB, PARN, Gld-2, and Maskin. Maskin binds to CPEB and eIF4E, blocking association of eIF4E with eIF4G. Polyadenylation–deadenylation cycles by GLD2 and PARN keep the polyA tail short. Phosphorylation of CPEB expels PARN, leading to polyA tail extension. Another factor released from the complex is ePAB, which now binds to the polyA tail and associates with eIF4E by competing out Maskin and recruits eIF4G leading to translation activation

(2013)]. For instance, in immature stage VI oocytes, PARN is expelled from the CPEB complex in response to CPEB phosphorylation; in the absence of a competing deadenylase activity, the polyadenylation activity of Gld2 results in polyA tail elongation (Barnard et al. 2004; Kim and Richter 2006). In addition, phosphorylation of CPEB by Cdc2 leads to the dissociation of embryonic polyA binding protein (ePAB) from the CPEB-containing mRNP complex, which competes out Maskin for binding to eIF4E allowing the recruitment of eIF4G to the initiation complex, which enables translational initiation by promoting the binding of the 40S ribosomal subunit to the 5′ cap (Fig. 6.7c) (Barnard et al. 2005; Cao et al. 2006; Cao and Richter 2002; Kim and Richter 2007).

An important aspect of the activation of early mRNAs is the operation of at least two positive feedback loops. In one, the initial association of Ringo-Cdc2 sets up a positive feedback loop leading to rapid increase in the levels of active cyclin B-Cdc2 (MPF). In this loop, Ringo-Cdc2 activates the Cdc25 phosphatase and inactivates the Myt1 kinase (Karaiskou et al. 1999). Since Cdc25 activates Cdc2 by removing a phosphate added by the Myt1 kinase, this leads to an increase in the active form of Cdc2. The second positive feedback loop consists of two components: phosphorylated Musashi activates the translation of its own mRNA; in addition, the MAPK pathway, activated by Musashi via Mos, also phosphorylates Musashi (Arumugam et al. 2012a, b).

Phosphorylation and activation of Musashi are thus essential for oocyte maturation, while activation of either the cyclin B-Cdc2 or the MAPK pathway is sufficient for this process (Haccard and Jessus 2006). This leads to the interesting conundrum: how is it possible for the upstream activator (Musashi) of pathway 1 (Mos-MAPK) to be essential for a given process, while that pathway itself is redundant with a second pathway (Cdc2), which in turn activates the upstream activator (Musashi) of the first pathway.

6.4 Concluding Remarks

Translational control clearly plays a predominant role in germ cell development, all the way from specification to gametogenesis, in diverse organisms. Functions of some of the factors involved, notably the Nanos and PUF family proteins, are well conserved pointing to the ancient origin of the central role for translational regulation in the germline. Recent work has identified several key factors that regulate translation in the germline. Some new mechanisms have emerged as well. For instance, the formation of mRNPs as a way to store mRNAs in translationally inactive state has emerged as a common theme in several species. These findings have opened up several new questions for the immediate future. The mRNA targets directly responsible for some of the key processes are not known for many of the newly identified RBPs. Although we did not discuss the role of small RNA molecules extensively in this chapter, their role in the germline is just emerging. The mechanistic details of how mRNPs assemble and what influences their dynamics are other central questions that need further investigation.

Acknowledgments Work in the KS laboratory is supported by grants from the Department of Science and Technology and the Department of Biotechnology, Government of India.

References

Ariz M, Mainpal R, Subramaniam K (2009) *C. elegans* RNA-binding proteins PUF-8 and MEX-3 function redundantly to promote germline stem cell mitosis. Dev Biol 326:295–304

Arumugam K, Macnicol MC, Macnicol AM (2012a) Autoregulation of Musashi1 mRNA translation during Xenopus oocyte maturation. Mol Reprod Dev 79:553–563

Arumugam K, MacNicol MC, Wang Y, Cragle CE, Tackett AJ, Hardy LL, MacNicol AM (2012b) Ringo/cyclin-dependent kinase and mitogen-activated protein kinase signaling pathways regulate the activity of the cell fate determinant Musashi to promote cell cycle re-entry in Xenopus oocytes. J Biol Chem 287:10639–10649

Austin J, Kimble J (1987) glp-1 is required in the germ line for regulation of the decision between mitosis and meiosis in *C. elegans*. Cell 51:589–599

Barnard DC, Ryan K, Manley JL, Richter JD (2004) Symplekin and xGLD-2 are required for CPEB-mediated cytoplasmic polyadenylation. Cell 119:641–651

Barnard DC, Cao Q, Richter JD (2005) Differential phosphorylation controls Maskin association with eukaryotic translation initiation factor 4E and localization on the mitotic apparatus. Mol Cell Biol 25:7605–7615

Barrios F, Filipponi D, Pellegrini M, Paronetto MP, Di Siena S, Geremia R, Rossi P, De Felici M, Jannini EA, Dolci S (2010) Opposing effects of retinoic acid and FGF9 on Nanos2 expression and meiotic entry of mouse germ cells. J Cell Sci 123:871–880

Batchelder C, Dunn MA, Choy B, Suh Y, Cassie C, Shim EY, Shin TH, Mello C, Seydoux G, Blackwell TK (1999) Transcriptional repression by the *Caenorhabditis elegans* germ-line protein PIE-1. Genes Dev 13:202–212

Biedermann B, Wright J, Senften M, Kalchhauser I, Sarathy G, Lee MH, Ciosk R (2009) Translational repression of cyclin E prevents precocious mitosis and embryonic gene activation during *C. elegans* meiosis. Dev Cell 17:355–364

Bontems F, Stein A, Marlow F, Lyautey J, Gupta T, Mullins MC, Dosch R (2009) Bucky ball organizes germ plasm assembly in zebrafish. Curr Biol 19:414–422

Bowerman B, Draper BW, Mello CC, Priess JR (1993) The maternal gene skn-1 encodes a protein that is distributed unequally in early *C. elegans* embryos. Cell 74:443–452

Bowles J, Feng CW, Spiller C, Davidson TL, Jackson A, Koopman P (2010) FGF9 suppresses meiosis and promotes male germ cell fate in mice. Dev Cell 19:440–449

Cao Q, Richter JD (2002) Dissolution of the maskin-eIF4E complex by cytoplasmic polyadenylation and poly(A)-binding protein controls cyclin B1 mRNA translation and oocyte maturation. EMBO J 21:3852–3862

Cao Q, Kim JH, Richter JD (2006) CDK1 and calcineurin regulate Maskin association with eIF4E and translational control of cell cycle progression. Nat Struct Mol Biol 13:1128–1134

Cao Q, Padmanabhan K, Richter JD (2010) Pumilio 2 controls translation by competing with eIF4E for 7-methyl guanosine cap recognition. RNA 16:221–227

Carreira-Rosario A, Bhargava V, Hillebrand J, Kollipara RK, Ramaswami M, Buszczak M (2016) Repression of Pumilio protein expression by Rbfox1 promotes germ cell differentiation. Dev Cell 36:562–571

Castagnetti S, Ephrussi A (2003) Orb and a long poly(A) tail are required for efficient oskar translation at the posterior pole of the Drosophila oocyte. Development 130:835–843

Chang JS, Tan L, Schedl P (1999) The Drosophila CPEB homolog, orb, is required for oskar protein expression in oocytes. Dev Biol 215:91–106

Charlesworth A, Meijer HA, de Moor CH (2013) Specificity factors in cytoplasmic polyadenylation. Wiley Interdiscip Rev RNA 4:437–461

Chekulaeva M, Hentze MW, Ephrussi A (2006) Bruno acts as a dual repressor of oskar translation, promoting mRNA oligomerization and formation of silencing particles. Cell 124:521–533

Chen D, McKearin D (2003) Dpp signaling silences bam transcription directly to establish asymmetric divisions of germline stem cells. Curr Biol 13:1786–1791

Chicoine J, Benoit P, Gamberi C, Paliouras M, Simonelig M, Lasko P (2007) Bicaudal-C recruits CCR4-NOT deadenylase to target mRNAs and regulates oogenesis, cytoskeletal organization, and its own expression. Dev Cell 13:691–704

Ciosk R, DePalma M, Priess JR (2006) Translational regulators maintain totipotency in the *Caenorhabditis elegans* germline. Science 311:851–853

Copeland PR, Wormington M (2001) The mechanism and regulation of deadenylation: identification and characterization of Xenopus PARN. RNA 7:875–886

Cragle C, MacNicol AM (2014) Musashi protein-directed translational activation of target mRNAs is mediated by the poly(A) polymerase, germ line development defective-2. J Biol Chem 289:14239–14251

Crist CG, Montarras D, Buckingham M (2012) Muscle satellite cells are primed for myogenesis but maintain quiescence with sequestration of Myf5 mRNA targeted by microRNA-31 in mRNP granules. Cell Stem Cell 11:118–126

Crittenden SL, Troemel ER, Evans TC, Kimble J (1994) GLP-1 is localized to the mitotic region of the *C. elegans* germ line. Development 120:2901–2911

Crittenden SL, Bernstein DS, Bachorik JL, Thompson BE, Gallegos M, Petcherski AG, Moulder G, Barstead R, Wickens M, Kimble J (2002) A conserved RNA-binding protein controls germline stem cells in *Caenorhabditis elegans*. Nature 417:660–663

Dahanukar A, Walker JA, Wharton RP (1999) Smaug, a novel RNA-binding protein that operates a translational switch in Drosophila. Mol Cell 4:209–218

Detwiler MR, Reuben M, Li X, Rogers E, Lin R (2001) Two zinc finger proteins, OMA-1 and OMA-2, are redundantly required for oocyte maturation in *C. elegans*. Dev Cell 1:187–199

Draper BW, Mello CC, Bowerman B, Hardin J, Priess JR (1996) MEX-3 is a KH domain protein that regulates blastomere identity in early *C. elegans* embryos. Cell 87:205–216

Eckmann CR, Kraemer B, Wickens M, Kimble J (2002) GLD-3, a bicaudal-C homolog that inhibits FBF to control germline sex determination in *C. elegans*. Dev Cell 3:697–710

Eckmann CR, Crittenden SL, Suh N, Kimble J (2004) GLD-3 and control of the mitosis/meiosis decision in the germline of *Caenorhabditis elegans*. Genetics 168:147–160

Extavour CG, Akam M (2003) Mechanisms of germ cell specification across the metazoans: epigenesis and preformation. Development 130:5869–5884

Forbes A, Lehmann R (1998) Nanos and Pumilio have critical roles in the development and function of Drosophila germline stem cells. Development 125:679–690

Forrest KM, Gavis ER (2003) Live imaging of endogenous RNA reveals a diffusion and entrapment mechanism for nanos mRNA localization in Drosophila. Curr Biol 13:1159–1168

Fox PM, Vought VE, Hanazawa M, Lee MH, Maine EM, Schedl T (2011) Cyclin E and CDK-2 regulate proliferative cell fate and cell cycle progression in the *C. elegans* germline. Development 138:2223–2234

Francis R, Barton MK, Kimble J, Schedl T (1995) gld-1, a tumor suppressor gene required for oocyte development in *Caenorhabditis elegans*. Genetics 139:579–606

Frise E, Hammonds AS, Celniker SE (2010) Systematic image-driven analysis of the spatial Drosophila embryonic expression landscape. Mol Syst Biol 6:345

Gallo CM, Wang JT, Motegi F, Seydoux G (2010) Cytoplasmic partitioning of P granule components is not required to specify the germline in *C. elegans*. Science 330:1685–1689

Gilboa L, Lehmann R (2004) How different is Venus from Mars? The genetics of germ-line stem cells in Drosophila females and males. Development 131:4895–4905

Govindan JA, Nadarajan S, Kim S, Starich TA, Greenstein D (2009) Somatic cAMP signaling regulates MSP-dependent oocyte growth and meiotic maturation in *C. elegans*. Development 136:2211–2221

Gunkel N, Yano T, Markussen FH, Olsen LC, Ephrussi A (1998) Localization-dependent translation requires a functional interaction between the 5′ and 3′ ends of oskar mRNA. Genes Dev 12:1652–1664

Guven-Ozkan T, Robertson SM, Nishi Y, Lin R (2010) zif-1 translational repression defines a second, mutually exclusive OMA function in germline transcriptional repression. Development 137:3373–3382

Haccard O, Jessus C (2006) Redundant pathways for Cdc2 activation in Xenopus oocyte: either cyclin B or Mos synthesis. EMBO Rep 7:321–325

Hansen D, Wilson-Berry L, Dang T, Schedl T (2004) Control of the proliferation versus meiotic development decision in the *C. elegans* germline through regulation of GLD-1 protein accumulation. Development 131:93–104

Hanyu-Nakamura K, Sonobe-Nojima H, Tanigawa A, Lasko P, Nakamura A (2008) Drosophila Pgc protein inhibits P-TEFb recruitment to chromatin in primordial germ cells. Nature 451:730–733

Harris AN, Macdonald PM (2001) Aubergine encodes a Drosophila polar granule component required for pole cell formation and related to eIF2C. Development 128:2823–2832

Harris JE, Govindan JA, Yamamoto I, Schwartz J, Kaverina I, Greenstein D (2006) Major sperm protein signaling promotes oocyte microtubule reorganization prior to fertilization in *Caenorhabditis elegans*. Dev Biol 299:105–121

Harris RE, Pargett M, Sutcliffe C, Umulis D, Ashe HL (2011) Brat promotes stem cell differentiation via control of a bistable switch that restricts BMP signaling. Dev Cell 20:72–83

Henderson ST, Gao D, Lambie EJ, Kimble J (1994) lag-2 may encode a signaling ligand for the GLP-1 and LIN-12 receptors of *C. elegans*. Development 120:2913–2924

Hubstenberger A, Cameron C, Shtofman R, Gutman S, Evans TC (2012) A network of PUF proteins and Ras signaling promote mRNA repression and oogenesis in *C. elegans*. Dev Biol 366:218–231

Hunter CP, Kenyon C (1996) Spatial and temporal controls target pal-1 blastomere-specification activity to a single blastomere lineage in *C. elegans* embryos. Cell 87:217–226

Jadhav S, Rana M, Subramaniam K (2008) Multiple maternal proteins coordinate to restrict the translation of *C. elegans* nanos-2 to primordial germ cells. Development 135:1803–1812

Jambor H, Mueller S, Bullock SL, Ephrussi A (2014) A stem-loop structure directs oskar mRNA to microtubule minus ends. RNA 20:429–439

Jan E, Motzny CK, Graves LE, Goodwin EB (1999) The STAR protein, GLD-1, is a translational regulator of sexual identity in *Caenorhabditis elegans*. EMBO J 18:258–269

Jeong J, Verheyden JM, Kimble J (2011) Cyclin E and Cdk2 control GLD-1, the mitosis/meiosis decision, and germline stem cells in *Caenorhabditis elegans*. PLoS Genet 7:e1001348

Joly W, Chartier A, Rojas-Rios P, Busseau I, Simonelig M (2013) The CCR4 deadenylase acts with Nanos and Pumilio in the fine-tuning of Mei-P26 expression to promote germline stem cell self-renewal. Stem Cell Rep 1:411–424

Kadyk LC, Kimble J (1998) Genetic regulation of entry into meiosis in *Caenorhabditis elegans*. Development 125:1803–1813

Kadyrova LY, Habara Y, Lee TH, Wharton RP (2007) Translational control of maternal Cyclin B mRNA by Nanos in the Drosophila germline. Development 134:1519–1527

Kalchhauser I, Farley BM, Pauli S, Ryder SP, Ciosk R (2011) FBF represses the Cip/Kip cell-cycle inhibitor CKI-2 to promote self-renewal of germline stem cells in *C. elegans*. EMBO J 30:3823–3829

Kalifa Y, Huang T, Rosen LN, Chatterjee S, Gavis ER (2006) Glorund, a Drosophila hnRNP F/H homolog, is an ovarian repressor of nanos translation. Dev Cell 10:291–301

Karaiskou A, Jessus C, Brassac T, Ozon R (1999) Phosphatase 2A and polo kinase, two antagonistic regulators of cdc25 activation and MPF auto-amplification. J Cell Sci 112 (Pt 21):3747–3756

Kaymak E, Ryder SP (2013) RNA recognition by the *Caenorhabditis elegans* oocyte maturation determinant OMA-1. J Biol Chem 288:30463–30472

Keady BT, Kuo P, Martinez SE, Yuan L, Hake LE (2007) MAPK interacts with XGef and is required for CPEB activation during meiosis in Xenopus oocytes. J Cell Sci 120:1093–1103

Kedde M, Strasser MJ, Boldajipour B, Oude Vrielink JA, Slanchev K, le Sage C, Nagel R, Voorhoeve PM, van Duijse J, Orom UA et al (2007) RNA-binding protein Dnd1 inhibits microRNA access to target mRNA. Cell 131:1273–1286

Kim JH, Richter JD (2006) Opposing polymerase-deadenylase activities regulate cytoplasmic polyadenylation. Mol Cell 24:173–183

Kim JH, Richter JD (2007) RINGO/cdk1 and CPEB mediate poly(A) tail stabilization and translational regulation by ePAB. Genes Dev 21:2571–2579

Kim KW, Wilson TL, Kimble J (2010) GLD-2/RNP-8 cytoplasmic poly(A) polymerase is a broad-spectrum regulator of the oogenesis program. Proc Natl Acad Sci USA 107:17445–17450

Kimble J (2011) Molecular regulation of the mitosis/meiosis decision in multicellular organisms. Cold Spring Harb Perspect Biol 3:a002683

Kim-Ha J, Smith JL, Macdonald PM (1991) oskar mRNA is localized to the posterior pole of the Drosophila oocyte. Cell 66:23–35

Koprunner M, Thisse C, Thisse B, Raz E (2001) A zebrafish nanos-related gene is essential for the development of primordial germ cells. Genes Dev 15:2877–2885

Korner CG, Wormington M, Muckenthaler M, Schneider S, Dehlin E, Wahle E (1998) The deadenylating nuclease (DAN) is involved in poly(A) tail removal during the meiotic maturation of Xenopus oocytes. EMBO J 17:5427–5437

Kugler JM, Lasko P (2009) Localization, anchoring and translational control of oskar, gurken, bicoid and nanos mRNA during Drosophila oogenesis. Fly (Austin) 3:15–28

Lamont LB, Crittenden SL, Bernstein D, Wickens M, Kimble J (2004) FBF-1 and FBF-2 regulate the size of the mitotic region in the *C. elegans* germline. Dev Cell 7:697–707

Leatherman JL, Levin L, Boero J, Jongens TA (2002) Germ cell-less acts to repress transcription during the establishment of the Drosophila germ cell lineage. Curr Biol 12:1681–1685

Lee MH, Schedl T (2001) Identification of in vivo mRNA targets of GLD-1, a maxi-KH motif containing protein required for *C. elegans* germ cell development. Genes Dev 15:2408–2420

Lehmann R (2012) Germline stem cells: origin and destiny. Cell Stem Cell 10:729–739

Li Y, Minor NT, Park JK, McKearin DM, Maines JZ (2009) Bam and Bgcn antagonize Nanos-dependent germ-line stem cell maintenance. Proc Natl Acad Sci USA 106:9304–9309

Li Y, Zhang Q, Carreira-Rosario A, Maines JZ, McKearin DM, Buszczak M (2013) Mei-p26 cooperates with Bam, Bgcn and Sxl to promote early germline development in the Drosophila ovary. PLoS One 8:e58301

Lin CL, Evans V, Shen S, Xing Y, Richter JD (2010) The nuclear experience of CPEB: implications for RNA processing and translational control. RNA 16:338–348

Lublin AL, Evans TC (2007) The RNA-binding proteins PUF-5, PUF-6, and PUF-7 reveal multiple systems for maternal mRNA regulation during *C. elegans* oogenesis. Dev Biol 303:635–649

Maduro MF, Meneghini MD, Bowerman B, Broitman-Maduro G, Rothman JH (2001) Restriction of mesendoderm to a single blastomere by the combined action of SKN-1 and a GSK-3beta homolog is mediated by MED-1 and -2 in *C. elegans*. Mol Cell 7:475–485

Mahowald AP (2001) Assembly of the Drosophila germ plasm. Int Rev Cytol 203:187–213

Marin VA, Evans TC (2003) Translational repression of a *C. elegans* Notch mRNA by the STAR/KH domain protein GLD-1. Development 130:2623–2632

Markussen FH, Michon AM, Breitwieser W, Ephrussi A (1995) Translational control of oskar generates short OSK, the isoform that induces pole plasma assembly. Development 121:3723–3732

Markussen FH, Breitwieser W, Ephrussi A (1997) Efficient translation and phosphorylation of Oskar require Oskar protein and the RNA helicase Vasa. Cold Spring Harb Symp Quant Biol 62:13–17

McLaren A (1999) Signaling for germ cells. Genes Dev 13:373–376

Mello CC, Draper BW, Krause M, Weintraub H, Priess JR (1992) The pie-1 and mex-1 genes and maternal control of blastomere identity in early *C. elegans* embryos. Cell 70:163–176

Mello CC, Schubert C, Draper B, Zhang W, Lobel R, Priess JR (1996) The PIE-1 protein and germline specification in *C. elegans* embryos. Nature 382:710–712

Mendez R, Barnard D, Richter JD (2002) Differential mRNA translation and meiotic progression require Cdc2-mediated CPEB destruction. EMBO J 21:1833–1844

Meng X, Lindahl M, Hyvonen ME, Parvinen M, de Rooij DG, Hess MW, Raatikainen-Ahokas A, Sainio K, Rauvala H, Lakso M et al (2000) Regulation of cell fate decision of undifferentiated spermatogonia by GDNF. Science 287:1489–1493

Merritt C, Seydoux G (2010) The Puf RNA-binding proteins FBF-1 and FBF-2 inhibit the expression of synaptonemal complex proteins in germline stem cells. Development 137:1787–1798

Merritt C, Rasoloson D, Ko D, Seydoux G (2008) 3′ UTRs are the primary regulators of gene expression in the *C. elegans* germline. Curr Biol 18:1476–1482

Micklem DR, Adams J, Grunert S, St Johnston D (2000) Distinct roles of two conserved Staufen domains in oskar mRNA localization and translation. EMBO J 19:1366–1377

Millonigg S, Minasaki R, Nousch M, Eckmann CR (2014) GLD-4-mediated translational activation regulates the size of the proliferative germ cell pool in the adult *C. elegans* germ line. PLoS Genet 10:e1004647

Mishima Y, Giraldez AJ, Takeda Y, Fujiwara T, Sakamoto H, Schier AF, Inoue K (2006) Differential regulation of germline mRNAs in soma and germ cells by zebrafish miR-430. Curr Biol 16:2135–2142

Mootz D, Ho DM, Hunter CP (2004) The STAR/Maxi-KH domain protein GLD-1 mediates a developmental switch in the translational control of *C. elegans* PAL-1. Development 131:3263–3272

Nakamura A, Sato K, Hanyu-Nakamura K (2004) Drosophila cup is an eIF4E binding protein that associates with Bruno and regulates oskar mRNA translation in oogenesis. Dev Cell 6:69–78

Nelson MR, Leidal AM, Smibert CA (2004) Drosophila Cup is an eIF4E-binding protein that functions in Smaug-mediated translational repression. EMBO J 23:150–159

Newton FG, Harris RE, Sutcliffe C, Ashe HL (2015) Coordinate post-transcriptional repression of Dpp-dependent transcription factors attenuates signal range during development. Development 142:3362–3373

Nie K, Gomez M, Landgraf P, Garcia JF, Liu Y, Tan LH, Chadburn A, Tuschl T, Knowles DM, Tam W (2008) MicroRNA-mediated down-regulation of PRDM1/Blimp-1 in Hodgkin/Reed-Sternberg cells: a potential pathogenetic lesion in Hodgkin lymphomas. Am J Pathol 173:242–252

Ogura K, Kishimoto N, Mitani S, Gengyo-Ando K, Kohara Y (2003) Translational control of maternal glp-1 mRNA by POS-1 and its interacting protein SPN-4 in *Caenorhabditis elegans*. Development 130:2495–2503

Oldenbroek M, Robertson SM, Guven-Ozkan T, Spike C, Greenstein D, Lin R (2013) Regulation of maternal Wnt mRNA translation in *C. elegans* embryos. Development 140:4614–4623

Olivieri G, Olivieri A (1965) Autoradiographic study of nucleic acid synthesis during spermatogenesis in *Drosophila melanogaster*. Mutat Res 2:366–380

Padmanabhan K, Richter JD (2006) Regulated Pumilio-2 binding controls RINGO/Spy mRNA translation and CPEB activation. Genes Dev 20:199–209

Pagano JM, Farley BM, Essien KI, Ryder SP (2009) RNA recognition by the embryonic cell fate determinant and germline totipotency factor MEX-3. Proc Natl Acad Sci USA 106:20252–20257

Posada J, Yew N, Ahn NG, Vande Woude GF, Cooper JA (1993) Mos stimulates MAP kinase in Xenopus oocytes and activates a MAP kinase kinase in vitro. Mol Cell Biol 13:2546–2553

Priti A, Subramaniam K (2015) PUF-8 functions redundantly with GLD-1 to promote the meiotic progression of spermatocytes in *Caenorhabditis elegans*. G3 (Bethesda)

Rongo C, Gavis ER, Lehmann R (1995) Localization of oskar RNA regulates oskar translation and requires Oskar protein. Development 121:2737–2746

Saffman EE, Styhler S, Rother K, Li W, Richard S, Lasko P (1998) Premature translation of oskar in oocytes lacking the RNA-binding protein bicaudal-C. Mol Cell Biol 18:4855–4862

Sagata N, Oskarsson M, Copeland T, Brumbaugh J, Vande Woude GF (1988) Function of c-mos proto-oncogene product in meiotic maturation in Xenopus oocytes. Nature 335:519–525

Sassone-Corsi P (2002) Unique chromatin remodeling and transcriptional regulation in spermatogenesis. Science 296:2176–2178

Schafer M, Nayernia K, Engel W, Schafer U (1995) Translational control in spermatogenesis. Dev Biol 172:344–352

Schmid M, Kuchler B, Eckmann CR (2009) Two conserved regulatory cytoplasmic poly (A) polymerases, GLD-4 and GLD-2, regulate meiotic progression in *C. elegans*. Genes Dev 23:824–836

Schumacher B, Hanazawa M, Lee MH, Nayak S, Volkmann K, Hofmann ER, Hengartner M, Schedl T, Gartner A (2005) Translational repression of *C. elegans* p53 by GLD-1 regulates DNA damage-induced apoptosis. Cell 120:357–368

Semotok JL, Cooperstock RL, Pinder BD, Vari HK, Lipshitz HD, Smibert CA (2005) Smaug recruits the CCR4/POP2/NOT deadenylase complex to trigger maternal transcript localization in the early Drosophila embryo. Curr Biol 15:284–294

Seydoux G, Dunn MA (1997) Transcriptionally repressed germ cells lack a subpopulation of phosphorylated RNA polymerase II in early embryos of *Caenorhabditis elegans* and *Drosophila melanogaster*. Development 124:2191–2201

Seydoux G, Mello CC, Pettitt J, Wood WB, Priess JR, Fire A (1996) Repression of gene expression in the embryonic germ lineage of *C. elegans*. Nature 382:713–716

Slaidina M, Lehmann R (2014) Translational control in germline stem cell development. J Cell Biol 207:13–21

Smibert CA, Wilson JE, Kerr K, Macdonald PM (1996) smaug protein represses translation of unlocalized nanos mRNA in the Drosophila embryo. Genes Dev 10:2600–2609

Song X, Wong MD, Kawase E, Xi R, Ding BC, McCarthy JJ, Xie T (2004) Bmp signals from niche cells directly repress transcription of a differentiation-promoting gene, bag of marbles, in germline stem cells in the Drosophila ovary. Development 131:1353–1364

Spike CA, Coetzee D, Eichten C, Wang X, Hansen D, Greenstein D (2014a) The TRIM-NHL protein LIN-41 and the OMA RNA-binding proteins antagonistically control the prophase-to-metaphase transition and growth of *Caenorhabditis elegans* oocytes. Genetics 198:1535–1558

Spike CA, Coetzee D, Nishi Y, Guven-Ozkan T, Oldenbroek M, Yamamoto I, Lin R, Greenstein D (2014b) Translational control of the oogenic program by components of OMA ribonucleoprotein particles in *Caenorhabditis elegans*. Genetics 198:1513–1533

Strome S, Wood WB (1982) Immunofluorescence visualization of germ-line-specific cytoplasmic granules in embryos, larvae, and adults of *Caenorhabditis elegans*. Proc Natl Acad Sci USA 79:1558–1562

Subramaniam K, Seydoux G (1999) *nos-1* and *nos-2*, two genes related to *Drosophila nanos*, regulate primordial germ cell development and survival in *Caenorhabditis elegans*. Development 126:4861–4871

Suh N, Jedamzik B, Eckmann CR, Wickens M, Kimble J (2006) The GLD-2 poly(A) polymerase activates gld-1 mRNA in the *Caenorhabditis elegans* germ line. Proc Natl Acad Sci USA 103:15108–15112

Sulston JE, Schierenberg E, White JG, Thomson JN (1983) The embryonic cell lineage of the nematode *Caenorhabditis elegans*. Dev Biol 100:64–119

Suzuki A, Saba R, Miyoshi K, Morita Y, Saga Y (2012) Interaction between NANOS2 and the CCR4-NOT deadenylation complex is essential for male germ cell development in mouse. PLoS One 7:e33558

Suzuki A, Niimi Y, Shinmyozu K, Zhou Z, Kiso M, Saga Y (2016) Dead end1 is an essential partner of NANOS2 for selective binding of target RNAs in male germ cell development. EMBO Rep 17:37–46

Tabara H, Hill RJ, Mello CC, Priess JR, Kohara Y (1999) pos-1 encodes a cytoplasmic zinc-finger protein essential for germline specification in *C. elegans*. Development 126:1–11

Tenenhaus C, Subramaniam K, Dunn MA, Seydoux G (2001) PIE-1 is a bifunctional protein that regulates maternal and zygotic gene expression in the embryonic germ line of *Caenorhabditis elegans*. Genes Dev 15:1031–1040

Trcek T, Grosch M, York A, Shroff H, Lionnet T, Lehmann R (2015) Drosophila germ granules are structured and contain homotypic mRNA clusters. Nat Commun 6:7962

Vaid S, Ariz M, Chaturbedi A, Kumar GA, Subramaniam K (2013) PUF-8 negatively regulates RAS/MAPK signalling to promote differentiation of *C. elegans* germ cells. Development 140:1645–1654

Vourekas A, Alexiou P, Vrettos N, Maragkakis M, Mourelatos Z (2016) Sequence-dependent but not sequence-specific piRNA adhesion traps mRNAs to the germ plasm. Nature 531:390–394

Walker AK, Boag PR, Blackwell TK (2007) Transcription reactivation steps stimulated by oocyte maturation in *C. elegans*. Dev Biol 304:382–393

Wang L, Eckmann CR, Kadyk LC, Wickens M, Kimble J (2002) A regulatory cytoplasmic poly (A) polymerase in *Caenorhabditis elegans*. Nature 419:312–316

Weidinger G, Stebler J, Slanchev K, Dumstrei K, Wise C, Lovell-Badge R, Thisse C, Thisse B, Raz E (2003) dead end, a novel vertebrate germ plasm component, is required for zebrafish primordial germ cell migration and survival. Curr Biol 13:1429–1434

West JA, Viswanathan SR, Yabuuchi A, Cunniff K, Takeuchi A, Park IH, Sero JE, Zhu H, Perez-Atayde A, Frazier AL et al (2009) A role for Lin28 in primordial germ-cell development and germ-cell malignancy. Nature 460:909–913

Wharton KA, Thomsen GH, Gelbart WM (1991) Drosophila 60A gene, another transforming growth factor beta family member, is closely related to human bone morphogenetic proteins. Proc Natl Acad Sci USA 88:9214–9218

Wilson JE, Connell JE, Macdonald PM (1996) aubergine enhances oskar translation in the Drosophila ovary. Development 122:1631–1639

Xie T, Spradling AC (1998) decapentaplegic is essential for the maintenance and division of germline stem cells in the Drosophila ovary. Cell 94:251–260

Yochem J, Greenwald I (1989) glp-1 and lin-12, genes implicated in distinct cell-cell interactions in *C. elegans*, encode similar transmembrane proteins. Cell 58:553–563

Youngren KK, Coveney D, Peng X, Bhattacharya C, Schmidt LS, Nickerson ML, Lamb BT, Deng JM, Behringer RR, Capel B et al (2005) The Ter mutation in the dead end gene causes germ cell loss and testicular germ cell tumours. Nature 435:360–364

Zaessinger S, Busseau I, Simonelig M (2006) Oskar allows nanos mRNA translation in Drosophila embryos by preventing its deadenylation by Smaug/CCR4. Development 133:4573–4583

Zhang F, Barboric M, Blackwell TK, Peterlin BM (2003) A model of repression: CTD analogs and PIE-1 inhibit transcriptional elongation by P-TEFb. Genes Dev 17:748–758

Zhou Z, Shirakawa T, Ohbo K, Sada A, Wu Q, Hasegawa K, Saba R, Saga Y (2015) RNA binding protein Nanos2 organizes post-transcriptional Buffering system to retain primitive state of mouse spermatogonial stem cells. Dev Cell 34:96–107

Chapter 7
Role of Cdc6 During Oogenesis and Early Embryo Development in Mouse and *Xenopus laevis*

Ewa Borsuk, Joanna Jachowicz, Malgorzata Kloc, Jean-Pierre Tassan, and Jacek Z. Kubiak

Abstract Cdc6 is an important player in cell cycle regulation. It is involved in the regulation of both S-phase and M-phase. Its role during oogenesis is crucial for repression of the S-phase between the first and the second meiotic M-phases, and it also regulates, via CDK1 inhibition, the M-phase entry and exit. This is of special importance for the reactivation of the major M-phase-regulating kinase CDK1 (Cyclin-Dependent Kinase 1) in oocytes entering metaphase II of meiosis and in embryo cleavage divisions, in which precise timing allows coordination between cell cycle events and developmental program of the embryo. In this chapter, we discuss the role of Cdc6 protein in oocytes and early embryos.

E. Borsuk • J. Jachowicz
Faculty of Biology, Institute of Zoology, Department of Embryology, University of Warsaw, Warsaw, Poland

M. Kloc
The Methodist Hospital Research Institute, 6670 Bertner Ave, Houston, TX 77030, USA

J.-P. Tassan
CNRS UMR6092, Cell Cycle Team, Rennes, France

Université Rennes 1, Institut de Génétique et Développement de Rennes (IGDR), Rennes, France

J.Z. Kubiak (✉)
CNRS UMR6092, Cell Cycle Team, Rennes, France

Université Rennes 1, Institut de Génétique et Développement de Rennes (IGDR), Rennes, France

Department of Regenerative Medicine, Military Institute of Hygiene and Epidemiology (WIHE), Warsaw, Poland
e-mail: jacek.kubiak@univ-rennes1.fr

S. Arur (ed.), *Signaling-Mediated Control of Cell Division*, Results and Problems in Cell Differentiation 59, DOI 10.1007/978-3-319-44820-6_7

7.1 Introduction

During oogenesis, the last round of DNA replication takes place just before entry into meiosis. The next S-phase will occur after fertilization in the pronuclei of the one-cell embryo. Proper control of the origins of replication (ORI) and assuring that oocytes are unable to restart replication are of the key importance for the control of oogenesis and the formation of female gametes with the right ploidy. The initiation of replication is possible only after chromatin licensing, i.e., after assembly of the pre-replication complexes (pre-RCs) on the origins of replication. The pre-RC forms by the sequential loading of ORI with the following proteins: ORC (Origin Recognition Complex), Cdc6 (Cell Division Cycle 6), and Cdt1 (Cycle-dependent transcript 1). The presence of Cdc6 is essential for loading of the last component of pre-RC, the MCM (Minichromosome Maintenance Complex), which has the helicase activity. Correct formation of pre-RCs is indispensable for S-phase initiation, and the absence of any component on the ORI prevents DNA replication. The maintenance of the chromatin in replication repressive state, after termination of DNA synthesis, is of equal importance in mitotic cell cycles and between the first and second meiotic division. This is achieved by gradual detachment of some of the replication-related proteins from the chromatin and maintaining them in chromatin unbound, phosphorylated state. Besides its role in controlling initiation of DNA synthesis, Cdc6 was recently also shown to be involved in M-phase regulation both during meiosis (oocyte maturation) and during the first embryonic mitosis in *Xenopus* and in the mouse. Therefore, this protein plays an essential role in cell cycle regulation in oocytes and embryos.

7.2 Homologs and Isoforms of Cdc6 in Eukaryotes

Cdc6 protein is ubiquitous in all eukaryotes and has a high similarity even in evolutionarily distant organisms. A gene similar to eukaryotic Cdc6 is also present in *Archeae*. The function of Cdc6 was first defined in yeasts *Saccharomyces cerevisiae* (Cdc6p) and *Schizosaccharomyces pombe* (*Cdc18* gene) (reviewed by Borlado and Mendez 2008). The proteins similar in structure and function to yeast Cdc6 were found in *Xenopus laevis* and mice (Berger et al. 1999) as well as in humans (Williams et al. 1997). Sequence analysis of Cdc6 in the NCBI GenBank database shows that Cdc6 exhibits very high sequence similarity between various organisms. For example, Cdc6 of *Mus musculus* is 60 % identical to *Xenopus laevis* and 40 % to *Drosophila melanogaster* protein. This similarity is even higher between human and mouse, where the homology is 80 %. The number of amino acids is also comparable, ranging from about 500 to 600. In the majority of organisms, only one isoform of Cdc6 is present; in *Xenopus* it consists of 554 amino acids and in *Drosophila* it has 662 amino acids. The only report demonstrating the existence of two isoforms of Cdc6 present simultaneously comes from the studies on mice and murine cell lines. Two isoforms of murine Cdc6 are different variants of transcripts resulting from alternative mRNA splicing. Variant 1 of Cdc6 gene transcript (# NM 011799.2 GenBank) was identified for the first time in the sequences of mouse ES cells originating from cDNA library in BAC (Bacteria Artificial Chromosome) (Berger et al. 1999). The variant 2 transcript (# NM 001025779.1 GenBank) was discovered by MGC Program (*Mammalian Gene Collection Program*) and identified for the first time in the brain tissue of mouse embryo at 12.5 E (Strausberg et al. 2002). Variant 1 encodes isoform a and variant 2 isoform b of Cdc6 protein. Both isoforms have high homology because the longer one (isoform a) includes the shorter one (isoform b). The function of each of the two isoforms is unknown. It is also not obvious why mouse is so far the only organism with two isoforms of Cdc6.

7.3 Motifs and Domains in Cdc6

All Cdc6 isoforms originating from different organisms have similar motifs and domains. The main domain of Cdc6 protein is localized in its C-terminus. Its presence is typical for replication-related proteins. The main function of this domain is binding to DNA and organization of three-dimensional (tertiary) structure of Cdc6 (Marchler-Bauer et al. 2011). The presence of AAA+ domain in Cdc6 locates this protein within the family of ATPase proteins with the chaperon activity. This activity allows Cdc6 to influence the formation of protein complexes (Borlado and Mendez 2008). Domain AAA+ is a part of PTZ000112 multidomain whose function is linked to the interactions with ORC proteins in the pre-replication complex (Marchler-Bauer et al. 2011). The N-terminal region of Cdc6 protein contains the binding sites for Cyclin-Dependent Kinases (CDKs) and the amino acids preferentially phosphorylated by CDKs (Mimura et al. 2004; Borlado and Mendez 2008).

7.4 The Presence and Function of Cdc6 Protein During Meiosis

7.4.1 Role of Cdc6 in Preventing S-Phase Initiation During Meiosis

During oogenesis, the meiotic division is highly extended in time, and its progression is physiologically blocked twice: for the first time at the diplotene of prophase of the first division and for the second time at the metaphase of second division (MII). Last round of DNA replication before meiosis takes place in oogonia. The next one will occur after fertilization in one-cell embryo. During the period of time separating these two rounds of replication, the oocyte grows in size, acquires competence to meiotic maturation, and represses DNA synthesis. It is known that in proliferating somatic cells the repression of DNA synthesis (which takes place in G2) occurs mainly through the maintenance of the pre-RC proteins in phosphorylated state (Kelly and Brown 2000). It is possible that similar mechanism controlling the repression of the S-phase between the first and second meiotic division also operates during oogenesis. Research performed on *Xenopus* oocytes revealed that the Cdc6 protein plays a key role in this mechanism. In fully grown *Xenopus* oocytes, blocked in the 1st prophase, the repression of replication is most probably controlled at the level of pre-RC formation. Because the proteins of ORC complex, such as Orc1, Orc2, Cdt1, and MCM, are present in the primary oocytes (Whitmire et al. 2002), Cdc6 was predicted to be a key to controlling the formation of pre-RCs. Indeed, the incubation of sperm chromatin in total, immature oocyte extract revealed that the complete pre-RCs assembly occurred only when the extract was supplemented with recombinant *Xenopus* Cdc6 protein. This proved that in immature oocytes the Cdc6 is the only missing pre-RCs component. Low levels of this protein are detected for the first time at GVBD (Whitmire et al. 2002) or soon after (Lemaître et al. 2002), and it persists during meiosis I. A significant increase in its expression occurs at meiosis II, At the same time, the level of other components of pre-RCs remains constant (Whitmire et al. 2002). Northern blot analysis showed that *cdc6* mRNA is stored as a maternal RNA in the primary oocytes and that accumulation of Cdc6 protein can be blocked by injection of antisense *cdc6* oligonucleotides into GV oocytes (Lemaître et al. 2002). The Cdc6 synthesis is regulated at the translational level by polyadenylation of maternal mRNA. A shift in electrophoretic mobility of *cdc6* mRNA observed soon after GVBD is consistent with its polyadenylation occurring during maturation. The repression of Cdc6 synthesis during prophase arrest and its reinitiation after GVBD was further confirmed by analysis of the behavior of exogenous vs. endogenous Cdc6 (Daldello et al. 2015). GV-intact oocytes were injected with mRNA encoding *Xenopus* histidine- or GST-tagged wild-type Cdc6 (his-WT-Cdc6 or GST-WT-Cdc6) and stimulated with progesterone to initiate maturation. Both types of proteins were translated at prophase I, but they were unstable and proteolytically degraded shortly before GVBD. The degradation mechanism relies on a proteolytic system dependent on Skp1-Cullins-F-box complex, in which Cdc4 is the F-box protein (SCF^{Cdc4}). This mechanism is controlled by initial activation of CDK1 provoking GVBD and the meiotic M-phase entry. In contrast, the absence of endogenous Cdc6 during prophase arrest does not depend on degradation but on the repression of its synthesis. Western blot analysis confirmed the lack of Cdc6 protein at various stages of oogenesis (from stage I to full-grown stage VI), and inhibition of CDK1 activation or inhibition of the proteasome proteolytic activity is not sufficient to promote Cdc6 stabilization. The endogenous Cdc6 protein is detected from 30 min after GVBD and is present at MII (Daldello et al. 2015).

These results confirm the role of Cdc6 as a key protein in maintaining the primary oocytes in replication repressive state. On the other hand, the timing of Cdc6 appearance and its slow accumulation in maturing oocytes suggests that the competence to pre-RC formation and DNA synthesis is achieved much earlier than in one-cell embryo. Although the DNA replication between meiosis I and meiosis II is repressed, oocytes that were artificially forced, by inhibition of protein synthesis with cycloheximide (CHX) treatment for 30–60 min after GVBD, to transit to interphase will initiate the S-phase (Furuno et al. 1994). Depletion of *cdc6* mRNA, by injection of antisense oligonucleotides before initiation of maturation, prevents this unscheduled replication. This effect can be rescued by injection of recombinant Cdc6 protein soon after GVBD (Lemaître et al. 2002). Treatment of oocytes with CHX at the moment of GVBD also results in their transition to interphase; however, in this case, the DNA synthesis does not occur (Furuno et al. 1994). This result mirrors the Cdc6 expression pattern. Addition of CHX at GVBD inhibits the first wave of Cdc6 synthesis (Lemaître et al. 2002; Whitmire et al. 2002), which prevents the formation of pre-RCs. This was further proven by rescuing DNA replication in maturing oocytes (treated with CHX at GVBD) by injection of Cdc6 protein. This experimental protocol induced DNA replication in a Cdc6 concentration-dependent manner. Moreover, replication induced by Cdc6 injection was aphidicoline sensitive suggesting that DNA replication rather than DNA repair was taking place (Whitmire et al. 2002). All these data indicate the existence of a mechanism controlling DNA replication competence in which Cdc6 plays a key role. According to this mechanism, the lack of Cdc6 protein in primary oocytes provides a way to prevent pre-RCs formation and initiation of S-phase during long period of oocyte growth. Cdc6 synthesis is initiated soon after GVBD and the protein accumulates gradually, leading to the increased competence for DNA replication during maturation. Similar mechanism, with Cdc6 protein as a key regulator, was described also in mouse oocytes and other Eukaryotes (*Schizosaccharomyces pombe* and *Drosophila melanogaster*) (Lemaître et al. 2004) suggesting that this mechanism is evolutionarily conserved from unicellular eukaryotes up to mammals.

The acquisition of a competence to replicate at the end of meiosis I was unexpected because the lack of DNA synthesis between the first and second meiotic division is crucial for the production of the haploid gametes. This is achieved by a precise regulation of Cdc6 turnover in maturing oocytes. As described for *Xenopus* oocytes, two mechanisms are involved in this regulation. The one is responsible for Cdc6 stabilization and the other for its degradation (Daldello et al. 2015). The stabilizing mechanism is set up when CDK1 activity is established in MI. The analysis of interaction of Cdc6 (both exogenous and endogenous) with CDK1–Cyclin B complexes suggests that the presence of active complexes and nondegradable Cyclin B accounts for Cdc6 stabilization in MI and during entry to MII (Daldello et al. 2015). However, during the first hour after GVBD, the accumulation of Cdc6 is slow due to the antagonistic effects of Cyclin B (which acts as a positive regulator) and Mos-MAPK pathway (which exerts negative effect on Cdc6 accumulation). Inhibition of Cyclin B by injection of antisense oligonucleotides into prophase-arrested oocytes, which were stimulated to initiate maturation by progesterone treatment, resulted in much delayed accumulation of Cdc6. In contrast, inhibition of Mos synthesis and, in consequence, the lack of MAPK phosphorylation/activation lead to earlier accumulation of Cdc6 (Daldello et al. 2015). These two antagonistic activities control the Cdc6 level. At the end of meiosis I, the levels of Cdc6 are not sufficient to inhibit CDK1 activation, although they are sufficient to promote pre-RC formation and DNA synthesis in oocytes experimentally forced to enter interphase. When the oocyte proceeds through meiosis and arrests at metaphase II the Cdc6 is stabilized by its interaction with newly synthesized Cyclin B. The negative effect of Mos-MAPK is bypassed and Cdc6 is able to accumulate. The competence to replicate DNA is already established in MII but is repressed by stable and highly active CDK1. Such a model, proposed by Daldello et al. (2015), puts Cdc6 in a position of a crucial player in the control of correct progression of events during subsequent meiotic divisions.

The fully grown GV oocytes of *Xenopus* and mouse lack Cdc6, which explains their incompetence for DNA replication. However, the nuclei of growing mouse oocytes are able to synthesize DNA when they are fused with parthenogenetic one-cell embryos or with single blastomeres from two-cell embryos derived from in vivo fertilized oocytes (Czołowska and Borsuk 2000). Immunofluorescent analysis of pre-RC proteins revealed that Cdc6 protein in its chromatin bound, insoluble form is present in the nuclei of growing mouse oocytes (Borsuk and Czolowska 2010; Fig. 7.1). Some other components of pre-RC proteins such as Mcm2, -6, and -7 (Swiech et al. 2007) and Orc2 (Czajkowska, personal information) were also detected. All these proteins, except Cdc6, undergo pronounced rearrangement in hybrid cells, which is probably the prerequisite for the formation of functional pre-RCs and initiation of replication (Borsuk and Czolowska 2010). However, in physiological conditions, the S-phase is efficiently repressed during the oocyte growth period, despite the presence of above-mentioned proteins. It is likely that (a) the accumulation of these proteins in insoluble granules, (b) the presence of

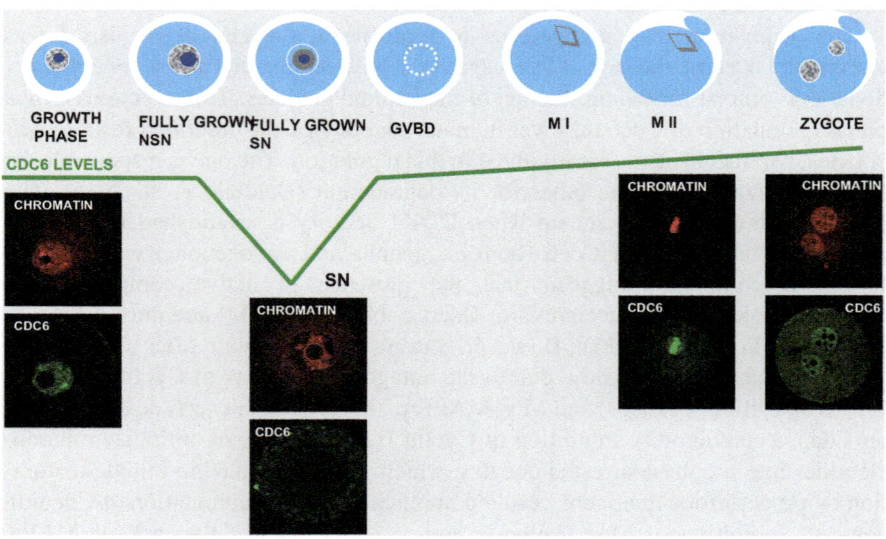

Fig. 7.1 Changes in Cdc6 levels and localization of its chromatin-bound form during oogenesis and at G1-phase of the first cell cycle of mouse embryo

Cdc6 in regions of constitutive heterochromatin, and (c) the lack of insoluble form of some other proteins (such as Mcm6) prevent the formation of pre-RCs (Swiech et al. 2007; Borsuk and Czolowska 2010). The rearrangement of chromatin from NSN (non-surrounded nucleolus) to SN (surrounded nucleolus) in fully grown oocytes (Debey et al. 1993; Zuccotti et al. 1995) is accompanied by detachment of replication-related proteins from chromatin (Fig. 7.1). After leaving the nucleus, they remain in oocyte cytoplasm in soluble form until G1 of the first cell cycle when they are transported into the pronuclei and attach to the chromatin (Swiech et al. 2007; Fig. 7.1).

7.4.2 Role of Cdc6 in the Progression of Meiotic Division

The early expression of Cdc6 during oocyte maturation, at the time when its function as a replication factor is not necessary, suggests that Cdc6 may play an additional role in meiosis, such as involvement in chromosome segregation or spindle assembly and function. Several pre-RC proteins exhibit such functions that are unrelated to DNA replication. For instance, Orc1, Orc2, and Orc6 have been shown to be involved in centrosome duplication and chromosome segregation in mitotic cells (Hemerly et al. 2009; Prasanth et al. 2004; Prasanth et al. 2002). The immunofluorescent staining of Cdc6 protein in maturing *Xenopus* oocytes revealed its concentration around barrel-shaped microtubule arrays, which surround condensed chromosomes 15 min after GVBD (Narasimhachar et al. 2012). Along with the formation of metaphase I spindle, the Cdc6 moves towards the spindle poles, forming, at first, two cup-like structures and eventually concentrating at the spindle poles. The animal pole of the spindle, which is attached to the animal cortex of the oocyte, shows higher intensity of Cdc6 staining than the vegetal pole-oriented one. This may suggest a potential involvement of Cdc6 protein in the attachment of spindle to the oocyte cortex. It has been shown that during telophase I/cytokinesis I, the fraction of Cdc6 was extruded in the first polar body along with the group of chromosomes and one spindle pole. During the second meiotic division, the distribution of Cdc6 within the meiotic spindle was similar to its distribution during meiosis I (Narasimhachar et al. 2012). Altogether, these data suggested that Cdc6 protein is also involved in meiotic spindle assembly and its attachment to the animal pole of the oocyte. To verify this hypothesis, both mRNA and Cdc6 protein were depleted from *Xenopus* oocytes by injection of antisense oligonucleotides and an antibody against Cdc6 prior the initiation of maturation. Depending on the concentration of antisense oligonucleotides, the formation of abnormal monopolar or multipolar meiotic spindles was observed. Even when Cdc6 protein was not completely depleted, as judged by the presence of weak protein staining at the spindle poles, the spindle assembly and its attachment to the cortex were defective. Moreover, the extrusion of the first polar body was never observed. Further increase of injected antisense oligonucleotides concentration resulted in complete depletion of Cdc6 and the lack of any microtubules associated with condensed chromosomes (Narasimhachar et al. 2012). Altogether, these data support the idea that in maturing *Xenopus* oocytes, the Cdc6 is indispensable for meiotic spindle assembly and its attachment to the cortex and therefore for chromosome segregation. Although it is not known how the Cdc6 functions in this capacity, the possible mechanism may be similar to the one involved in the formation of MTOC-TMA (microtubule-organizing center-transient microtubule array), the acentriolar centrosome that gathers and transports condensed chromosomes to the animal cortex. The localization of Cdc6 is reminiscent of localization of NuMa, one of the centrosomal proteins that accumulates at spindle poles both during meiosis and mitosis (Sun and Schatten 2006) and whose presence is critical for the establishment of functional spindle poles. It seems probable that Cdc6 may interact with NuMa and participate in this process.

Surprisingly, in MII mouse oocytes, the Cdc6 is present on chromosomes instead of the spindle poles (Anger et al. 2005; Fig. 7.1), with higher concentration at the kinetochores (our unpublished data). Depletion of the Cdc6 mRNA from GV-intact oocytes by injection of specific dsRNA resulted in the lack of maturation-associated increase in Cdc6 protein in the oocytes, which were allowed to resume meiosis. Although these oocytes underwent GVBD, they failed to reach metaphase I. The morphological analysis revealed the lack of proper meiotic spindle or any microtubules connected with chromosomes. The chromosomes were condensed, but they formed a hollow sphere instead of distinguished bivalents organized in metaphase plate. Because the MPF and MAPK were normally and fully activated in oocytes injected with Cdc6 dsRNA, it is obvious that the failure to form bivalents and spindle was a direct effect of lack of Cdc6.

Despite the differences in the localization of Cdc6 between maturing oocytes of *Xenopus* and mouse, the role of this protein in organization of metaphase plate and chromosome segregation is unquestionable. The molecular mechanism underlying this function remains unknown and needs to be further investigated.

7.5 Inhibition of CDK1 by Cdc6 and Its Role in Regulation of Timing of Mitotic Divisions in Embryos

It has been shown in HeLa cells that Cdc6 is actively involved in inhibition of the major M-phase kinase CDK1 upon mitotic exit (Yim and Erikson 2010). The involvement of Cdc6 in CDK1 inhibition before and during mitotic M-phase was postulated by several studies (Calzada et al. 2001; Clay-Farrace et al. 2003; Greenwood et al. 1998). Using cell-free extract of *Xenopus laevis* embryos obtained just before the first embryonic mitosis, we have shown that Cdc6 indeed inhibits CDK1 and determines both the timing of M-phase entry and the level of CDK1 activation during mitosis (El Dika et al. 2014). Recombinant Cdc6 added to the extract before the entry into M-phase delayed or abolished CDK1 activation in a dose-dependent manner. Alternatively, the reduction in Cdc6 level via (a) degradation of Cdc6 through addition of norcantharidin (Li et al. 2006), (b) addition of blocking antibodies against *Xenopus* Cdc6 thus neutralizing its function, or (c) immunodepletion using the same antibody results in accelerated M-phase

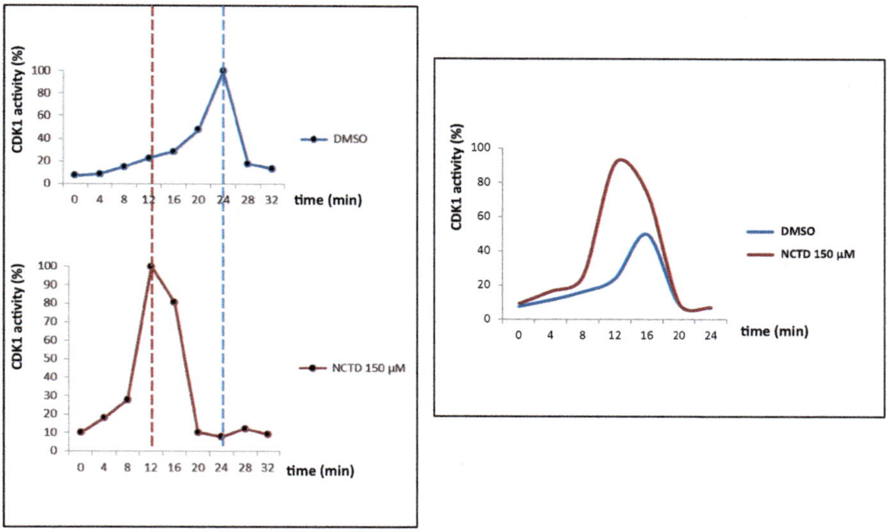

Fig. 7.2 Profiles of CDK1 activity during M-phase in *Xenopus laevis* embryos cell-free extract upon control conditions (DMSO) and when Cdc6 levels are drastically lowered by addition of 150 μM norcantharidin (NCTD). On the *left*: two profiles aligned in time; on the *right*: two profiles superimposed to show modification of the dynamics and amplitude of CDK1 activation

entry and an increase in the level of CDK1 above normal physiological levels. Also the profile of CDK1 activation was altered: the slow phase of activation of CDK1 was highly reduced or abolished. In consequence, CDK1 activation proceeded immediately from the minimum to the non-physiologically high maximum (see Fig. 7.2).

Importantly, the addition of recombinant Cdc6 completely rescued the phenotype: the timing of M-phase entry came back to normal, CDK1 activation changed profile, the slow phase of activation was reestablished, and the final maximal level was also rescued. These results show that Cdc6 behaves like CDK1 inhibitor, and the experimental modifications of the timing and the mode of this kinase activation were solely due to the modifications of Cdc6 levels before and during the M-phase.

To verify whether Cdc6 plays the same role in intact embryos, the timing of M-phase entry in mouse zygotes was studied (El Dika et al. 2014). Norcantharidin treatment as well as anti-MmCdc6 injection clearly accelerated mitotic initiation showing that interfering with Cdc6 level or activity speeds up mitosis. This shows that Cdc6 plays its mitotic role by controlling CDK1 activity and the dynamics of its activation not only in *Xenopus* embryo cell-free extract but also in intact mammalian cells.

Most probably, Cdc6 does not inhibit CDK1 directly because it has very different structure from all known *bona fide* CDK inhibitor proteins. It is therefore plausible that Cdc6 allows association of CDK1 with its inhibitor protein. It has been shown that during S-phase, the Cdc6 has an opposite activity to another CDK enzyme, the CDK2 (Kan et al. 2008; Uranbileg et al. 2012). In this case, Cdc6 promotes separation of p21WAF1/CIP1 and p27KIP1 inhibitors from CDK2 allowing its activation. It is therefore tempting to speculate that Cdc6 may act as an intermediary between CDK enzymes and their inhibitors and activates some of them (e.g., CDK2) and inactivates others (CDK1). Because the CDK2-activating process is regulated by phosphorylation (ibid.), one can speculate that during M-phase the specific phosphorylation events may regulate the opposite action of Cdc6, i.e., association of p21WAF1/CIP1 or p27KIP1 with CDK1. If this hypothesis is true, then Cdc6 could act as a platform regulating association of CDK enzymes with their inhibitors.

7.6 Conclusions

Recent studies have shown that Cdc6 plays essential roles in oocytes and embryos. By controlling both the S-phase and M-phase, the Cdc6 is an important regulator of the cell cycle during oogenesis and early embryo development. Specific modifications of Cdc6 metabolism/accumulation regulate the ability of oocytes to replicate DNA during meiotic divisions. Cdc6 is also indispensable for DNA replication in embryos and somatic cells. The involvement of Cdc6 in M-phase regulation in zygotes suggests that it also controls the timing of embryonic mitoses.

References

Anger M, Stein P, Schultz RM (2005) CDC6 requirement for spindle formation during maturation of mouse oocytes. Biol Reprod 72:188–194

Berger C, Strub A, Staib C, Lepke M, Zisimopoulou P, Hoehn K, Nanda I, Schmid M, Grummt F (1999) Identification and characterization of a mouse homolog to yeast Cdc6p. Cytogenet Cell Genet 86:307–316

Borlado LR, Mendez J (2008) Cdc6: From DNA replication to cell cycle checkpoints and oncogenesis. Carcinogenesis 29:237–243

Borsuk E, Czolowska R (2010) Factors engaged in reactivation of DNA replication in the nuclei of growing mouse oocytes introduced into the cytoplasm of parthenogenetic one-cell embryos. Int J Dev Biol 54:21–31

Calzada A, Sacristan M, Sanchez E, Bueno A (2001) Cdc6 cooperates with Sic1 and Hct1 to inactivate mitotic cyclin-dependent kinases. Nature 412:355–358

Clay-Farrace L, Pelizon C, Santamaria D, Pines J, Laskey RA (2003) Human replication protein Cdc6 prevents mitosis through a checkpoint mechanism that implicates Chk1. EMBO J 22:704–712

Czołowska R, Borsuk E (2000) Induction of DNA replication in the germinal vesicle of the growing mouse oocyte. Dev Biol 223:205–215

Daldello EM, Le T, Poulhe R, Jessus C, Haccard O, Dupré A (2015) Control of Cdc6 accumulation by Cdk1 and MAPK is essential for completion of oocyte meiotic divisions i *Xenopus*. J Cell Sci 128:2482–2496

Debey P, Szöllösi MS, Szöllösi D, Vautier D, Girousse A, Besombes D (1993) Competent mouse oocytes isolated from antral follicles exhibit different chromatin organization and follow different maturation dynamics. Mol Reprod Dev 36:59–74

El Dika M, Laskowska-Kaszub K, Koryto M, Dudka D, Prigent C, Tassan JP, Kloc M, Polanski Z, Borsuk E, Kubiak JZ (2014) CDC6 controls dynamics of the first embryonic M-phase entry and progression via CDK1 inhibition. Dev Biol 396:67–80

Furuno N, Nishizawa M, Okazaki K, Tanaka H, Iwashita J, Nakajo N, Ogawa Y, Sagata N (1994) Suppression of DNA replication via Mos function during meiotic divisions in *Xenopus* oocytes. EMBO J 13:2399–2410

Greenwood E, Nishitani H, Nurse P (1998) Cdc18p can block mitosis by two independent mechanisms. J Cell Sci 111:3101–3108

Hemerly AS, Prasanth SG, Siddiqui K, Stillman B (2009) Orc1 controls centriole and centrosome copy number in human cells. Science 323:789–793

Kan Q, Jinno S, Yamamoto H, Kobayashi K, Okayama H (2008) ATP-dependent activation of p21WAF1/CIP1-associated Cdk2 by Cdc6. Proc Natl Acad Sci U S A 105:4757–4762

Kelly TJ, Brown GW (2000) Regulation of chromosome replication. Annu Rev Biochem 69:829–880

Lemaître JM, Bocquet S, Mechali M (2002) Competence to replicate in the unfertilized egg is conferred by Cdc6 during meiotic maturation. Nature 419:718–722

Lemaître JM, Bocquet S, Terret M, Namdar M, Ait-Ahmed O, Kearsey S, Verlah M, Mechali M (2004) The regulation of competence to replicate in meiosis by Cdc6 is conserved during evolution. Mol Reprod Dev 69:94–100

Li JL, Cai YC, Liu XH, Xian LJ (2006) Norcantharidin inhibits DNA replication and induces apoptosis with the cleavage of initiation protein Cdc6 in HL-60 cells. Anticancer Drugs 17:307–314

Marchler-Bauer A, Lu S, Anderson JB, Chitsaz F, Derbyshire MK, DeWeese-Scott C, Fong JH, Geer LY, Geer RC, Gonzales NR, Gwadz M, Hurwitz DI, Jackson JD, Ke Z, Lanczycki CJ, Lu F, Marchler GH, Mullokandov M, Omelchenko MV, Robertson CL, Song JS, Thanki N, Yamashita RA, Zhang D, Zhang N, Zheng C, Bryant SH (2011) CDD: A Conserved Domain Database for the functional annotation of proteins. Nucleic Acids Res 39:225–229

Mimura S, Seki T, Tanaka S, Diffley JFX (2004) phosphorylation depenedent binding of mitotic cyclins to Cdc6 contributes to DNA replication control. Nature 431:1118–1123

Narasimhachar Y, Webster DR, Gard DL, Coué M (2012) Cdc6 is required for meiotic spindle assembly in *Xenopus* oocytes. Cell Cycle 11:1–8

Prasanth SG, Prasanth KV, Siddiqui K, Spector DL, Stillman B (2004) Human Orc2 localizes to centrosomes, centromeres and heterochromatin during chromosome inheritance. EMBO J 23:2651–2663

Prasanth SG, Prasanth KV, Stillman B (2002) Orc6 involved in DNA replication, chromosome segregation and cytokinesis. Science 297:1026–1031

Strausberg RL et al (2002) Generation and initial analysis of more than 15,000 full-length human and mouse cDNA sequences. Proc Natl Acad Sci U S A 99:16899–16903

Sun QY, Schatten H (2006) Role of NuMA in vertebrate cells: Review of an intriguing multifunctional protein. Front Biosci 11:1137–1146

Swiech L, Kisiel K, Czolowska R, Zientarski M, Borsuk E (2007) Accumulation and dynamics of proteins of MCM family during mouse oogenesis and first embryonic cell cycle. Int J Dev Biol 51:283–295

Uranbileg B, Yamamoto H, Park JH, Mohanty AR, Arakawa-Takeuchi S, Jinno S, Okayama H (2012) Cdc6 protein activates p27KIP1-bound Cdk2 protein only after the bound p27 protein undergoes C-terminal phosphorylation. J Biol Chem 287:6275–6283

Williams RS, Shohet RV, Stillman B (1997) A human protein related to yeast Cdc6p. Proc Natl Acad Sci U S A 94:142–147

Whitmire E, Khan B, Coué M (2002) Cdc6 sybnthesis regulates replication competence in *Xenopus* oocytes. Nature 419:722–725

Yim H, Erikson RL (2010) Cell division cycle 6, a mitotic substrate of polo-like kinase 1, regulates chromosomal segregation mediated by cyclin-dependent kinase 1 and separase. Proc Natl Acad Sci U S A 107:19742–19747

Zuccotti M, Piccinelli A, Giorgio Rossi P, Garagna S, Redi CA (1995) Chromatin organization during mouse oocyte growth. Mol Reprod Dev 41:479–485

Chapter 8
Oocyte Activation and Fertilisation: Crucial Contributors from the Sperm and Oocyte

Marc Yeste, Celine Jones, Siti Nornadhirah Amdani, and Kevin Coward

Abstract This chapter intends to summarise the importance of sperm- and oocyte-derived factors in the processes of sperm–oocyte binding and oocyte activation. First, we describe the initial interaction between sperm and the zona pellucida, with particular regard to acrosome exocytosis. We then describe how sperm and oocyte membranes fuse, with special reference to the discovery of the sperm protein IZUMO1 and its interaction with the oocyte membrane receptor JUNO. We then focus specifically upon oocyte activation, the fundamental process by which the oocyte is alleviated from metaphase II arrest by a sperm-soluble factor. The identity of this sperm factor has been the source of much debate recently, although mounting evidence, from several different laboratories, provides strong support for phospholipase C ζ (PLCζ), a sperm-specific phospholipase. Herein, we discuss the evidence in support of PLCζ and evaluate the potential role of other candidate proteins, such as post-acrosomal WW-binding domain protein (PAWP/WBP2NL). Since the cascade of downstream events triggered by the sperm-borne oocyte activation factor heavily relies upon specialised cellular machinery within the oocyte, we also discuss the critical role of oocyte-borne factors, such as the inositol trisphosphate receptor (IP_3R), protein kinase C (PKC), store-operated calcium entry (SOCE) and calcium/calmodulin-dependent protein kinase II (CaMKII), during the process of oocyte activation. In order to place the implications of these various factors and processes into a clinical context, we proceed to describe their potential association with oocyte activation failure and discuss how clinical techniques such as the in vitro maturation of oocytes may affect oocyte activation ability. Finally, we contemplate the role of artificial oocyte activating agents in the clinical rescue of oocyte activation deficiency and discuss options for more endogenous alternatives.

M. Yeste (✉)
Unit of Cell Biology, Department of Biology, Faculty of Sciences, University of Girona, C/ Maria Aurèlia Campany, 69, Campus Montilivi, E-17071 Girona, Spain

Nuffield Department of Obstetrics and Gynaecology, University of Oxford, Level 3, Women's Centre, John Radcliffe Hospital, Headington, Oxford, UK
e-mail: marc.yeste@udg.edu

C. Jones • S.N. Amdani • K. Coward
Nuffield Department of Obstetrics and Gynaecology, University of Oxford, Level 3, Women's Centre, John Radcliffe Hospital, Headington, Oxford, UK

S. Arur (ed.), *Signaling-Mediated Control of Cell Division*, Results and Problems in Cell Differentiation 59, DOI 10.1007/978-3-319-44820-6_8

8.1 Introduction: A Global View of Fertilisation

In mammals, fertilisation is defined by the process in which a capacitated spermatozoon binds to the zona pellucida (ZP), fuses with the oocyte membrane and activates the oocyte, a series of events which subsequently results in the start of embryogenesis (Yeste 2013a). In mammals, this process occurs internally, within the fallopian tube (Suarez and Pacey 2006) which provides the optimal environment for sperm–oocyte interaction (Yeste 2013b).

In order for a spermatozoon to interact with the oocyte, it must have previously been capacitated (Yeste 2013b). Sperm capacitation consists of a wide array of changes, including Ca^{2+} influx, phosphorylation of crucial sperm proteins, changes in sperm motility patterns and an increase in membrane lipid disorder (Aitken and Nixon 2013). As a result of these changes, both sperm plasma and the outer acrosome membranes become more fusogenic, which ultimately provides the sperm with the ability to trigger the acrosome reaction (Yeste 2013b). While recent data indicate that acrosomal exocytosis arises before the sperm interact with the ZP (La Spina et al. 2016), the actual binding of the sperm head with ZP is known to be mediated by glycoproteins (Töpfer-Petersen et al. 2008), a feature which has been demonstrated in several mammalian species.

Following the successful passage of sperm through the ZP, the equatorial region of the acrosome-reacted spermatozoon is exposed, allowing for the sperm protein IZUMO1 to interact with its oocyte receptor JUNO (Bianchi et al. 2014). As a result of this interaction, membrane fusion occurs in a process that is yet to be fully understood and the spermatozoon is engulfed by the cytoplasm. Upon membrane fusion, the oocyte is activated, triggering a series of Ca^{2+} oscillations which will result in the cortical reaction, release of the oocyte from meiosis II (MII) arrest and eventually the induction of embryogenesis (Williams 2002; Kashir et al. 2010). The following sections will describe each of these distinct steps in detail.

8.2 Interactions Between Sperm and the Zona Pellucida

Precise interaction between sperm and the zona pellucida (ZP) involves glycoproteins from the sperm head which recognise specific oligosaccharide ligands on ZP-glycoproteins.

8.2.1 ZP-glycoproteins

In mammals, the ZP is made up of more than one glycoprotein, and the exact number of different glycoproteins that together constitute the ZP varies across species (Harris et al. 1994). While the ZP is composed of four glycoproteins in humans (Lefièvre et al. 2004; Conner et al. 2005), non-human primates (Goudet et al. 2008), cats (Stetson et al. 2015), rats (Hoodbhoy et al. 2005) and hamsters (Izquierdo-Rico et al. 2009), only three glycoproteins are required to constitute the ZP in mice, cattle and pigs (Noguchi et al. 1994; Goudet et al. 2008; Mugnier et al. 2009).

The function of these different ZP-glycoproteins differs. ZPB/ZP4 and ZPC/ZP3 are involved in sperm recognition and primary binding (Kudo et al. 1998; Yurewicz et al. 1998), whereas ZPA/ZP2 is involved in secondary binding with the inner acrosomal sperm membrane. Binding of ZP with sperm is mediated by N- and O-glycans attached to ZP-glycoproteins. Particularly, sialyl-Lewis^x sequence [NeuAcα2-3Galβ1-4(Fucα1-3)GlcNAc] is the dominant antenna sequence of these N- and O-glycans in humans (Pang et al. 2011). Moreover, following secondary binding, the cortical reaction is triggered and the released proteases cleave ZPA/ZP2 causing the ZP to harden and thus prevent polyspermia (Gupta et al. 2012).

8.2.2 Do ZP-glycoproteins Induce the Acrosome Reaction?

Acrosome-reacted sperm are able to penetrate the ZP. This takes place by alternate cycles of sperm–ZP binding and limited proteolytic digestion through the ZP matrix. The force exerted by the forward motility of capacitated sperm also contributes to ZP penetration (O'Rand et al. 1986; Green 1987). During the acrosome reaction, proacrosin (the inactive form) is converted into acrosin, an active serine protease. Across mammalian species, proacrosin shows high affinity for sulphated oligosaccharide chains on ZP-glycoproteins, which also appear to be involved in the process of converting proacrosin to acrosin (Töpfer-Petersen et al. 2008). These activation properties and the regulation of acrosin activity rely heavily upon spatial proximity between polysulphate-binding sites and the localisation of proacrosin/acrosin (Tranter et al. 2000).

The acrosome reaction involves phospholipases, such as phospholipase A2 (PLA2), although the exact role of PLA2 isoforms has yet to be completely addressed. While it has been reported that an intracellular PLA2 isoform, known as iPLA2β, is involved in spontaneous acrosome reaction, it appears that both iPLA2β and a secreted PLA2 (mGX sPLA2) are involved in the acrosome reaction and are induced by progesterone, at least when this hormone is at low concentrations. Indeed, when progesterone is at a concentration similar to that found in cumulus cells, the involvement of these two isoforms relies upon the moment at which the acrosome reaction is triggered. Therefore, the iPLA2β isoform appears to be involved in early progesterone-induced acrosome reaction (0–5 min), while mGX sPLA2 participates in late acrosome exocytosis (20–30 min; Abi Nahed et al. 2016). In contrast, when progesterone is at high concentrations (10 μM), either other phospholipases are involved (PLA1, PLB, PLD1) or another pathway is triggered by progesterone.

Currently, it is not fully understood how, and at what point, the acrosome reaction is triggered, although it is clear that only capacitated sperm may undergo acrosome exocytosis. As reviewed by Buffone et al. (2014), three different scenarios have been proposed in order to explain how the acrosome reaction occurs in the fallopian tube.

The classical view is that the acrosome reaction takes place upon the ZP surface and involves specific interactions between key molecules. Following primary binding, proacrosin is activated to acrosin, which initiates the acrosome reaction and permits secondary binding. In this instance, acrosome-intact sperm would interact with specific ZP proteins such as ZPC/ZP3 in the mouse and ZP1, ZPC/ZP3 and ZPB/ZP4 in the human. This view was based upon the observation that samples of solubilised ZP triggers the acrosome reaction in sperm from humans and other mammalian species (Cross et al. 1988). Although this classical view was predominant over many decades, other factors in the fallopian tube, such as the presence of progesterone in the follicular fluid, were reported to exert a priming role in the ZP-mediated induction of the acrosome reaction (Schuffner et al. 2002; Munuce et al. 2006), and this led to researchers to put forward alternative models. Therefore, a second model emerged which stated that the acrosome reaction occurs at the level of the cumulus cells and is triggered by factors released by these cells. A third model, however, hypothesises that the acrosome reaction arises in the fallopian tube before reaching the cumulus cells, and it is triggered by factors present in the oviduct or released by the cumulus cells (Buffone et al. 2014).

While there is a significant consensus of opinion in support of the acrosome reaction being triggered when sperm interact with ZP-glycoproteins, the current view indicates that sperm binding to the ZP is not sufficient to induce this reaction. Indeed, in vitro fertilisation (IVF) experiments have shown that acrosome exocytosis occurs prior to sperm interaction with the ZP, indicating that proteins within the ZP are not the main inducers of the acrosome reaction (reviewed by Buffone et al. 2014). In addition, although some authors suggested that contact of sperm with cumulus cells could trigger the acrosome reaction, recent studies in mice indicate that acrosome exocytosis takes place in the upper isthmus of the fallopian tube, so that only few sperm are still acrosome intact when they reach the ampulla (La Spina et al. 2016). This recent view suggests that, at least in murine species, the acrosome reaction is not triggered when sperm bind to, or is in the vicinity of, the ZP, but rather that progesterone released from the cumulus cells, and flowing downstream of the oviduct, is able to initiate the acrosome reaction earlier on.

Another interesting issue is how sperm degrade the ZP matrix. Although, as aforementioned, sperm serine proteases (mainly acrosin) have been thought to play a major role in such degradation; recent data indicate that metalloproteases and the sperm proteasome also participate in the degradation of ZP proteins (Saldívar-Hernández et al. 2015).

8.2.3 Sperm Receptors for ZP-proteins in Mammals

The interaction between sperm and the ZP takes place in two steps: primary and secondary binding. However, whether one or more than one sperm protein acts as a receptor for ZP-glycoproteins is still being investigated (Barroso et al. 2009). Thus far, separate studies, conducted in different species, have suggested the following primary sperm receptors for ZP: a 95 KDa protein kinase, the sperm receptor for ZPC/ZP3 (also known as sp56), zonadhesin, β1,4-galactosyltransferase, SED1 and ADAM3 (Bookbinder et al. 1995; Gao and Garbers 1998; Bi et al. 2003; Tardif and Cormier 2011; Gupta 2014). With regard to secondary binding, proacrosin is known to play a crucial role (Honda et al. 2002).

A recent study by Tanphaichitr et al. (2015) attempted to shed light on these points and has reported that proteins in the sperm head form high molecular weight complexes which bind to the ZP. These complexes contain acrosomal proteins, such as zonadhesin, proacrosin and acrosin binding protein (ACRBP). On the other hand, evidence exists to indicate that other proteins are also involved in sperm–ZP interaction. For example, sperm devoid of Heat Shock Protein A2 (HSPA2) are unable to interact with the ZP. In this context, it is important to note that BCL2-associated athanogene 6 protein (BAG6) has recently been identified as an interacting protein involved in the regulation of HSPA2 function in human sperm (Bromfield et al. 2015). Furthermore, it has recently been reported that the epididymal protein Cysteine-Rich Secretory Protein 1 (CRISP1), which associates with the sperm membrane during epididymal maturation and is involved in the sperm–oocyte fusion (Cohen et al. 2001), also interacts with ZPC/ZP3 (Maldera et al. 2014). These results concur with previous studies conducted in rodent species (Busso et al. 2007). Other research has shown that glutathione S-transferase M3 interacts with ZBP/ZP4 and that voltage-dependent anion channel 2 (VDAC2) interacts with ZPA/ZP2 and ZPC/ZP3 (Petit et al. 2013).

At present, no molecule has been unequivocally established as a primary sperm receptor for ZP-glycoproteins. We know that different sperm proteins interact with the ZP, although we have yet to address whether such proteins act synergistically. Another relevant issue that warrants further research relates to acrosomal proteins which exhibit affinity for the ZP; it remains to be elucidated whether these proteins bind to the ZP before or after the acrosome reaction.

Finally, one should note that many of the existing studies of interaction between ZP-glycoproteins and sperm have been conducted in mice, and care must be taken when extrapolating these data to the human because of the reasons illustrated below. Firstly, because the mouse ZP is composed of three glycoproteins, while the human ZP is made up of four. Secondly, the ZPC/ZP3 and ZPA/ZP2 have been identified as primary and secondary receptors, respectively, in mouse sperm, but in humans, primary binding is mediated by ZP1, ZPC/ZP3 and ZBP/ZP4 (reviewed by Gupta 2015). In addition, since some studies have used recombinant ZPs, more research is required in order to confirm whether interactions between recombinant ZPs and sperm proteins also occur in native forms.

8.2.4 Sperm Passing Through the ZP and ZP Hardening

The precise way in which proacrosin acts during fertilisation has also become the source of much debate over the last decade or so. Indeed, it is generally accepted that binding first occurs to the acrosome, and then secondary binding occurs to the ZP after which the ZP digests (Honda et al. 2002). However, a knockout mouse null for acrosin showed delayed fertilisation but normal fecundity and viable offspring (Adham et al. 1997). For this reason, the model has now been refined and has indicated that the crucial role of acrosin is related to ZP binding, rather than ZP hydrolysis. Mao and Yang (2013) suggested that sperm only require a smooth pathway and not the complete hydrolysis of ZP in order to reach the oolemma. In addition, since acrosin quickly disappears following the acrosome reaction, it has been suggested that its direct role in ZP hydrolysis is not as crucial as previously thought. However, it is possible that acrosin is involved in the activation of other acrosomal proteins rather than in the digestion of ZP itself. This hypothesis deserves further research but requires the identification of other acrosin substrates (reviewed by Mao and Yang 2013).

Furthermore, Dietzel et al. (2013) found that Fetuin-B (FETUB) is involved in the regulation of ZP hardening via the inhibition of ovastacin (ASTL1), an oocyte-specific member of the astacin family of metalloendoproteases in cortical granules, which cleaves ZPA/ZP2 (Burkart et al. 2012). This indicates that beyond the direct binding of ZP and sperm proteins, there are other factors that also intervene in the precise regulation of this interaction.

8.3 Fusion of Sperm and Oocyte Membranes

It has been established that the equatorial/post-acrosomal region of an acrosome-reacted spermatozoon adheres to and fuses with the oolemma because the plasma membrane of this region remains intact following acrosome exocytosis (Stein et al. 2004). A range of different candidate proteins have been suggested to be involved in the specific binding/fusion of sperm and oocyte membranes, although the specific role of some of these molecules remains to be validated (Vjugina and Evans 2008).

Two cysteine-rich secretory proteins (CRISP) isolated from sperm have been found to play a role in key interactions: the epididymal protein CRISP1 is suggested to interact with the ZP (mainly with ZP3) and the oolemma, while the testicular protein TPX1 (CRISP2) also interacts with the oocyte membrane (Busso et al. 2005; Vadnais et al. 2008; Cohen et al. 2011; Maldera et al. 2014). Other sperm proteins involved in the binding of sperm and oocyte membranes include equatorin, a protein localised to the inner acrosome membrane, SPACA6, ADAM-1, ADAM-2 and IZUMO1 (Toshimori et al. 1998; Inoue et al. 2005, 2008). Given the pivotal role of IZUMO1, its role will be discussed later in this chapter.

With regard to oocyte proteins, previous studies in mammals have clearly emphasised the role of integrins ($\alpha v\beta_1$ and β_1), and integrinβ_1 was suggested to recognise sperm protein ADAM-2 (Blobel 2000; Linfor and Berger 2000). However, the involvement of integrins in sperm and oocyte fusion has been dismissed in humans (Sengoku et al. 2004). Another oocyte protein suggested to be involved in the binding of sperm and oocyte membranes is P-selectin (Fusi et al. 1996). Moreover, bioinformatic tools have further suggested that the oocyte proteins CD151 and CD9 interact with the sperm protein CD49 and also led to hypothesise that oocyte protein CD81 interacts with sperm protein ITGA4 (Sabetian et al. 2014). Notwithstanding, this approach strongly indicated that the oocyte protein CD9, known to complex with integrins on the oocyte membrane (Kaji et al. 2000; Miyado et al. 2000), interacts with ADAM2 (Sabetian et al. 2014). However, there are some inconsistencies with regard to the role of integrins in gamete fusion and further research is warranted.

The most important contribution to gamete fusion involves IZUMO1 and JUNO proteins. IZUMO1 is a family member of immunoglobulin sperm proteins, whose function does not seem to rely on glycosylation (Inoue et al. 2005, 2008). In 2014, Bianchi et al. (2014) made a crucial discovery when they found that folate receptor 4 (FOLR4), renamed as JUNO, was the oocyte interactor for IZUMO1. Binding between IZUMO1 and JUNO is conserved across animal species and not only plays a crucial role for oocyte–sperm membrane adhesion but also for blocking polyspermy (Bianchi et al. 2014; Gupta 2014). It appears that a fragment of the N-terminus (Asp^5 to Leu^{113}) of IZUMO1, known as 'IZUMO1 putative functional fragment', is the specific region that interacts with the surface of the oocyte (Inoue et al. 2013). Following binding of monomeric IZUMO1 with JUNO, there is IZUMO1 dimerisation (Inoue et al. 2015).

The ability of Juno protein extracts from hamster oocytes to bind to human IZUMO1 may explain why zona-free hamster oocytes are able to interact and fuse with human sperm and why this has been used as a heterologous penetration assay in assisted reproductive technology (ART; Bianchi and Wright 2015). Finally, a question that remains to be elucidated is whether JUNO interacts with other proteins, such as $\alpha6\beta1$ and CD9 (Gupta 2014).

8.4 Oocyte Activation (I): Sperm Factors

Upon the fusion of sperm and oocyte membranes, the whole spermatozoon, except the plasma membrane, is engulfed by the oocyte cytoplasm (reviewed in Yanagimachi 1998). Then, a process known as 'oocyte activation' occurs and allows the oocyte, arrested at metaphase II, to complete the second meiotic division (Jones 2007; Horner and Wolfner 2008; Dale et al. 2010). Under normal circumstances, the presence of a sperm in the oocyte cytoplasm evokes a characteristic pattern of intracellular calcium (Ca^{2+}) oscillations (Swann 1990; Kline and Kline 1992) which orchestrates a series of key events, such as the resumption of meiosis,

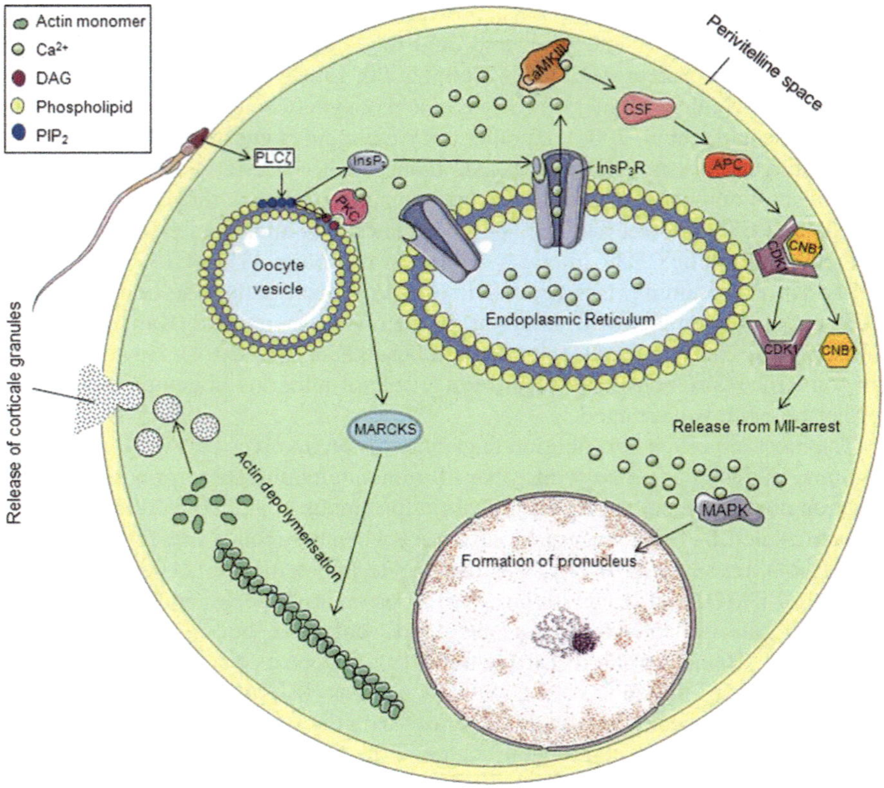

Fig. 8.1 Oocyte activation triggered by PLCζ, a sperm-specific protein factor. Reproduced with permission from Yeste et al. (2016)

polar body extrusion, exocytosis of cortical granules, cytoskeletal rearrangements, recruitment of maternal mRNA and the formation of pronuclei (Swann et al. 2004; Ajduk et al. 2006; Swann and Yu 2008; Swann and Lai 2016; Fig. 8.1). These periodic Ca^{2+} oscillations, which occur over a specific temporal window, are critical for both oocyte activation and the onset of embryogenesis (Amdani et al. 2016).

The specific identity of the sperm protein responsible for activating the oocyte has been the source of much debate over recent years. Three theories were proposed to explain how Ca^{2+} is released upon game fusion: (1) the 'sperm factor theory' which hypothesised that a catalytic substance in the sperm head triggered Ca^{2+} release in the oocyte following gamete fusion (Dale et al. 1985); (2) the 'receptor theory' which proposed that a signal transduction pathway was initiated upon the interaction of an oocyte membrane receptor and a sperm protein (Kline et al. 1988) and (3) the 'calcium bomb' theory which suggested that Ca^{2+} entered the oocyte upon sperm–oocyte fusion, either from stores in the sperm or via channels on the sperm plasma membrane (Jaffe 1983). The 'receptor' and 'calcium bomb' theories have both now been dismissed, despite the receptor theory having gained some credence initially (Williams et al. 1998). In this latter case, the main reason to reject this theory was that no oocyte membrane protein is involved in oocyte activation following intracytoplasmic sperm injection (ICSI) (Neri et al. 2014), a technique clinically proven to cause fertilisation. The 'sperm factor theory' became established when sperm-soluble extracts injected into the ooplasm were found to induce Ca^{2+} oscillations (Jones et al. 1998). Oscillin was the first protein proposed as the sperm-borne oocyte activation factor (SOAF; Parrington et al. 1996), but this hypothesis was later discounted when injections of recombinant oscillin into mouse oocytes failed to induce Ca^{2+} release. Subsequent studies successfully identified another sperm-specific cytosolic protein, known as phospholipase c zeta (PLCζ), as being a strong candidate for the SOAF (Saunders et al. 2002, 2007; Kashir et al. 2010; Amdani et al. 2016).

Apart from PLCζ, several other proteins have been suggested to be involved in oocyte activation in mammalian and non-mammalian species, including citrate synthase, truncated c-Kit tyrosine kinase and post-acrosomal WW-domain binding protein (PAWP/WBP2NL) (Sette et al. 2002; Harada et al. 2007; Wu et al. 2007a, b; Aarabi et al. 2014a, b). However, while mounting evidence supports the role of PLCζ, the relevance of some of the other proteins remains unclear, particularly with regard to inconsistencies in the literature relating to PAWP/WBP2NL (Amdani et al. 2015, 2016), with studies from separate laboratories reporting conflicting results (Wu et al. 2007a, b: Aarabi et al. 2014a, b; Nomikos et al. 2014, 2015; Satouh et al. 2015; Escoffier et al. 2016).

8.4.1 Phospholipase Cζ

As with all phospholipases C (PLC), PLCζ hydrolyses phosphatidylinositol 4,5-bisphosphate (PIP_2) into inositol 1,4,5-trisphosphate (IP_3) and diacylglycerol (DAG) (Yang et al. 2013). While IP_3 leads to the release of Ca^{2+} from the endoplasmic reticulum via interaction with its receptor (IP_3R), DAG acts together with Ca^{2+} and activates protein kinase C (PKC; Taylor et al. 2014). In spite of this, PLCζ, which is the smallest PLC isoform, is very different from the other PLCs in that it is not localised to the plasma membrane linked to G protein-coupled receptors (GPCRs) and does not possess a PH domain. Thus, rather than hydrolysing PIP_2 present in the plasma membrane, PLCζ targets PIP_2-containing vesicles distributed across the ooplasm via its X–Y linker region (Fig. 8.1; Yu et al. 2012; Theodoridou et al. 2013; Swann and Lai 2013). In this context, it is worth mentioning that the X–Y linker is composed of positively charged residues which interact electrostatically with PIP_2, which is negatively charged (Nomikos et al. 2011a). However, it appears that not only does the X–Y linker play a role for PIP_2 hydrolysis (Nomikos et al. 2011b) but also the EF-hand domain Nomikos et al. (2015).

Thus far, PLCζ has been found in the sperm of several non-mammalian (reviewed in Kashir et al. 2013) and mammalian species, including humans, rodents, pigs and monkeys (Cox et al. 2002; Saunders et al. 2002; Yoneda et al. 2006; Cooney et al. 2010), thus indicating that the role of PLCζ is conserved throughout the animal kingdom. Although PLCζ has been suggested to be involved in events other than oocyte activation (Young et al. 2009), data from patients with certain types of infertility clearly indicate that abnormal localisation, genetic mutations or the absence of PLCζ can result in oocyte activation deficiency and total fertilisation failure (Yoon et al. 2008; Heytens et al. 2009, 2010; Kashir et al. 2011, 2012a, b; Yelumalai et al. 2015). In some cases, the presence of sperm devoid of PLCζ occurs together with other major sperm deficiencies, such as globozoospermia (Taylor et al. 2010; Yassine et al. 2015).

In spite of all these lines of evidence, the precise mechanism through which PLCζ activates the oocyte has yet to be fully elucidated. Related to this, Swann and Lai (2013) hypothesised the existence of a specific receptor on the surface of the PIP_2-containing vesicles distributed within the oocyte cytoplasm. Since only oocytes have been reported to present such vesicles, this could explain why PLCζ is able to trigger Ca^{2+} in oocytes but not in other somatic cells.

8.4.2 PAWP/WBP2NL

Over the last few years, PAWP/WBP2NL has been considered to be the main competitor of PLCζ as the SOAF. The name of this protein is derived from its localisation, as it is found in the post-acrosomal sheath of the sperm perinuclear theca (Wu et al. 2007a, b). The perinuclear theca is a compartment in sperm thought to be involved in oocyte activation (Sutovsky et al. 2003). With regard to the function of this protein, there are some reports relating to its ability to trigger Ca^{2+} oscillations in both mammalian and non-mammalian oocytes (Wu et al. 2007a; Aarabi et al. 2014a), the mechanism of action of this protein. However, other research groups have not observed the same results (Nomikos et al. 2014, 2015). While the proportions of sperm exhibiting PAWP/WBP2NL have been related to reproductive outcomes in humans and cattle (Aarabi et al. 2014b; Kennedy et al. 2014), recent data using a knockout mice model do not support the role of PAWP/WBP2NL as a SOAF candidate, since sperm from such null mice are able to trigger normal Ca^{2+} oscillations and give normal embryo development (Satouh et al. 2015). Therefore, further research is warranted to address these inconsistencies and to clarify whether there is an unknown role for this sperm protein.

8.5 Oocyte Activation (II): Oocyte Factors

Aside from sperm contributions, the process of oocyte activation relies heavily upon cellular machinery involved in the signal transduction pathway initiated by the SOAF. Therefore, in order to discuss this process further, it is imperative to also refer to the oocyte factors involved in the process. While the quality of the oocyte is one major reason for total failure in fertilisation following ICSI (Miyara et al. 2003; Kilani and Chapman 2014), there are other reports that suggest that there may be a relationship between total fertilisation failure and the expression of different genes during oocyte maturation, including growth differentiation factor 9 (GDF9), bone morphogenetic protein 15 (BMP15), BCL2-associated transcription factor 1 (BCLAF1), Leiomodin 3 (LMOD3) and F-box protein 5 (FBXO5) (Gasca et al. 2008; Grøndahl et al. 2013; Li et al. 2014).

8.5.1 *PIP$_2$*

Phospholipase C ζ initiates a PLC pathway in which PIP_2, DAG and IP_3R are involved (Fig. 8.1). Previous studies have indicated that oocytes present a significant number of PIP_2-containing vesicles distributed across the cytoplasm, which are specifically targeted by PLCζ (Yu et al. 2008, 2012). For this reason, it was also suggested that if such vesicles contain reduced amounts of PIP_2 or are low in number, it may hamper the transduction pathway cascade triggered by sperm (Yeste et al. 2016).

8.5.2 *IP$_3$ Receptor (IP$_3$R)*

Following PIP_2 hydrolysis, IP_3 binds to IP_3R, causing the channel to open and Ca^{2+} to be released from the endoplasmic reticulum (Bhanumathy et al. 2012). The activity of IP_3R in somatic cells is regulated by other kinase PKA and phosphatase proteins PP1 and PP2A (DeSouza et al. 2002), cAMP (Taylor et al. 2014) and calmodulin (CaM) (Kasri et al. 2006). While it remains to be established how this receptor is regulated in mammalian oocytes, our current knowledge of other cells suggests that this mechanism is highly complex with precise regulation. In addition to this, one should note that the amounts of IP_3R are known to be correlated with Ca^{2+} levels and that, upon fertilisation, there is a downregulation of IP_3Rs resulting from the increase and sustained levels of IP_3 (Ross et al. 2008; Cooney et al. 2010; Lee et al. 2010). Despite the crucial role of IP_3R in triggering Ca^{2+} oscillations, it appears that Ca^{2+} in mammalian oocytes could also be mediated either by other mechanisms or that other calcium stores could also be involved (Malcuit et al. 2005).

Besides its vital function during oocyte activation, IP_3R is also important during oocyte maturation (Pesty et al. 1998; Vanderheyden et al. 2009; Verbert et al. 2008). Indeed, while IP_3R1, one of the IP_3R isoforms, presents a diffuse, irregular and granular distribution in germinal vesicle (GV) human oocytes, its localisation changes to a reticular and peripheral distribution when the MII stage is reached. Upon fertilisation, IP_3R1 is present in the centre and periphery at the two- to four-cell stage embryos, but redistributes to a pattern across the entire cytosol in six- to eight-cell embryos. These localisation changes occur concomitant with an increase in the levels of this receptor (Goud et al. 1999). Therefore, these data collectively show the relevance of this receptor in the signalling processes underlying oocyte maturation, activation and embryo development.

8.5.3 Protein Kinase C

Protein kinase Cs (PKCs) are involved in the downstream cascade triggered by Ca^{2+} oscillations, and different PKC isoforms (conventional, non-classic and atypical) have been identified in both GV and MII oocytes (Baluch et al. 2004). The distribution of conventional PKC isoform changes following oocyte maturation and upon sperm–oocyte fusion (Chen et al. 2014), as PKCα translocates from the ooplasm to the oocyte cortex (Luria et al. 2000; Halet et al. 2004). This translocation is related to the exocytosis of cortical granules, as it allows phosphorylation of myristoylated alanine-rich C-kinase substrate (MARCKS) proteins, which lead the actin network to disassemble and allow cortical granules to exocytose (Gallicano et al. 1997; Eliyahu and Shalgi 2002; Eliyahu et al. 2005, 2006; Tsaadon et al. 2008; Fig. 8.1).

8.5.4 The Vital Role of Ca^{2+}

Considering that Ca^{2+} oscillations have a major role during oocyte maturation and activation (Machaca 2007), the mechanisms involved in maintaining its homeostasis in oocytes are crucial. Although this mechanism is not completely understood, the present section intends to summarise our current state of knowledge (Fig. 8.2).

Firstly, the store-operated calcium entry system is crucial for maintaining Ca^{2+} homeostasis, especially in cells that are non-electrochemically excitable. This system consists of a series of channel proteins that provide regulation across internal Ca^{2+} stores, cytosol and the extracellular zone (Putney 2011; Smyth et al. 2010). The main components of this system include stromal interaction molecule 1 (STIM1), Ca^{2+} release-activated Ca^{2+} channel protein 1 (ORAI1) and sarco/endoplasmic reticulum Ca^{2+} ATPase (SERCA; Fig. 8.2).

STIM1 and ORAI1 are important for oocyte maturation, when there is an increasing influx of Ca^{2+} to the oocyte, which mainly accumulates in the

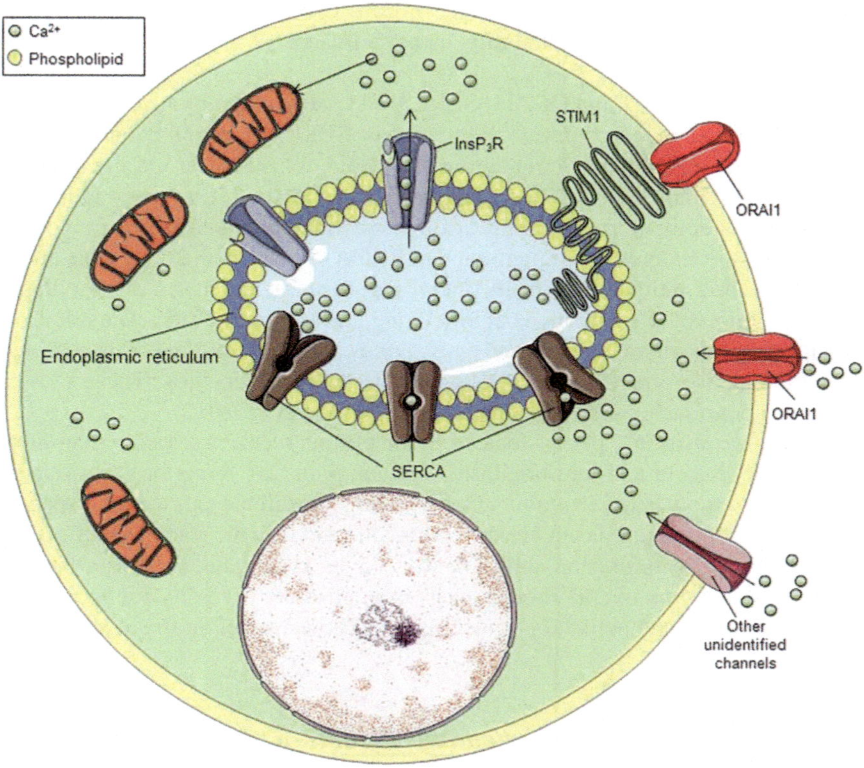

Fig. 8.2 Oocyte mechanisms involved in the regulation of Ca^{2+}. It is important to note the relevance of SOCE (SERCA, ORAI1) for Ca^{2+} homeostasis, as well as the role of mitochondria. Reproduced with permission from Yeste et al. (2016)

endoplasmic reticulum (Cheon et al. 2013) and is likely to be related to its role during oocyte activation. As shown in Fig. 8.2, STIM1 is a transmembrane protein present in the endoplasmic reticulum and plasma membranes that acts as a sensor of low Ca^{2+} levels (Zhang et al. 2005; Roos et al. 2005). For this reason, alterations/deficiencies of STIM1 could underlie oocyte activation deficiency, as its role is crucial for the generation of Ca^{2+} oscillations triggered by the SOAF (Yeste et al. 2016).

STIM1 is strongly associated with ORAI1. Upon hydrolysis of PIP_2 into IP_3 and DAG, IP_3 interacts with its receptor in the membrane of the endoplasmic reticulum (Fig. 8.1). This leads the pore to open, so that Ca^{2+} may flow out from the lumen of the reticulum. Specific domains of STIM1, located in the lumen of the endoplasmic reticulum, act as a sensor of Ca^{2+} levels, and this causes the relocation of STIM1 from the plasma membrane where it interacts directly with ORAI1 channel (Fig. 8.2). This channel subsequently opens and allows Ca^{2+} to enter the cell (Feske et al. 2006; Putney 2009; Gudlur et al. 2014). It has been suggested that deficiencies in ORAI1 could render the oocyte unable to influx Ca^{2+} properly, which could in turn affect its ability to be activated (Wang et al. 2012; Yeste et al. 2016).

In addition to STIM1 and ORAI1, SERCA and plasma membrane Ca^{2+} ATPases (PMCAs) are also involved in Ca^{2+} homeostasis. SERCA is an ion pump located in the membrane of endoplasmic reticulum and pump cytosolic Ca^{2+} to the lumen of the reticulum (Ullah et al. 2007). As proposed in the Fig. 8.2, Ca^{2+} passing through ORAI1 and other membrane channels from the extracellular space to the cytosol is pumped into the endoplasmic reticulum by SERCA (Wakai et al. 2013). This allows the endoplasmic reticulum to refill, thereby generating another Ca^{2+} oscillation upon interaction of IP_3 with its receptor (Wang and Machaty 2013). The relevance of SERCA for oocyte maturation and activation is apparent from the organisation of one particular isoform (SERCA2b) into clusters in MII oocytes. These clusters disappear upon reaching the pronuclear stage (Wakai et al. 2013).

PMCAs are also Ca^{2+} pumps located in the plasma membrane rather than in the endoplasmic reticulum. Therefore, instead of pumping Ca^{2+} from the cytosol to the lumen of the endoplasmic reticulum, Ca^{2+} is pumped from the extracellular space to the cytosol. At present, it is not clear whether alterations in the role of SERCAs and PMCAs may compromise the ability of the oocyte to activate. However, because the precise regulation of Ca^{2+} homeostasis is an absolute necessity for appropriate oocyte maturation and activation, much research is warranted in this realm.

8.5.5 *Ca^{2+}/Calmodulin-dependent Protein Kinase II*

The cascade of downstream events triggered by Ca^{2+} oscillations involves multiple proteins, some of which have yet to be identified. One of these proteins is Ca^{2+}/calmodulin-dependent protein kinase II (CaMKII) which effectively contributes to MII resumption (Von Stetina and Orr-Weaver 2011). One of the phosphorylation substrates of CaMKII is F-box only protein 43 (FBXO43, also known as EMI2 and ERP1). FBXO43 prevents the degradation of CCNB1, which together with CDK1 forms the M phase/maturation-promoting factor (MPF) heterodimer (Jones 2007). Phosphorylation of FBXO43 by CaMKII leads to its degradation, so that it is no longer able to prevent CCNB1 degradation. MPF is thus also degraded, and this allows the oocyte to resume meiosis II (Madgwick et al. 2005; Ducibella et al. 2006; Knott et al. 2006; Fig. 8.1).

8.6 Infertility and Oocyte Activation Deficiency

To conclude this chapter, it is necessary to refer to the clinical link between oocyte activation deficiency and infertility. Infertility affects around 15 % of couples worldwide (Ombelet et al. 2008; Louis et al. 2013; Thoma et al. 2013). Between

20 and 30 % of cases are due to a male factor, 20–35 % of cases to a female factor and 25–40 % of cases relate to problems associated with both male and female factors. The remaining 10–25 % of cases is unexplained (Yeste et al. 2016). Since the birth of the first in vitro baby in 1978, ART has emerged as a powerful tool for those couples seeking treatment for infertility (Johnson 2011). In spite of this, normal IVF cycles fail in some cases (Bhattacharya et al. 2013) and intracytoplasmic sperm injection (ICSI) may be advised. Despite ICSI proving to be very successful, complete fertilisation failure still occurs in 1–5 % of ICSI cycles (reviewed in Yeste et al. 2016). Although ICSI failure can occur for a variety of reasons, oocyte activation deficiency (OAD) is considered to be one of the main factors for ICSI failure and occurs in 40 % of cases (Mahutte and Arici 2003; Vanden Meerschaut et al. 2013, 2014a; Beck-Fruchter et al. 2014). In most cases, this appears to be related to the inability of sperm to activate the oocyte. However, the involvement of oocyte factors should not be dismissed, as oocyte quality is crucial for successful fertilisation (Kashir et al. 2010; Neri et al. 2014). Thus far, this chapter has provided a global picture of fertilisation and oocyte activation and has focused upon the pivotal role of sperm protein PLCζ. In the next section, we discuss how two different ART techniques, in vitro maturation and oocyte activation deficiency, relate to OAD and contemplate our current options for rescuing fertility in patient with OAD.

8.7 Links Between Oocyte Activation Deficiency and In Vitro Maturation and the Use of Artificial Oocyte Activation as a Therapeutic Option

8.7.1 In Vitro Maturation (IVM)

As previously discussed, some oocyte proteins are crucial for both oocyte maturation and activation. The in vitro maturation (IVM) of human oocytes is a powerful clinical technique that may be advised in cases when women require fertility preservation, such as young women suffering from cancer that cannot delay chemo- or radiotherapy or women who cannot receive treatment by ovarian hyperstimulation (Chian et al. 2013). While there have been recent advances in the clinical application of this technique in humans, fertilisation and implantation rates for IVM oocytes are lower than those observed when in vivo oocytes are used (Son et al. 2008). Furthermore, there are clear differences in terms of the pattern of Ca^{2+} released at fertilisation when in vivo and in vitro matured oocytes are compared (Nikiforaki et al. 2014). This emphasises the idea that cytoplasmic changes taking place during oocyte maturation are crucial for the proteins involved in Ca^{2+} homeostasis and that this may underlie the observed differences between oocytes matured in vitro and in vivo. In addition to this, non-synchronisation between cytoplasmic and nuclear maturation during IVM has been highlighted as one of the main reasons for poor developmental competence in IVM oocytes (Combelles et al. 2002; Moor et al. 1998; De Vincentiis et al. 2013). All of these findings deserve further attention and may contribute to a better understanding of the links between IVM and OAD.

8.7.2 *Artificial Oocyte Activation (AOA)*

Artificial oocyte activation (AOA) is one of the more recent ART therapies and is used to rescue fertility in couples presenting with a fertilisation rate below 30 % following ICSI (Heindryckx et al. 2005, 2008; Montag et al. 2012). In this technique, mechanical, electric or chemical stimuli are used to help alleviate oocytes from MII arrest (Nasr-Esfahani et al. 2010; Ebner et al. 2012, 2015; Vanden Meerschaut et al. 2014a). Of all these stimuli, chemical activation is the most commonly used procedure and consists of exposing oocytes to chemical agents which increase permeability of the oocyte's plasma membrane to Ca^{2+} contained in the surrounding media. The most commonly used chemical agents are calcium ionophores (calcimycin and ionomycin).

With regard to the safety of AOA, one should note that this technique appears to be safe, with no chromosomal abnormalities or concerns over child health (Heindryckx et al. 2005; Ebner et al. 2012, 2015; Montag et al. 2012; Vanden Meerschaut et al. 2014b; Deemeh et al. 2014; Kim et al. 2015). However, safety concerns still exist, and as Van Blerkom et al. (2015) have recently pointed out, there are basic questions that are yet to be answered. For example, whether, following AOA, miscarriage rates are higher or incidence of minor and major congenital malformations falls within the acceptable limits still remains to be addressed. Because of all these concerns, AOA has been authorised in the UK only recently by the British Human Fertilisation and Embryo Authority (HFEA), the potential risks for embryo viability having been highlighted (Van Blerkom et al. 2015; Thompson 2015). In addition, this authority has emphasised that AOA should only be restricted to selected patients, such as those with PLCζ deficiency (Thompson 2015).

Regarding the aforementioned concerns, one should note that patterns of Ca^{2+} transients differ between normal fertilisation and AOA, which could have an impact upon downstream events (Yanagida et al. 2008). Indeed, while Ca^{2+} oscillations are evoked in normal fertilisation, there is a single Ca^{2+} transient when artificial oocyte activators are used. Furthermore, there is a single transient increase if mouse oocytes are injected with a non-mammalian PLCζ, which negatively affects embryogenesis (Yu et al. 2008). Studies in rabbits revealed that alterations in the pattern of Ca^{2+} oscillations during oocyte activation may be detrimental for organogenesis (Ozil and Huneau 2001). In this scenario, replacing artificial activators with a recombinant PLCζ protein would lead the oocytes to show a Ca^{2+} pattern more similar to that observed in normal fertilisation. This strategy could benefit those cases of OAD related to PLCζ deficiency and represents a far more endogenous and safer alternative (Kashir et al. 2010). In this regard, it is worth noting that such strategy has already been proven to be successful, as Kashir et al. (2011) were able to produce the first recombinant human PLCζ protein, and recombinant PLCζ forms have been found to be able to evoke Ca^{2+} oscillations and start embryogenesis (Kashir et al. 2011; Yoon et al. 2012).

8.8 Conclusions

This chapter intends to summarise the main steps involving sperm and oocyte proteins during the fertilisation and activation of mammalian oocytes. In this aspect, one should note the recent advances being made with regard to the moment at which the acrosome reaction is triggered in the fallopian tube and the precise role of ZP-glycoproteins in this process. In this context, we have also emphasised the relevance of the recent discovery concerning the interaction between the sperm protein IZUMO1 and its interactor JUNO upon the oocyte membrane. In addition to this, we have discussed the sperm proteins involved in the activation of the oocyte, especially PLCζ, although we have included the debate surrounding other putative sperm factors, such as PAWP/WBP2NL. Because the cascade triggered by the SOAF depends heavily upon cellular machinery inside the oocyte, this chapter has also focused upon store-operated Ca^{2+} mechanisms and proteins involved in the regulation of Ca^{2+} homeostasis. To conclude, we contemplate the relevance of OAD in the context of infertility and discuss the potential relationship between IVM and OAD. Finally, we provide a brief description of AOA, the only method available at present which can be used to rescue fertility in OAD and proceed to describe the limitations of this technique and the need to provide a safer and more endogenous alternative.

Authors' Contribution M.Y. wrote the chapter. C.J., S.N.A. and K.C. critically revised the chapter and approved the final version.

Funding M.Y. was funded by the European Commission, FP7-People Programme, Marie Curie-IEF (Grant Number: 626061). S.N.A. is funded by the Ministry of Education (Brunei Darussalam) and the Chancellor's Scholarship, Universiti Brunei Darussalam.

Conflict of Interest The authors declare that there is no conflict of interest in regards to the contents of this chapter.

References

Aarabi M, Balakier H, Bashar S, Moskovtsev SI, Sutovsky P, Librach CL, Oko R (2014a) Sperm-derived WW domain-binding protein, PAWP, elicits calcium oscillations and oocyte activation in humans and mice. FASEB J 28:4434–4440

Aarabi M, Balakier H, Bashar S, Moskovtsev SI, Sutovsky P, Librach CL, Oko R (2014b) Sperm content of postacrosomal WW binding protein is related to fertilization outcomes in patients undergoing assisted reproductive technology. Fertil Steril 102:440–447

Abi Nahed R, Martinez G, Escoffier J, Yassine S, Karaouzène T, Hograindleur JP, Turk J, Kokotos G, Ray PF, Bottari S, Lambeau G, Hennebicq S, Arnoult C (2016) Progesterone-induced acrosome exocytosis requires sequential involvement of calcium-independent phospholipase A2β (iPLA2β) and group X secreted phospholipase A2 (sPLA2). J Biol Chem 291: 3076–3089

Adham IM, Nayernia K, Engel W (1997) Spermatozoa lacking acrosin protein show delayed fertilization. Mol Reprod Dev 46:370–376

Aitken RJ, Nixon B (2013) Sperm capacitation: a distant landscape glimpsed but unexplored. Mol Hum Reprod 19:785–793

Ajduk A, Yamauchi Y, Ward MA (2006) Sperm chromatin remodeling after intra-cytoplasmic sperm injection differs from that of in vitro fertilization. Biol Reprod 75:442–451

Amdani SN, Yeste M, Jones C, Coward K (2015) Sperm factors and oocyte activation: current controversies and considerations. Biol Reprod 93:50. doi:10.1095/biolreprod.115.130609

Amdani SN, Yeste M, Jones C, Coward K (2016) Phospholipase C zeta (PLCζ) and male infertility: clinical update and topical developments. Adv Biol Regul 61:58–67. doi:10.1016/j.jbior.2015.11.009

Baluch DP, Koeneman BA, Hatch KR, McGaughey RW, Capco DG (2004) PKC isotypes in post-activated and fertilized mouse eggs: association with the meiotic spindle. Dev Biol 274:45–55

Barroso G, Valdespin C, Vega E, Kershenovich R, Avila R, Avendaño C, Oehninger S (2009) Developmental sperm contributions: fertilization and beyond. Fertil Steril 92:835–848

Beck-Fruchter R, Lavee M, Weiss A, Geslevich Y, Shalev E (2014) Rescue intracytoplasmic sperm injection: a systematic review. Fertil Steril 101:690–698

Bhanumathy C, da Fonseca PC, Morris EP, Joseph SK (2012) Identification of functionally critical residues in the channel domain of inositol trisphosphate receptors. J Biol Chem 287: 43674–43684

Bhattacharya S, Maheshwari A, Mollison J (2013) Factors associated with failed treatment: an analysis of 121,744 women embarking on their first IVF cycles. PLoS One 8:e82249. doi:10.1371/journal.pone.0082249

Bi M, Hickox R, Winfrey P, Olson E, Hardy M (2003) Processing, localization and binding activity of zonadhesin suggest a function in sperm adhesion to the zona pellucida during exocytosis of the acrosome. Biochem J 375:477–488

Bianchi E, Wright GJ (2015) Cross-species fertilization: the hamster egg receptor, Juno, binds the human sperm ligand, Izumo1. Philos Trans R Soc Lond B Biol Sci 370:20140101. doi:10.1098/rstb.2014.0101

Bianchi E, Doe B, Goulding D, Wright GJ (2014) Juno is the egg Izumo receptor and is essential for mammalian fertilization. Nature 508:483–487

Blobel CP (2000) Functional processing of fertilin: evidence for a critical role of proteolysis in sperm maturation and activation. Rev Reprod 5:75–83

Bookbinder H, Cheng A, Bleil D (1995) Tissue- and species-specific expression of SP56, a mouse sperm fertilization protein. Science 269:86–89

Bromfield E, Aitken RJ, Nixon B (2015) Novel characterization of the HSPA2-stabilizing protein BAG6 in human spermatozoa. Mol Hum Reprod 21:755–769

Buffone MG, Hirohashi N, Gerton GL (2014) Unresolved questions concerning mammalian sperm acrosomal exocytosis. Biol Reprod 90:112. doi:10.1095/biolreprod.114.117911

Burkart AD, Xiong B, Baibakov B, Jiménez-Movilla M, Dean J (2012) Ovastacin, a cortical granule protease, cleaves ZP2 in the zona pellucida to prevent polyspermy. J Cell Biol 197: 37–44

Busso D, Cohen DJ, Hayashi M, Kasahara M, Cuasnicú PS (2005) Human testicular protein TPX1/CRISP-2: localization in spermatozoa, fate after capacitation and relevance for gamete interaction. Mol Hum Reprod 11:299–305

Busso D, Cohen DJ, Maldera JA, Dematteis A, Cuasnicu PS (2007) A novel function for CRISP1 in rodent fertilization: involvement in sperm-zona pellucida interaction. Biol Reprod 77: 848–854

Chen Q, Zhang W, Ran H, Feng L, Yan H, Mu X, Han Y, Liu W, Xia G, Wang C (2014) PKCδ and θ possibly mediate FSH-induced mouse oocyte maturation via NOX-ROS-TACE cascade signaling pathway. PLoS One 9:e111423. doi:10.1371/journal.pone.0111423

Cheon B, Lee HC, Wakai T, Fissore RA (2013) Ca2+ influx and the store-operated Ca2+ entry pathway undergo regulation during mouse oocyte maturation. Mol Biol Cell 24:1396–1410

Chian RC, Uzelac PS, Nargund G (2013) In vitro maturation of human immature oocytes for fertility preservation. Fertil Steril 99:1173–1181

Cohen DJ, Ellerman DA, Busso D, Morgenfeld MM, Piazza AD, Hayashi M, Young ET, Kasahara M, Cuasnicu PS (2001) Evidence that human epididymal protein ARP plays a role in gamete fusion through complementary sites on the surface of the human egg. Biol Reprod 65:1000–1005

Cohen DJ, Maldera JA, Vasen G, Ernesto JI, Muñoz MW, Battistone MA, Cuasnicú PS (2011) Epididymal protein CRISP1 plays different roles during the fertilization process. J Androl 32: 672–678

Combelles CM, Cekleniak NA, Racowsky C, Albertini DF (2002) Assessment of nuclear and cytoplasmic maturation in in-vitro matured human oocytes. Hum Reprod 17:1006–1016

Conner J, Lefievre L, Hughes C, Barratt L (2005) Cracking the egg: increased complexity in the zona pellucida. Hum Reprod 20:1148–1152

Cooney MA, Malcuit C, Cheon B, Holland MK, Fissore RA, D'Cruz NT (2010) Species-specific differences in the activity and nuclear localization of murine and bovine phospholipase C zeta 1. Biol Reprod 83:92–101

Cox LJ, Larman MG, Saunders CM, Hashimoto K, Swann K, Lai FA (2002) Sperm phospholipase Czeta from humans and cynomolgus monkeys triggers Ca2+ oscillations, activation and development of mouse oocytes. Reproduction 124:611–623

Cross L, Morales P, Overstreet W, Hanson W (1988) Induction of acrosome reactions by the human zona pellucida. Biol Reprod 38:235–244

Dale B, DeFelice LJ, Ehrenstein G (1985) Injection of a soluble sperm extract into sea urchin eggs triggers the cortical reaction. Experientia 41:1068–1070

Dale B, Wilding M, Coppola G, Tosti E (2010) How do spermatozoa activate oocytes? Reprod Biomed Online 21:1–3

De Vincentiis S, De Martino E, Buffone MG, Brugo-Olmedo S (2013) Use of metaphase I oocytes matured in vitro is associated with embryo multinucleation. Fertil Steril 99:414–421

Deemeh MR, Tavalaee M, Nasr-Esfahani MH (2014) Health of children born through artificial oocyte activation: a pilot study. Reprod Sci 22:322–328

DeSouza N, Reiken S, Ondrias K, Yang YM, Matkovich S, Marks AR (2002) Protein kinase A and two phosphatases are components of the inositol 1,4,5-trisphosphate receptor macromolecular signaling complex. J Biol Chem 277:39397–39400

Dietzel E, Wessling J, Floehr J, Schäfer C, Ensslen S, Denecke B, Rösing B, Neulen J, Veitinger T, Spehr M, Tropartz T, Tolba R, Renné T, Egert A, Schorle H, Gottenbusch Y, Hildebrand A, Yiallouros I, Stöcker W, Weiskirchen R, Jahnen-Dechent W (2013) Fetuin-B, a liver-derived plasma protein is essential for fertilization. Dev Cell 25:106–112

Ducibella T, Schultz RM, Ozil JP (2006) Role of calcium signals in early development. Semin Cell Dev Biol 17:324–332

Ebner T, Köster M, Shebl O, Moser M, Van der Ven H, Tews G, Montag M (2012) Application of a ready-to-use calcium ionophore increases rates of fertilization and pregnancy in severe male factor infertility. Fertil Steril 98:1432–1437

Ebner T, Montag M, Oocyte Activation Study Group, Montag M, Van der Ven K, Van der Ven H, Ebner T, Shebl O, Oppelt P, Hirchenhain J, Krüssel J, Maxrath B, Gnoth C, Friol K, Tigges J, Wünsch E, Luckhaus J, Beerkotte A, Weiss D, Grunwald K, Struller D, Etien C (2015) Live birth after artificial oocyte activation using a ready-to-use ionophore: a prospective multicentre study. Reprod Biomed Online 30:359–365

Eliyahu E, Shalgi R (2002) A role for protein kinase C during rat egg activation. Biol Reprod 67: 189–195

Eliyahu E, Tsaadon A, Shtraizent N, Shalgi R (2005) The involvement of protein kinase C and actin filaments in cortical granule exocytosis in the rat. Reproduction 129:161–170

Eliyahu E, Shtraizent N, Tsaadon A, Shalgi R (2006) Association between myristoylated alanin-rich C kinase substrate (MARCKS) translocation and cortical granule exocytosis in rat eggs. Reproduction 131:221–231

Escoffier J, Lee HC, Yassine S, Zouari R, Martinez G, Karaouzène T, Coutton C, Kherraf ZE, Halouani L, Triki C, Nef S, Thierry-Mieg N, Savinov SN, Fissore R, Ray PF, Arnoult C (2016) Homozygous mutation of PLCZ1 leads to defective human oocyte activation and infertility that is not rescued by the WW-binding protein PAWP. Hum Mol Genet 25:878–891

Feske S, Gwack Y, Prakriya M, Srikanth S, Puppel SH, Tanasa B, Hogan PG, Lewis RS, Daly M, Rao A (2006) A mutation in Orai1 causes immune deficiency by abrogating CRAC channel function. Nature 441:179–185

Fusi F, Montesano M, Bernocchi N, Panzeri C, Ferrara F, Villa A, Bronson R (1996) P-selectin is expressed on the oolemma of human and hamster oocytes following sperm adhesion and is also detected on the equatorial region of acrosome-reacted human spermatozoa. Mol Hum Reprod 2:341–347

Gallicano GI, McGaughey RW, Capco DG (1997) Activation of protein kinase C after fertilization is required for remodeling the mouse egg into the zygote. Mol Reprod Dev 46:587–601

Gao Z, Garbers DL (1998) Species diversity in the structure of zonadhesin, a sperm-specific membrane protein containing multiple cell adhesion molecule-like domains. J Biol Chem 273:3415–3421

Gasca S, Reyftmann L, Pellestor F, Rème T, Assou S, Anahory T, Dechaud H, Klein B, De Vos J, Hamamah S (2008) Total fertilization failure and molecular abnormalities in metaphase II oocytes. Reprod Biomed Online 17:772–781

Goud PT, Goud AP, Van Oostveldt P, Dhont M (1999) Presence and dynamic redistribution of type I inositol 1,4,5-trisphosphate receptors in human oocytes and embryos during in-vitro maturation, fertilization and early cleavage divisions. Mol Hum Reprod 5:441–451

Goudet G, Mugnier S, Callebaut I, Monget P (2008) Phylogenetic analysis and identification of pseudogenes reveal a progressive loss of zona pellucida genes during evolution of vertebrates. Biol Reprod 78:796–806

Green DPL (1987) Mammalian sperm can not penetrate the zona pellcuida soley by force. Exp Cell Res 169:31–38

Grøndahl ML, Borup R, Vikeså J, Ernst E, Andersen CY, Lykke-Hartmann K (2013) The dormant and the fully competent oocyte: comparing the transcriptome of human oocytes from primordial follicles and in metaphase II. Mol Hum Reprod 19:600–617

Gudlur A, Quintana A, Zhou Y, Hirve N, Mahapatra S, Hogan PG (2014) STIM1 triggers a gating rearrangement at the extracellular mouth of the ORAI1 channel. Nat Commun 5:5164

Gupta SK (2014) Unraveling the intricacies of mammalian fertilization. Asian J Androl 16: 801–802

Gupta SK (2015) Role of zona pellucida glycoproteins during fertilization in humans. J Reprod Immunol 108:90–97

Gupta SK, Bhandari B, Shrestha A, Biswal BK, Palaniappan C, Malhotra SS, Gupta N (2012) Mammalian zona pellucida glycoproteins: structure and function during fertilization. Cell Tissue Res 349:665–678

Halet G, Tunwell R, Parkinson SJ, Carroll J (2004) Conventional PKCs regulate the temporal pattern of Ca2+ oscillations at fertilization in mouse eggs. J Cell Biol 164:1033–1044

Harada Y, Matsumoto T, Hirahara S, Nakashima A, Ueno S, Oda S, Miyazaki S, Iwao Y (2007) Characterization of a sperm factor for egg activation at fertilization of the newt Cynops pyrrhogaster. Dev Biol 306:797–808

Harris JD, Hibler DW, Fontenot GK, Hsu KT, Yurewicz EC, Sacco AG (1994) Cloning and characterization of zona pellucida genes and cDNAs from a variety of mammalian species: the ZPA, ZPB and ZPC gene families. DNA Seq 4:361–393

Heindryckx B, Van der Elst J, De Sutter P, Dhont M (2005) Treatment option for sperm- or oocyte-related fertilization failure: assisted oocyte activation following diagnostic heterologous ICSI. Hum Reprod 20:2237–2241

Heindryckx B, De Gheselle S, Gerris J, Dhont M, De Sutter P (2008) Efficiency of assisted oocyte activation as a solution for failed intracytoplasmic sperm injection. Reprod Biomed Online 17:662–668

Heytens E, Parrington J, Coward K, Young C, Lambrecht S, Yoon SY, Fissore RA, Hamer R, Deane CM, Ruas M, Grasa P, Soleimani R, Cuvelier CA, Gerris J, Dhont M, Deforce D, Leybaert L, De Sutter P (2009) Reduced amounts and abnormal forms of phospholipase C zeta (PLCzeta) in spermatozoa from infertile men. Hum Reprod 24:2417–2428

Heytens E, Schmitt-John T, Moser JM, Jensen NM, Soleimani R, Young C, Coward K, Parrington J, De Sutter P (2010) Reduced fertilization after ICSI and abnormal phospholipase C zeta presence in spermatozoa from the wobbler mouse. Reprod Biomed Online 21:742–749

Honda A, Siruntawinwti J, Baba T (2002) Role of acrosomal matrix proteinases in sperm-zona pellucida interactions. Hum Reprod Update 5:405–412

Hoodbhoy T, Joshi S, Boja ES, Williams SA, Stanley P, Dean J (2005) Human sperm do not bind to rat zonae pellucidae despite the presence of four homologous glycoproteins. J Biol Chem 280:12721–12731

Horner VL, Wolfner MF (2008) Transitioning from egg to embryo: triggers and mechanisms of egg activation. Dev Dyn 237:527–544

Inoue N, Ikawa M, Isotani A, Okabe M (2005) The immunoglobulin superfamily protein Izumo is required for sperm to fuse with eggs. Nature 434:234–238

Inoue N, Ikawa M, Okabe M (2008) Putative sperm fusion protein IZUMO and the role of N-glycosylation. Biochem Biophys Res Comm 377:910–914

Inoue N, Hamada D, Kamikubo H, Hirata K, Kataoka M, Yamamoto M, Ikawa M, Okabe M, Hagihara Y (2013) Molecular dissection of IZUMO1, a sperm protein essential for sperm-egg fusion. Development 140:3221–3229

Inoue N, Hagihara Y, Wright D, Suzuki T, Wada I (2015) Oocyte-triggered dimerization of sperm IZUMO1 promotes sperm-egg fusion in mice. Nat Commun 6:8858. doi:10.1038/ncomms9858

Izquierdo-Rico MJ, Jiménez-Movilla M, Llop E, Pérez-Oliva AB, Ballesta J, Gutiérrez-Gallego R, Jiménez-Cervantes C, Avilés M (2009) Hamster zona pellucida is formed by four glycoproteins: ZP1, ZP2, ZP3, and ZP4. J Proteome Res 8:926–941

Jaffe F (1983) Sources of calcium in egg activation: a review and hypothesis. Dev Biol 99:265–276

Johnson MH (2011) Robert Edwards: the path to IVF. Reprod Biomed Online 23:245–262

Jones KT (2007) Intracellular calcium in the fertilization and development of mammalian eggs. Clin Exp Pharmacol Physiol 34:1084–1089

Jones KT, Soeller C, Cannell MB (1998) The passage of $Ca2+$ and fluorescent markers between the sperm and egg after fusion in the mouse. Development 125:4627–4635

Kaji K, Oda S, Shikano T, Ohnuki T, Uematsu Y, Sakagami J, Tada N, Miyazaki S, Kudo A (2000) The gamete fusion process is defective in eggs of CD9-deficient mice. Nat Genet 24:279–282

Kashir J, Heindryckx B, Jones C, De Sutter P, Parrington J, Coward K (2010) Oocyte activation, phospholipase C zeta and human infertility. Hum Reprod Update 16:690–703

Kashir J, Jones C, Lee HC, Rietdorf K, Nikiforaki D, Durrans C, Ruas M, Tee ST, Heindryckx B, Galione A, De Sutter P, Fissore RA, Parrington J, Coward K (2011) Loss of activity mutations in phospholipase C zeta (PLCζ) abolishes calcium oscillatory ability of human recombinant protein in mouse oocytes. Hum Reprod 26:3372–3387

Kashir J, Konstantinidis M, Jones C, Heindryckx B, De Sutter P, Parrington J, Wells D, Coward K (2012a) Characterization of two heterozygous mutations of the oocyte activation factor phospholipase C zeta (PLCζ) from an infertile man by use of minisequencing of individual sperm and expression in somatic cells. Fertil Steril 98:423–431

Kashir J, Konstantinidis M, Jones C, Lemmon B, Lee HC, Hamer R, Heindryckx B, Deane CM, De Sutter P, Fissore RA, Parrington J, Wells D, Coward K (2012b) A maternally inherited autosomal point mutation in human phospholipase C zeta (PLCζ) leads to male infertility. Hum Reprod 27:222–231

Kashir J, Deguchi R, Jones C, Coward K, Stricker SA (2013) Comparative biology of sperm factors and fertilization-induced calcium signals across the animal kingdom. Mol Reprod Dev 80:787–815

Kasri NN, Török K, Galione A, Garnham C, Callewaert G, Missiaen L, Parys JB, De Smedt H (2006) Endogenously bound calmodulin is essential for the function of the inositol 1,4,5-trisphosphate receptor. J Biol Chem 281:8332–8338

Kennedy CE, Krieger KB, Sutovsky M, Xu W, Vargovič P, Didion BA, Ellersieck MR, Hennessy ME, Verstegen J, Oko R, Sutovsky P (2014) Protein expression pattern of PAWP in bull spermatozoa is associated with sperm quality and fertility following artificial insemination. Mol Reprod Dev 81:436–449

Kilani S, Chapman MG (2014) Meiotic spindle normality predicts live birth in patients with recurrent in vitro fertilization failure. Fertil Steril 101:403–406

Kim JW, Yang SH, Yoon SH, Kim SD, Jung JH, Lim JH (2015) Successful pregnancy and delivery after ICSI with artificial oocyte activation by calcium ionophore in in-vitro matured oocytes: a case report. Reprod Biomed Online 30:373–377

Kline D, Kline JT (1992) Repetitive calcium transients and the role of calcium in exocytosis and cell cycle activation in the mouse egg. Dev Biol 149:80–89

Kline D, Simoncini L, Mandel G, Maue RA, Kado RT, Jaffe LA (1988) Fertilization events induced by neurotransmitters after injection of mRNA in xenopus eggs. Science 241:464–467

Knott JG, Gardner AJ, Madgwick S, Jones KT, Williams CJ, Schultz RM (2006) Calmodulin-dependent protein kinase II triggers mouse egg activation and embryo development in the absence of Ca^{2+} oscillations. Dev Biol 296:388–395

Kudo K, Yonezawy N, Katsumata T, Aoki H, Nakano M (1998) Localization of carbohydrate chains of pig sperm ligand in the glycoprotein ZPB of egg zona pellucida. Eur J Biol 252: 492–499

La Spina FA, Puga Molina LC, Romarowski A, Vitale AM, Falzone TL, Krapf D, Hirohashi N, Buffone MG (2016) Mouse sperm begin to undergo acrosomal exocytosis in the upper isthmus of the oviduct. Dev Biol 411:172–182

Lee B, Yoon SY, Malcuit C, Parys JB, Fissore RA (2010) Inositol 1,4,5-trisphosphate receptor 1 degradation in mouse eggs and impact on [Ca2+]i oscillations. J Cell Physiol 222:238–247

Lefièvre L, Conner SJ, Salpekar A, Olufowobi O, Ashton P, Pavlovic B, Lenton W, Afnan M, Brewis IA, Monk M, Hughes DC, Barratt CL (2004) Four zona pellucida glycoproteins are expressed in the human. Hum Reprod 19:1580–1586

Li Y, Li RQ, Ou SB, Zhang NF, Ren L, Wei LN, Zhang QX, Yang DZ (2014) Increased GDF9 and BMP15 mRNA levels in cumulus granulosa cells correlate with oocyte maturation, fertilization, and embryo quality in humans. Reprod Biol Endocrinol 12:81. doi:10.1186/1477-7827-12-81

Linfor J, Berger T (2000) Potential role of αv and β1 integrins as oocyte adhesion molecules during fertilization in pigs. J Reprod Fertil 120:65–72

Louis JF, Thoma ME, Sørensen DN, McLain AC, King RB, Sundaram R, Keiding N, Buck Louis GM (2013) The prevalence of couple infertility in the United States from a male perspective: evidence from a nationally representative sample. Andrology 1:741–748

Luria A, Tennenbaum T, Sun QY, Rubinstein S, Breitbart H (2000) Differential localization of conventional protein kinase C isoforms during mouse oocyte development. Biol Reprod 62: 1564–1570

Machaca K (2007) Ca2+ signaling differentiation during oocyte maturation. J Cell Physiol 213: 331–340

Madgwick S, Levasseur M, Jones KT (2005) Calmodulin-dependent protein kinase II, and not protein kinase C, is sufficient for triggering cell-cycle resumption in mammalian eggs. J Cell Sci 118:3849–3859

Mahutte NG, Arici A (2003) Failed fertilization: is it predictable? Curr Opin Obstet Gynecol 15: 211–218

Malcuit C, Knott JG, He C, Wainwright T, Parys JB, Robl JM, Fissore RA (2005) Fertilization and inositol 1,4,5-trisphosphate (IP3)-induced calcium release in type-1 inositol 1,4,5-trisphosphate receptor down-regulated bovine eggs. Biol Reprod 73:2–13

Maldera JA, Weigel Muñoz M, Chirinos M, Busso D, GE Raffo F, Battistone MA, Blaquier JA, Larrea F, Cuasnicu PS (2014) Human fertilization: epididymal hCRISP1 mediates sperm-zona pellucida binding through its interaction with ZP3. Mol Hum Reprod 20:341–349

Mao HT, Yang WX (2013) Modes of acrosin functioning during fertilization. Gene 526:75–79

Miyado K, Yamada G, Yamada S, Hasuwa H, Nakamura Y, Ryu F, Suzuki K, Kosai K, Inoue K, Ogura A, Okabe M, Mekada E (2000) Requirement of CD9 on the egg plasma membrane for fertilization. Science 287:321–324

Miyara F, Aubriot FX, Glissant A, Nathan C, Douard S, Stanovici A, Herve F, Dumont-Hassan M, LeMeur A, Cohen-Bacrie P, Debey P (2003) Multiparameter analysis of human oocytes at metaphase II stage after IVF failure in non-male infertility. Hum Reprod 18:1494–1503

Montag M, Köster M, van der Ven K, Bohlen U, van der Ven H (2012) The benefit of artificial oocyte activation is dependent on the fertilization rate in a previous treatment cycle. Reprod Biomed Online 24:521–526

Moor RM, Dai Y, Lee C, Fulka J Jr (1998) Oocyte maturation and embryonic failure. Hum Reprod Update 4:223–236

Mugnier S, Dell'Aquila ME, Pelaez J, Douet C, Ambruosi B, De Santis T, Lacalandra GM, Lebos C, Sizaret PY, Delaleu B, Monget P, Mermillod P, Magistrini M, Meyers SA, Goudet G (2009) New insights into the mechanisms of fertilization: comparison of the fertilization steps, composition, and structure of the zona pellucida between horses and pigs. Biol Reprod 81:856–870

Munuce MJ, Quintero I, Caille AM, Ghersevich S, Berta CL (2006) Comparative concentrations of steroid hormones and proteins in human peri-ovulatory peritoneal and follicular fluids. Reprod Biomed Online 13:202–207

Nasr-Esfahani MH, Deemeh MR, Tavalaee M (2010) Artificial oocyte activation and intracytoplasmic sperm injection. Fertil Steril 94:520–526

Neri QV, Lee B, Rosenwaks Z, Machaca K, Palermo GD (2014) Understanding fertilization through intracytoplasmic sperm injection (ICSI). Cell Calcium 55:24–37

Nikiforaki D, Vanden Meerschaut F, Qian C, De Croo I, Lu Y, Deroo T, Van den Abbeel E, Heindryckx B, De Sutter P (2014) Oocyte cryopreservation and in vitro culture affect calcium signalling during human fertilization. Hum Reprod 29:29–40

Noguchi S, Yonezawa N, Katsumata T, Hashizume K, Kuwayama M, Hamano S, Watanabe S, Nakano M (1994) Characterization of the zona pellucida glycoproteins from bovine ovarian and fertilized eggs. Biochim Biophys Acta 1201:7–14

Nomikos M, Elgmati K, Theodoridou M, Calver BL, Cumbes B, Nounesis G, Swann K, Lai FA (2011a) Male infertility-linked point mutation disrupts the Ca2+ oscillation-inducing and PIP_2 hydrolysis activity of sperm PLCζ. Biochem J 434:211–217

Nomikos M, Elgmati K, Theodoridou M, Georgilis A, Gonzalez-Garcia JR, Nounesis G, Swann K, Lai FA (2011b) Novel regulation of PLCζ activity via its XY-linker. Biochem J 438:427–432

Nomikos M, Sanders JR, Theodoridou M, Kashir J, Matthews E, Nounesis G, Lai FA, Swann K (2014) Sperm-specific post-acrosomal WW-domain binding protein (PAWP) does not cause Ca2+ release in mouse oocytes. Mol Hum Reprod 20:938–947

Nomikos M, Sanders JR, Kashir J, Sanusi R, Buntwal L, Love D, Ashley P, Sanders D, Knaggs P, Bunkheila A, Swann K, Lai FA (2015) Functional disparity between human PAWP and PLCζ in the generation of Ca2+ oscillations for oocyte activation. Mol Hum Reprod 21:702–710

O'Rand M, Welch JE, Fisher SJ (1986) Sperm membrane and zona pellucida interaction during fertilization. In: Dhinsa DS, Bahl O (eds) Molecular and cellular aspects of reproduction. Plenum Press, New York, pp 131–144

Ombelet W, Cooke I, Dyer S, Serour G, Devroey P (2008) Infertility and the provision of infertility medical services in developing countries. Hum Reprod Update 14:605–621

Ozil JP, Huneau D (2001) Activation of rabbit oocytes: the impact of the Ca2+ signal regime on development. Development 128:917–928

Pang PC, Chiu PC, Lee CL, Chang LY, Panico M, Morris HR, Haslam SM, Khoo KH, Clark GF, Yeung WS, Dell A (2011) Human sperm binding is mediated by the sialyl-Lewis(x) oligosaccharide on the zona pellucida. Science 333:1761–1764

Parrington J, Swann K, Shevchenko VI, Sesay AK, Lai FA (1996) Calcium oscillations in mammalian eggs triggered by a soluble sperm protein. Nature 379:364–368

Pesty A, Avazeri N, Lefèvre B (1998) Nuclear calcium release by InsP3-receptor channels plays a role in meiosis reinitiation in the mouse oocyte. Cell Calcium 24:239–251

Petit FM, Serres C, Bourgeon F, Pineau C, Auer J (2013) Identification of sperm head proteins involved in zona pellucida binding. Hum Reprod 28:852–865

Putney JW (2009) Capacitative calcium entry: from concept to molecules. Immunol Rev 231: 10–22

Putney JW (2011) Origins of the concept of store-operated calcium entry. Front Biosci 3:980–984

Roos J, DiGregorio PJ, Yeromin AV, Ohlsen K, Lioudyno M, Zhang S, Safrina O, Kozak JA, Wagner SL, Cahalan MD, Veliçelebi G, Stauderman KA (2005) STIM1, an essential and conserved component of store-operated Ca2+ channel function. J Cell Biol 169:435–445

Ross PJ, Beyhan Z, Iager AE, Yoon SY, Malcuit C, Schellander K, Fissore RA, Cibelli JB (2008) Parthenogenetic activation of bovine oocytes using bovine and murine phospholipase C zeta. BMC Dev Biol 8:16. doi:10.1186/1471-213X-8-16

Sabetian S, Shamsir MS, Abu Naser M (2014) Functional features and protein network of human sperm-egg interaction. Syst Biol Reprod Med 60:329–337

Saldívar-Hernández A, González-González ME, Sánchez-Tusié A, Maldonado-Rosas I, López P, Treviño CL, Larrea F, Chirinos M (2015) Human sperm degradation of zona pellucida proteins contributes to fertilization. Reprod Biol Endocrinol 13:99

Satouh Y, Nozawa K, Ikawa M (2015) Sperm postacrosomal WW domain-binding protein is not required for mouse egg activation. Biol Reprod 93:94

Saunders CM, Larman MG, Parrington J, Cox LJ, Royse J, Blayney LM, Swann K, Lai FA (2002) PLC zeta: a sperm-specific trigger of Ca(2+) oscillations in eggs and embryo development. Development 129:3533–3544

Saunders CM, Swann K, Lai FA (2007) PLCzeta, a sperm-specific PLC and its potential role in fertilization. Biochem Soc Symp 74:23–36

Schuffner AA, Bastiaan HS, Duran HE, Lin ZY, Morshedi M, Franken DR, Oehninger S (2002) Zona pellucida-induced acrosome reaction in human sperm: dependency on activation of pertussis toxin-sensitive Gi protein and extracellular calcium, and priming effect of progesterone and follicular fluid. Mol Hum Reprod 8:722–727

Sengoku K, Takuma N, Miyamoto T, Horikawa M, Ishikawa M (2004) Integrins are not involved in the process of human sperm-oolemmal fusion. Hum Reprod 19:639–644

Sette C, Paronetto MP, Barchi M, Bevilacqua A, Geremia R, Rossi P (2002) Tr-kit-induced resumption of the cell cycle in mouse eggs requires activation of a Src-like kinase. EMBO J 21:5386–5395

Smyth JT, Hwang SY, Tomita T, DeHaven WI, Mercer JC, Putney JW (2010) Activation and regulation of store-operated calcium entry. J Cell Mol Med 14:2337–2349

Son WY, Chung JT, Demirtas E, Holzer H, Sylvestre C, Buckett W, Chian RC, Tan SL (2008) Comparison of in-vitro maturation cycles with and without in-vivo matured oocytes retrieved. Reprod Biomed Online 17:59–67

Stein KK, Primafoff P, Myles D (2004) Sperm-egg fusion: events at the plasma membrane. J Cell Sci 117:6269–6274

Stetson I, Avilés M, Moros C, García-Vázquez FA, Gimeno L, Torrecillas A, Aliaga C, Bernardo-Pisa MV, Ballesta J, Izquierdo-Rico MJ (2015) Four glycoproteins are expressed in the cat zona pellucida. Theriogenology 83:1162–1173

Suarez SS, Pacey AA (2006) Sperm transport in the female reproductive tract. Hum Reprod Update 12:23–37

Sutovsky P, Manandhar G, Wu A, Oko R (2003) Interactions of sperm perinuclear theca with the oocyte: implications for oocyte activation, anti-polyspermy defense, and assisted reproduction. Microsc Res Tech 61:362–378

Swann K (1990) A cytosolic sperm factor stimulates repetitive calcium increases and mimics fertilization in hamster eggs. Development 110:1295–1302

Swann K, Lai FA (2013) PLCζ and the initiation of Ca(2+) oscillations in fertilizing mammalian eggs. Cell Calcium 53:55–62

Swann K, Lai FA (2016) Egg activation at fertilization by a soluble sperm protein. Physiol Rev 96:127–149

Swann K, Yu Y (2008) The dynamics of calcium oscillations that activate mammalian eggs. Int J Dev Biol 52:585–594

Swann K, Larman MG, Saunders CM, Lai FA (2004) The cytosolic sperm factor that triggers Ca2+ oscillations and egg activation in mammals is a novel phospholipase C: PLCzeta. Reproduction 127:431–439

Tanphaichitr N, Kongmanas K, Kruevaisayawan H, Saewu A, Sugeng C, Fernandes J, Souda P, Angel JB, Faull KF, Aitken RJ, Whitelegge J, Hardy D, Berger T, Baker M (2015) Remodeling of the plasma membrane in preparation for sperm-egg recognition: roles of acrosomal proteins. Asian J Androl 17:574–582

Tardif S, Cormier N (2011) Role of zonadhesin during sperm-egg interaction: a species-specific acrosomal molecule with multiple functions. Mol Hum Reprod 17:661–668

Taylor SL, Yoon SY, Morshedi MS, Lacey DR, Jellerette T, Fissore RA, Oehninger S (2010) Complete globozoospermia associated with PLCζ deficiency treated with calcium ionophore and ICSI results in pregnancy. Reprod Biomed Online 20:559–564

Taylor CW, Tovey SC, Rossi AM, Lopez Sanjurjo CI, Prole DL, Rahman T (2014) Structural organization of signalling to and from IP3 receptors. Biochem Soc Trans 42:63–70

Theodoridou M, Nomikos M, Parthimos D, Gonzalez-Garcia JR, Elgmati K, Calver BL, Sideratou Z, Nounesis G, Swann K, Lai FA (2013) Chimeras of sperm PLCζ reveal disparate protein domain functions in the generation of intracellular Ca2+ oscillations in mammalian eggs at fertilization. Mol Hum Reprod 19:852–864

Thoma ME, McLain AC, Louis JF, King RB, Trumble AC, Sundaram R, Buck Louis GM (2013) Prevalence of infertility in the United States as estimated by the current duration approach and a traditional constructed approach. Fertil Steril 99:1324–1331

Thompson P (2015) HFEA response to 'A plea for caution and more research in the "experimental" use of ionophores in ICSI'. Reprod Biomed Online 31:829–830

Töpfer-Petersen E, Ekhlasi-Hundrieser M, Tsolova M (2008) Glycobiology of fertilization in the pig. Int J Dev Biol 52:717–736

Toshimori K, Saxena DK, Tanii I, Yoshinaga K (1998) An MN9 antigenic molecule, equatorin, is required for successful sperm-oocyte fusion in mice. Biol Reprod 59:22–29

Tranter R, Read JA, Jones R, Brady RL (2000) Effector sites in the three-dimensional structure of mammalian sperm β-acrosin. Structure 8:1179–1188

Tsaadon L, Kaplan-Kraicer R, Shalgi R (2008) Myristoylated alanine-rich C kinase substrate, but not Ca2+/calmodulin-dependent protein kinase II, is the mediator in cortical granules exocytosis. Reproduction 135:613–624

Ullah G, Jung P, Machaca K (2007) Modeling Ca2+ signaling differentiation during oocyte maturation. Cell Calcium 42:556–564

Vadnais ML, Foster DN, Roberts KP (2008) Molecular cloning and expression of the CRISP family of proteins in the boar. Biol Reprod 79:1129–1134

Van Blerkom J, Cohen J, Johnson M (2015) A plea for caution and more research in the 'experimental' use of ionophores in ICSI. Reprod Biomed Online 30:323–324

Vanden Meerschaut F, Leybaert L, Nikiforaki D, Qian C, Heindryckx B, De Sutter P (2013) Diagnostic and prognostic value of calcium oscillatory pattern analysis for patients with ICSI fertilization failure. Hum Reprod 28:87–98

Vanden Meerschaut F, Nikiforaki D, Heindryckx B, De Sutter P (2014a) Assisted oocyte activation following ICSI fertilization failure. Reprod Biomed Online 28:560–571

Vanden Meerschaut F, D'Haeseleer E, Gysels H, Thienpont Y, Dewitte G, Heindryckx B, Oostra A, Roeyers H, Van Lierde K, De Sutter P (2014b) Neonatal and neurodevelopmental outcome of children aged 3-10 years born following assisted oocyte activation. Reprod Biomed Online 28:54–63

Vanderheyden V, Wakai T, Bultynck G, De Smedt H, Parys JB, Fissore RA (2009) Regulation of inositol 1,4,5-trisphosphate receptor type 1 function during oocyte maturation by MPM-2 phosphorylation. Cell Calcium 46:56–64

Verbert L, Lee B, Kocks SL, Assefa Z, Parys JB, Missiaen L, Callewaert G, Fissore RA, De Smedt H, Bultynck G (2008) Caspase-3-truncated type 1 inositol 1,4,5-trisphosphate receptor enhances intracellular Ca2+ leak and disturbs Ca2+ signalling. Biol Cell 100:39–49

Vjugina U, Evans JP (2008) New insights into the molecular basis of mammalian sperm-egg membrane interactions. Front Biosci 13:462–476

Von Stetina JR, Orr-Weaver TL (2011) Developmental control of oocyte maturation and egg activation in metazoan models. Cold Spring Harb Perspect Biol 3:a005553

Wakai T, Zhang N, Vangheluwe P, Fissore RA (2013) Regulation of endoplasmic reticulum Ca(2+) oscillations in mammalian eggs. J Cell Sci 126:5714–5724

Wang C, Machaty Z (2013) Calcium influx in mammalian eggs. Reproduction 145:97–105

Wang C, Lee K, Gajdócsi E, Papp AB, Machaty Z (2012) Orai1 mediates store-operated Ca2+ entry during fertilization in mammalian oocytes. Dev Biol 365:414–423

Williams CJ, Mehlmann LM, Jaffe LA, Kopf GS, Schultz RM (1998) Evidence that Gq family G proteins do not function in mouse egg activation at fertilization. Dev Biol 198:116–127

Williams CJ (2002) Signaling mechanisms of mammalian oocyte activation. Hum Reprod Update 8:313–321

Wu AT, Sutovsky P, Manandhar G, Xu W, Katayama M, Day BN, Park KW, Yi YJ, Xi YW, Prather RS, Oko R (2007a) PAWP, a sperm-specific WW domain-binding protein, promotes meiotic resumption and pronuclear development during fertilization. J Biol Chem 282: 12164–12175

Wu AT, Sutovsky P, Xu W, van der Spoel AC, Platt FM, Oko R (2007b) The postacrosomal assembly of sperm head protein, PAWP, is independent of acrosome formation and dependent on microtubular manchette transport. Dev Biol 312:471–483

Yanagida K, Fujikura Y, Katayose H (2008) The present status of artificial oocyte activation in assisted reproductive technology. Reprod Med Biol 7:133–142

Yanagimachi R (1998) Intracytoplasmic sperm injection experiments using the mouse as a model. Hum Reprod 13(Suppl 1):87–98

Yang YR, Follo MY, Cocco L, Suh PG (2013) The physiological roles of primary phospholipase C. Adv Biol Regul 53:232–241

Yassine S, Escoffier J, Martinez G, Coutton C, Karaouzène T, Zouari R, Ravanat JL, Metzler-Guillemain C, Lee HC, Fissore R, Hennebicq S, Ray PF, Arnoult C (2015) Dpy19l2-deficient globozoospermic sperm display altered genome packaging and DNA damage that compromises the initiation of embryo development. Mol Hum Reprod 21:169–185

Yelumalai S, Yeste M, Jones C, Amdani SN, Kashir J, Mounce G, Fatum M, Barratt C, McVeigh E, Coward K (2015) Total levels and proportions of sperm exhibiting phospholipase C Zeta (PLCζ) are significantly correlated with fertilization rates following intracytoplasmic sperm injection. Fertil Steril 104:561–568

Yeste M (2013a) Boar spermatozoa within the oviductal environment (III): fertilisation. In: Bonet S, Casas I, Holt WV, Yeste M (eds) Boar reproduction: fundamentals and new biotechnological trends. Springer, Berlin, pp 343–406

Yeste M (2013b) Boar spermatozoa within the oviductal environment (II): sperm capacitation. In: Bonet S, Casas I, Holt WV, Yeste M (eds) Boar reproduction: fundamentals and new biotechnological trends. Springer, Berlin, pp 281–342

Yeste M, Jones C, Amdani SN, Patel S, Coward K (2016) Oocyte activation deficiency: a role for an oocyte contribution? Hum Reprod Update 22:23–47

Yoneda A, Kashima M, Yoshida S, Terada K, Nakagawa S, Sakamoto A, Hayakawa K, Suzuki K, Ueda J, Watanabe T (2006) Molecular cloning, testicular postnatal expression, and oocyte-activating potential of porcine phospholipase Czeta. Reproduction 132:393–401

Yoon SY, Jellerette T, Salicioni AM, Lee HC, Yoo MS, Coward K, Parrington J, Grow D, Cibelli JB, Visconti PE, Mager J, Fissore RA (2008) Human sperm devoid of PLC, zeta 1 fail to induce Ca(2+) release and are unable to initiate the first step of embryo development. J Clin Invest 118: 3671–3681

Yoon SY, Eum JH, Lee JE, Lee HC, Kim YS, Han JE, Won HJ, Park SH, Shim SH, Lee WS, Fissore RA, Lee DR, Yoon TK (2012) Recombinant human phospholipase C zeta 1 induces intracellular calcium oscillations and oocyte activation in mouse and human oocytes. Hum Reprod 27:1768–1780

Young C, Grasa P, Coward K, Davis LC, Parrington J (2009) Phospholipase C zeta undergoes dynamic changes in its pattern of localization in sperm during capacitation and the acrosome reaction. Fertil Steril 91:2230–2242

Yu Y, Halet G, Lai FA, Swann K (2008) Regulation of diacylglycerol production and protein kinase C stimulation during sperm- and PLCzeta-mediated mouse egg activation. Biol Cell 100:633–643

Yu Y, Nomikos M, Theodoridou M, Nounesis G, Lai FA, Swann K (2012) PLCζ causes Ca(2+) oscillations in mouse eggs by targeting intracellular and not plasma membrane PI(4,5)P(2). Mol Biol Cell 23:371–380

Yurewicz EC, Sacco AG, Ghupta SK, Xu N, Gage DA (1998) Heterooligomerisation dependent binding of pig oocyte zona pellucida glycoproteins ZPB and ZPC to boar sperm membrane vesicles. J Biol Chem 273:7488–7494

Zhang SL, Yu Y, Roos J, Kozak JA, Deerinck TJ, Ellisman MH, Stauderman KA, Cahalan MD (2005) STIM1 is a Ca2+ sensor that activates CRAC channels and migrates from the Ca^{2+} store to the plasma membrane. Nature 437:902–905

Chapter 9
Sperm Navigation Mechanisms in the Female Reproductive Tract

Hieu D. Hoang and Michael A. Miller

Abstract Fertilization, the union of an oocyte and a sperm, is a fundamental process that restores the diploid genome and initiates embryonic development. For the sperm, fertilization is the end of a long journey, one that starts in the male testis before transitioning to the female reproductive tract's convoluted tubule architecture. Historically, motile sperm were thought to complete this journey using luck and numbers. A different picture of sperm has emerged recently as cells that integrate complex sensory information for navigation. Chemical, physical, and thermal cues have been proposed to help guide sperm to the waiting oocyte. Molecular mechanisms are being delineated in animal models and humans, revealing common features, as well as important differences. Exposure to pheromones and nutritional signals can modulate guidance mechanisms, indirectly impacting sperm motility performance and fertility. These studies highlight the importance of sensory information and signal transduction in fertilization.

9.1 Introduction

Sexual reproduction is essential for the survival and evolution of animal species (Hoekstra 1987; Czaran and Hoekstra 2004). A key process is meiosis, which generates haploid (1N) gametes that fuse during fertilization to restore the diploid (2N) chromosomal content (Hunt and LeMaire-Adkins 1998; Hillers et al. 2015). Females generate nutrient-rich, immotile gametes called oocytes, whereas males generate compact, motile spermatozoa (hereinafter referred to as sperm) (Matova and Cooley 2001). Sperm complete the meiotic divisions during spermatogenesis. Oocytes, on the other hand, are arrested in the meiotic prophase while growth occurs by synthesis and provision of nutrients, RNAs, lipids, and proteins. Meiosis resumes during oocyte maturation, a process that prepares the oocyte for

H.D. Hoang • M.A. Miller (✉)
Department of Cell, Developmental and Integrative Biology, University of Alabama at Birmingham School of Medicine, Birmingham, AL, USA
e-mail: mamiller@uab.edu

S. Arur (ed.), *Signaling-Mediated Control of Cell Division*, Results and Problems in Cell Differentiation 59, DOI 10.1007/978-3-319-44820-6_9

fertilization and embryogenesis (Masui 1985; Dekel 2005). Oocyte maturation is coupled to ovulation, which expulses the oocyte from the ovary into the oviduct. After ovulation, oocytes are arrested again in meiosis, most frequently in metaphase I or metaphase II, depending on the species (Marcello et al. 2013; Runft et al. 2002). Sperm must fertilize the oocyte during this second meiotic arrest. The fertilization window is short, ranging from a few minutes in species like the nematode *C. elegans* to a few days in some mammals. The sperm and oocyte must meet at the right place and time to generate a viable embryo. Sexually reproducing animals are left with a choice: either rely on chance or control the meeting.

Animal reproduction falls into two general classes, those species where fertilization occurs in the external environment and those where fertilization occurs in the female reproductive tract. External fertilizers include aquatic species, such as hydrozoans, sea urchins, and frogs (Miller et al. 2000; Kaupp et al. 2008; Olson et al. 2001). In these animals, males and females (or hermaphrodites) spawn their gametes into the aquatic environment, often in response to environmental triggers and other signals. Oocytes secrete positional cues called chemoattractants designed to attract sperm from conspecific males. Internal fertilizers include many terrestrial species, such as nematodes, fruit flies, and mammals (Han et al. 2010; Marcello et al. 2013; Loppin et al. 2015; Evans and Florman 2002). During mating, males provide an ejaculate containing sperm and other molecules into the female's uterus. Sperm then swim, or crawl in the case of nematodes, to the fertilization site, which can be a long distance away. In some species, females have sperm storage sites that protect sperm and house sperm battles from competing males (Neubaum and Wolfner 1999; Holt and Fazeli 2016). Fertilization and storage sites can be the same or different tissues. A confounding issue in internally fertilizing animals is that mating can occur when oocytes are in the wrong stage of meiosis. For instance, mammalian females undergo estrus, a hormone-driven cycle that coordinates oocyte maturation with other processes and behaviors important for reproduction (Boden et al. 2013). The length of estrus varies from days to weeks. To fertilize an oocyte, sperm have a formidable task: find a maturing oocyte within the female reproductive tract's convoluted architecture. The tract is a hostile environment, equipped to keep out pathogens, support early embryonic development, and expel embryos.

Several mechanisms have been proposed to control the sperm and oocyte encounter in the oviduct, the tube through which ovulated oocytes pass (Fig. 9.1). The first is the competitive race model (Eisenbach and Giojalas 2006; Eisenbach and Tur-Kaspa 1999). In this model, males inseminate large numbers of sperm that race to the fertilization site but are blind to its whereabouts. It is inherently inefficient, as most sperm fail to find a maturing oocyte and are instead expelled from the uterus or killed internally. As methods to evaluate sperm motility improve, there is little support for the competitive race model. Other proposed mechanisms include chemotaxis, rheotaxis, contractile forces, and thermotaxis (Fig. 9.1) (Perez-Cerezales et al. 2015b; Miki and Clapham 2013; Avila et al. 2011; Avila and Wolfner 2009). Chemotaxis is best understood in externally fertilizing species. Oocytes secrete chemical signals that diffuse away, forming a concentration

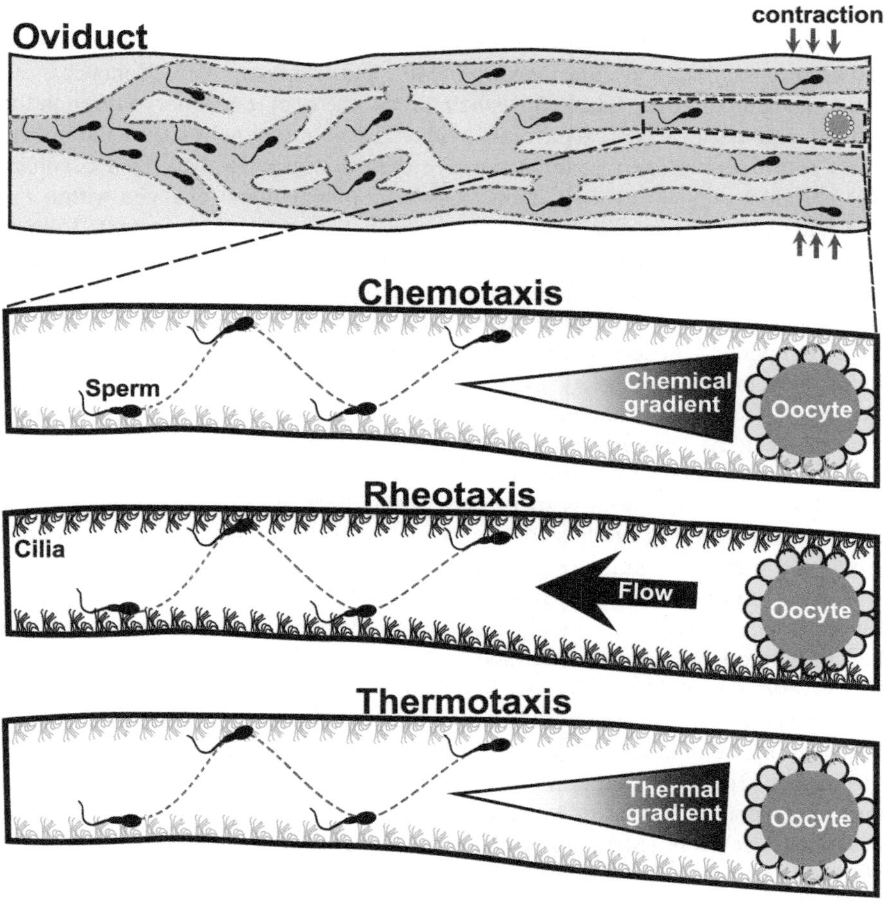

Fig. 9.1 Proposed in vivo sperm guidance mechanisms. Anatomical landmarks are based on the mammalian oviduct, but the principles apply to most animal species. Muscle contractions and cilia lining the oviduct generate fluid flow important for rheotaxis

gradient. Sperm interpret this gradient, altering their flagellar beating pattern to direct swimming toward the source (Kaupp et al. 2008). The most detailed molecular mechanisms have been delineated in the sea urchin *Arbacia punctulata*. The chemoattractant peptide resact binds to a receptor-type guanylyl cyclase, triggering a series of events that open Ca^{2+} channels in the flagellar membrane (Ward et al. 1985; Kaupp et al. 2008). Ca^{2+} influx changes the flagellar beat. By analogy, oocytes or surrounding cells have been proposed to secrete chemoattractants into the oviduct, thereby directing sperm to their location (Fig. 9.1). There is evidence for a chemotaxis-like mechanism in the *C. elegans* hermaphrodite uterus (Han et al. 2010; Hoang et al. 2013).

The ability of sperm to orient and swim against fluid flow is called positive rheotaxis (Fig. 9.1). Fluid flows in the oviduct are produced from cell secretions, muscle contraction, and ciliary beating (Miki and Clapham 2013; Kantsler et al. 2014; Tung et al. 2015). These flows help propel the oocyte or embryo through the oviduct and uterus. Rheotaxis against oviductal flow might help sperm and oocyte cross paths. Mouse sperm exhibit positive rheotaxis in vitro (Miki and Clapham 2013; Kantsler et al. 2014). Contractions or conformational changes within the reproductive tract can also help move sperm by physical displacement. Uterine conformational changes are thought to help sperm target the sperm storage site in *Drosophila* females (Avila and Wolfner 2009). Finally, thermotaxis is directed movement of cells up a temperature gradient (Perez-Cerezales et al. 2015a). Small temperature differences have been documented in regions of the mammalian oviduct. A low percentage of mammalian sperm appear capable of thermotaxis in vitro. The emerging picture is that sperm integrate complex information from their surrounding environment, translating this information into motility behavior to enhance the chance for fertilization.

In this chapter, we discuss sperm navigation mechanisms in the female reproductive tract, focusing on those delineated in genetically tractable animal models. Abnormal sperm navigation or guidance is likely to play a major role in infertility. It is also a model for understanding how motile cells find their target destinations, a fundamental question in developmental biology. Mammals are considered together because there is not a consensus model species. We also describe mechanisms in two invertebrate species, *Drosophila melanogaster* and *Caenorhabditis elegans*. Each chapter section starts with a description of relevant reproductive anatomy, followed by summaries of the experimental data. The focus is on molecular mechanisms that control sperm movement prior to fertilization. Excellent reviews on gametogenesis, fertilization, and other processes important for reproduction are described elsewhere (Marcello et al. 2013; Han et al. 2010; Matova and Cooley 2001; Hunt and LeMaire-Adkins 1998; Loppin et al. 2015; Evans and Florman 2002; Griswold 2016).

9.2 Mammals

The mammalian female reproductive tract is among the most complex of all animals. Although there are important differences between species, here we consider general features of the most well-studied tracts, including mouse, rabbit, bovine, and human (Eisenbach and Giojalas 2006). During mating, sperm and seminal fluid are ejaculated into the vagina or uterus (Fig. 9.2). The typical mammalian ejaculate contains millions of sperm, yet only several hundred make it to the sperm storage site just beyond the uterotubule junction (Chang and Suarez 2012; Suarez and Pacey 2006; Williams et al. 1993). Sperm must swim against fluid flows within the cervix and oviduct, generated in part by ciliary beating from epithelial cells (Tung et al. 2015; Miki and Clapham 2013; Kantsler et al. 2014).

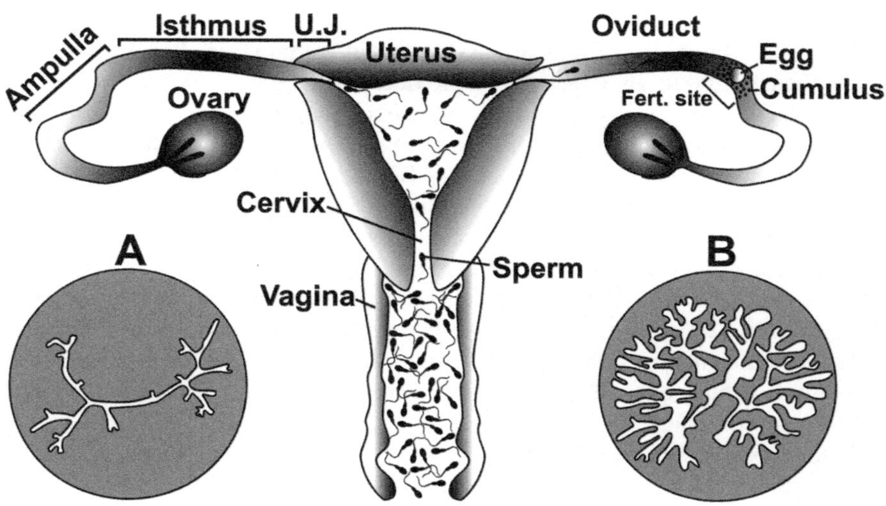

Fig. 9.2 Mammalian female reproductive tract. The illustration is adapted from Eisenbach and Giojalas (2006) and Suarez (2008). *U. J.* uterotubal junction, *Fert* fertilization. (**a**) Cross-sectional illustration of the U. J. showing a convoluted lumen. (**b**) Cross-sectional illustration of the ampulla

Moreover, the cervix and oviduct are not simply empty tubes. Their lumens have convoluted architectures composed of mucosal folds and microgrooves (Fig. 9.2) (Suarez 2008). Sperm are often found within these microgrooves (Suarez 2016). In vitro studies suggest that sperm swim within grooves, against gentle fluid flow, and in contrast to pathogens that are swept away in fluid (Tung et al. 2015; Denissenko et al. 2012; Kantsler et al. 2014; Miki and Clapham 2013). The uterotubule junction is thought to restrict entry of infectious organisms and leukocytes from the uterus while permitting entry of sperm. How this occurs is not well understood but may involve sperm surface proteins such as ADAM3 (Yamaguchi et al. 2009).

After passing through the uterotubule junction, sperm bind the epithelial surface within mucosal folds, forming a storage reservoir (Suarez 2008). Sperm are stored until ovulation approaches, when small numbers are gradually released. The storage reservoir is thought to help preserve sperm over time and prevent polyspermy (Suarez 2016). The sperm number near the egg is low at any given time, presumably due to controlled release from the reservoir (Chung et al. 2014; Miki and Clapham 2013; Chang and Suarez 2012). Using fluorescently labeled transgenic mouse sperm in dissected oviducts, Chang and Suarez observed that sperm attach and reattach to the oviductal epithelium after capacitation (Chang and Suarez 2012), a critical process that modulates sperm motility and adherence behaviors. Capacitation involves a suite of membrane changes and signaling events that enable sperm to fertilize oocytes (Aitken and Nixon 2013; Chang 1951; Austin 1952; Navarrete et al. 2015). In rabbits and mice, capacitation occurs in the sperm storage reservoir before ovulation, aiding in sperm release from the reservoir (Aitken and Nixon 2013; Suarez and Pacey 2006).

Capacitation triggers changes in sperm motility called hyperactivation, which is characterized by high amplitude, asymmetrical patterns in flagellar beating (Yanagimachi 2009). Hyperactivation provides sperm greater thrusting force to swim through oviductal mucus and to penetrate the cumulus matrix surrounding the oocyte (Suarez 2008, 2016). Capacitation involves a sperm ion channel called CatSper, encoded by at least seven genes (Carlson et al. 2003; Qi et al. 2007). CatSper is a pH-sensitive and weakly voltage-dependent Ca^{2+} channel that predominantly localizes to the flagellar principal piece (Chung et al. 2014). In mice and humans, loss of function mutations in *CatSper* subunits causes infertility (Qi et al. 2007; Avenarius et al. 2009). *CatSper* null sperm can swim but fail to hyperactivate (Carlson et al. 2003). A primary cause of infertility in *CatSper*-deficient mouse sperm is inefficient migration past the uterotubule junction to the fertilization site in the ampulla (Chung et al. 2014). Thus, CatSper is critical for sperm guidance to the oocyte–cumulus mass.

Chemotaxis, rheotaxis, and thermotaxis are proposed to help guide mammalian sperm through the oviduct (Eisenbach and Giojalas 2006; Perez-Cerezales et al. 2015b; Miki and Clapham 2013). Much of what is understood about these mechanisms comes from in vitro studies. A fraction of mouse, rabbit, and human spermatozoa (~1/10) exhibit chemotactic behavior in chemotaxis chambers (Perez-Cerezales et al. 2015b). Video microscopy and computer image analysis is consistent with increasing chemoattractant concentration suppressing turning (Publicover et al. 2008; Teves et al. 2006). Conditioned media application to chemotaxis chambers suggests that oocytes and cumulus cells secrete sperm chemoattractants (Eisenbach and Giojalas 2006; Oren-Benaroya et al. 2008; Sun et al. 2005). Several chemoattractant molecules have been proposed, including lipids, such as progesterone (Oren-Benaroya et al. 2008), as well as proteins, such as natriuretic peptide precursor A and CRISP1 (Ernesto et al. 2015; Bian et al. 2012).

The most compelling data is on progesterone, which is secreted by oocytes and cumulus cells. Sperm bias their swimming direction in response to picomolar concentration progesterone gradients (Guidobaldi et al. 2008; Teves et al. 2006; Publicover et al. 2008). Nanomolar to micromolar concentrations increase the number of hyperactivated sperm but do not affect directional motility. Progesterone has been shown to activate CatSper, promoting Ca^{2+} influx into the sperm flagellum (Lishko et al. 2011; Strunker et al. 2011). Ca^{2+} is implicated in sperm chemotaxis in diverse animals, including externally fertilizing species (Publicover et al. 2008; Kaupp et al. 2008). In sea urchin sperm, CatSper promotes Ca^{2+} influx and chemotactic steering (Seifert et al. 2015). Molecules like progesterone that activate CatSper are attractive chemoattractants for mammalian sperm. There are several outstanding questions for future investigations. How are progesterone gradients established in the oviduct? How do these gradients control sperm motility behavior? What is the effect of progesterone depletion in oocyte–cumulus complexes on fertilization? Answers to these questions may resolve apparent inconsistencies. For instance, the concentrations over which progesterone promotes sperm chemotactic behavior are too low to activate CatSper (Teves et al. 2006; Lishko et al. 2011;

Strunker et al. 2011). Furthermore, progesterone promotes mouse sperm chemotaxis (Perez-Cerezales et al. 2016), but it does not activate mouse CatSper, at least under the experimental conditions tested (Lishko et al. 2011). Current data on progesterone are provocative, but further studies are needed to conclude that it has chemoattractant function in the mammalian oviduct.

Several other molecules implicated as sperm chemoattractants also activate CatSper (Ernesto et al. 2015; Brenker et al. 2012). Only progesterone and prostaglandins promote Ca^{2+} influx in the low nanomolar concentration range (Brenker et al. 2012; Lishko et al. 2011). Other molecules, such as bourgeonal and CRISP1 activate CatSper at much higher concentrations. Such high concentrations are unlikely to be found in the oviduct. It is possible that multiple molecules act synergistically to regulate CatSper, either through direct binding to the channel or through an accessory protein(s). In this context, progesterone and prostaglandins appear to bind different sites (Lishko et al. 2011). Another possibility is that the endogenous chemoattractant(s) is yet to be identified. To date, it has not been possible to knockout potential chemoattractants genetically and examine sperm motility behavior within the oviduct. Developing this technology in a mammalian system should help test models of sperm guidance.

Fluid viscosity and flow affect sperm motility, assessed in vitro (Miki and Clapham 2013; Denissenko et al. 2012; Kantsler et al. 2014). Fluid viscosity in the oviduct varies during the estrus cycle (Miki and Clapham 2013; Jansen 1978). Ciliated epithelial cells in the isthmus are covered by a dense mucous layer that disappears after ovulation, when secretions of oviductal fluids increase. In mice, mating increases fluid flow from the oviduct to the uterus, a response requiring the hormone prolactin (Miki and Clapham 2013). Ciliary beating and muscle contraction in the isthmus are thought to push oviductal fluid to the uterus, washing away mucus. Mammalian sperm are proposed to exhibit positive rheotaxis against this flow. Roughly half of sperm, irrespective of capacitation status, swim against gentle fluid flow in vitro (Kantsler et al. 2014; Miki and Clapham 2013). Altering the viscosity of the medium reveals differences in motility behavior between capacitated and noncapacitated spermatozoa (Miki and Clapham 2013). In a current, sperm encounter tangential forces that reorient them against the flow. Sperm tail rotation and CatSper are required for rheotaxis (Miki and Clapham 2013). Ca^{2+} influx is proposed to trigger increased amplitude of the tail, thereby increasing tangential force. Increasing angular speed stabilizes swimming orientation. Although these behaviors are studied in vitro, there is evidence that rheotaxis is important in vivo. Using dissected mouse oviducts and transgenic fluorescent sperm, Miki and Clapham found that sperm reach the ampulla less efficiently when oviducts were ligated near the uterotubule junction to block fluid flow (Miki and Clapham 2013). These studies suggest that rheotaxis plays a key role in guiding sperm to the ampulla, where chemoattractants may provide further positional information.

Another potential mechanism for sperm guidance is thermotaxis (Bahat et al. 2003; Perez-Cerezales et al. 2015b). In rabbits and pigs, a temperature drop at ovulation creates a slight gradient in the oviduct with the ampulla 1–2 °C warmer than the sperm storage site (Hunter and Nichol 1986; Bahat et al. 2005). Human, rabbit, and mouse sperm respond in vitro to temperature changes by swimming up the gradient (Perez-Cerezales et al. 2015b). The percent of responsive sperm is low, between 1 and 10 % depending on the species. Recent studies implicate opsins and downstream phospholipase C and cyclic nucleotide pathways in thermotaxis (Perez-Cerezales et al. 2015a). In *Drosophila* larvae, rhodopsin is involved in temperature discrimination, independent of light (Shen et al. 2011). Rhodopsin-knockout mouse sperm have impaired themotaxis in vitro, although fertility in vivo is unaffected (Perez-Cerezales et al. 2015a). One possibility is that the thermotactic mechanism is redundant with other sperm guidance pathways. Alternatively, in vitro sperm behavior in response to temperature gradients has been proposed to reflect rheotaxis due to convection currents (Miki and Clapham 2013). The extent to which chemotaxis, rheotaxis, and thermotaxis contribute to sperm guidance is not clear. These studies highlight strengths and limitations of mammalian model systems, which have largely inaccessible and opaque reproductive tracts, for studying sperm navigation mechanisms.

9.3 *Drosophila melanogaster*

The *Drosophila* female reproductive tract consists of a uterine tube at the posterior end and a pair of ovaries at the anterior end (Schnakenberg et al. 2012; Heifetz and Rivlin 2010) (Fig. 9.3). In the middle lie two types of sperm storage organs, a single seminal receptacle and two mushroom-shaped spermathecae. The seminal receptacle is a long, narrow tube that attaches to the uterus at one end with the other end encircling it (Pattarini et al. 2006; Heifetz and Rivlin 2010; Schnakenberg et al. 2012). Each spermatheca consists of a duct connecting the uterus to a lumen lined by secretory cells. A thin layer of muscle and epithelial cells surround the spermathecal ducts, which are adjacent to two accessory glands of largely unknown function. Mature eggs descend from the ovaries through the oviduct to the uterus. Eggs are activated in the oviduct before entering the uterus, where fertilization occurs (Fig. 9.3). Sperm penetrate the egg's micropyle, a specialized opening that passes adjacent to the seminal receptacle and spermathecae ducts (Loppin et al. 2015). Mature spermatozoa are released from the sperm storage organs as the eggs pass (Heifetz and Rivlin 2010; Lefevre and Jonsson 1962).

Drosophila males deposit sperm and seminal fluid into the female's posterior uterus during mating (Avila et al. 2011). Secretions from the female tract are thought to mix with the ejaculate to form a sperm sac and facilitate hyperactivation (Alonso-Pimentel et al. 1994; Kottgen et al. 2011; Yang and Lu 2011). Sperm swim predominantly tail first in the uterus (Yang and Lu 2011; Kottgen et al. 2011). A series of uterine conformational changes help move sperm from the posterior to the anterior uterus, adjacent to the sperm storage organ ducts (Avila and Wolfner 2009). Females store roughly 600 sperm; the remainder are expelled from the uterus during egg laying (Lefevre and Jonsson 1962; Ohsako and Yamamoto 2011). Stored sperm are highly efficient at fertilizing eggs.

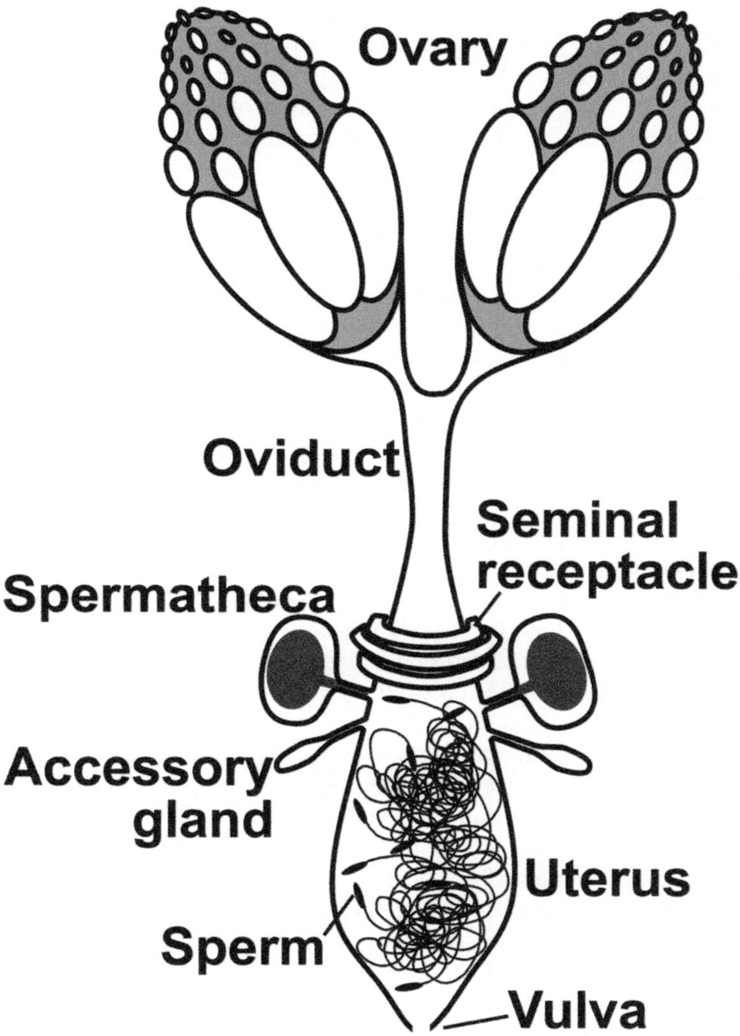

Fig. 9.3 *Drosophila melanogaster* female reproductive tract. The illustration is adapted from Schnakenberg et al. (2012). Drosophila sperm have long tails and swim tail first

How sperm target the storage organs is not well understood. The female nervous system is important for sperm storage (Arthur et al. 1998). Additionally, the neuromodulators octopamine and tyramine are required for sperm release (Avila et al. 2012). Males provide proteins in seminal fluid that are important for sperm storage, sperm retention in storage organs, increased ovulation rates, and postmating behaviors (Avila et al. 2011; Schnakenberg et al. 2012). The seminal fluid protein Acp36DE mediates uterine conformational changes or contractions that are critical for sperm storage (Avila and Wolfner 2009). During these changes, the uterus transitions from an S-shaped structure with a closed lumen to a more distended conformation, exposing the sperm storage organ ducts. As conformational changes occur, a sperm mass appears near the entryways of the sperm storage organs. Acp36DE knockdown prevents formation of sperm masses adjacent to sperm storage organ ducts (Avila and Wolfner 2009). At present, it is not known whether uterine conformational changes alone are sufficient to target sperm to the storage organs or whether guidance cues are also involved.

Amo, the *Drosophila* homolog of the human transient receptor potential Ca^{2+} channel TRPP2 (also known as PKD2), is essential for sperm storage. *Drosophila* males with *amo* mutations are sterile (Watnick et al. 2003; Gao et al. 2003). The mutant males produce motile sperm that are transferred to the uterus but fail to reach the storage organs. In mature sperm, Amo is expressed in the flagellum (Kottgen et al. 2011). Wild-type sperm exhibit hyperactive motility upon transfer to the uterus and swim backwards (i.e., tail first) into the sperm storage organs (Kottgen et al. 2011; Yang and Lu 2011). *Amo* mutant sperm are capable of swimming backwards but fail to exhibit hyperactivation and increased swimming speed. In some cases, the mutant sperm enter the seminal receptacle in the wrong orientation with abnormal flagellar movements that impede further progress (Yang and Lu 2011). These results indicate that swimming per se is not sufficient to localize sperm to the storage organs. What, if anything, triggers Amo activation and presumably Ca^{2+} influx into the flagellum is not known.

Female proteins influence sperm targeting to the storage organs. Upon mating, glucose dehydrogenase (Gld) expression increases in the spermathecal ducts (Iida and Cavener 2004; McGraw et al. 2004). Gld is not expressed in the seminal receptacle, however. *Gld* mutant females store normal numbers of sperm in the seminal receptacle but fewer sperm in the spermathecae. Often the mutant females have highly asymmetrical sperm distributions in the two spermathecae (Iida and Cavener 2004). Gld is thought to be a secreted protein, but its biochemical role in the female tract is uncertain. Other genes expressed in the spermathecae secretory cells are also likely to be important for sperm recruitment. Secretory cell ablation causes decreased sperm accumulation in the spermathecae but not in the seminal receptacle (Schnakenberg et al. 2011).

There are interesting differences between sperm recruitment to the seminal receptacle versus the two spermathecae. These organs have different tissue organizations, compositions, and evolutionary histories. Sperm stored in the seminal receptacle are the primary source used for fertilization (Manier et al. 2010). After mating, most sperm target the seminal receptacle first, as sperm accumulate in the spermathecae more slowly (Gilbert and Richmond 1981; Manier et al. 2010). Masculinizing the female fly nervous system, and consequently disrupting signaling from the nervous system to sperm storage organs, results in reduced sperm accumulation (Arthur et al. 1998). However, the spermathecae are more affected than the seminal receptacle. These data suggest that there may be different mechanisms important for targeting the seminal receptacle and spermathecae. Besides the physical role of uterine conformational changes, other mechanisms that contribute to sperm guidance have yet to be identified.

9.4 *Caenorhabditis elegans*

C. elegans is a free-living nematode species that reproduces as facultative hermaphrodites (Frezal and Felix 2015). In nature and laboratory conditions, wild-type *C. elegans* propagate primarily as hermaphrodites (Chasnov 2013). The natural male frequency varies from less than 0.1 % to 3.2 % (Hodgkin and Doniach 1997). Both hermaphrodites and males have five autosomal chromosomes but differ in sex chromosome content. Hermaphrodites have two X chromosomes, whereas males have only one X (Zarkower 2006; Meyer 2005). Males are generated from selfing hermaphrodites by a rare X chromosome nondisjunction event (Zarkower 2006; Hodgkin et al. 1979). Hermaphrodites also produce males following mating to a male. Male sperm are competitively superior to hermaphrodite sperm and have an equal chance of bearing an X, resulting in broods containing approximately 50 % males (Singson 2001; L'Hernault 2009).

C. elegans adult hermaphrodites are nearly indistinguishable anatomically from mated females in gonochoristic (female/male) *Caenorhabditis* species. The hermaphrodite gonad consists of two U-shaped arms connected to a common uterus (Fig. 9.4a) (Hirsh et al. 1976; Klass et al. 1976). At the distal end of each gonad arm lies a germline stem cell population, which is surrounded by the somatic distal tip cell (DTC) (Byrd and Kimble 2009; Hubbard and Greenstein 2000). The DTC controls stem cell proliferation and meiotic entry (Kimble and Crittenden 2005).

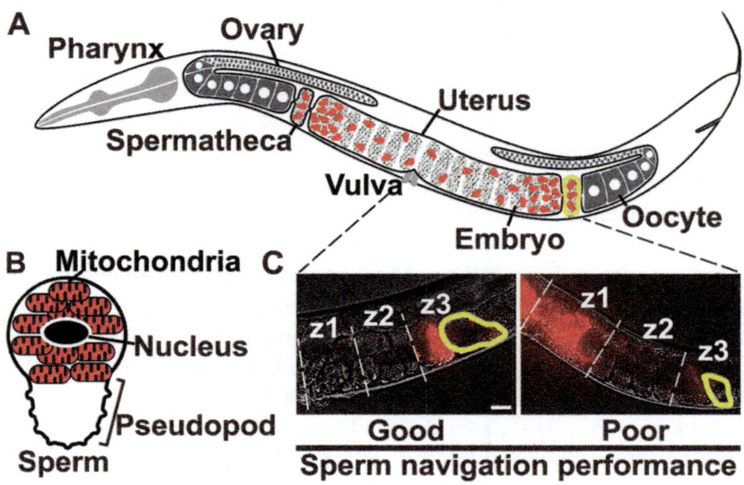

Fig. 9.4 *Caenorhabditis elegans* hermaphrodite reproductive tract. (**a**) Adult hermaphrodite anatomy, focusing on the gonad. Sperm fluoresce red due to vital dye labelling and the spermatheca is outlined in *yellow*. (**b**) Sperm illustration showing key organelles and structures. Spermatids extend a pseudopod during spermiogenesis, which occurs upon spermathecal entry (self sperm) or insemination (male sperm). (**c**) Representative images of sperm (*red*) distributions in wild-type and mutant hermaphrodite uteri one hour after mating. The uterus is divided into three zones (z1–z3) for quantification

Germ cells differentiate as they are displaced proximally toward the spermatheca, a sac-like structure that separates the ovary from the uterus. The spermatheca is the site of sperm storage and fertilization (Marcello et al. 2013) (Fig. 9.4a). During the L4 stage, germ cells differentiate as spermatids before they switch to oogenesis in adults (Klass et al. 1976; Hirsh et al. 1976; Ellis and Schedl 2007).

Hermaphrodite spermatids enter the spermatheca together with the first ovulating oocyte. Spermatids then activate for motility by extending a single pseudopod largely composed of major sperm protein (MSP) filaments (Fig. 9.4b) (Roberts and Stewart 2000, 2012; Ward et al. 1983). Hermaphrodite, but not male spermiogenesis, requires the *spe-8* group of genes (L'Hernault 2006; Nance et al. 1999; Shakes and Ward 1989). A single sperm rapidly fertilizes the oocyte, while the remaining sperm compete for position within the spermatheca (L'Hernault 2006; Han et al. 2010). In addition to sperm cytoskeletal elements, MSPs also function as secreted ligands that stimulate oogenesis, oocyte maturation, and ovulation (Han et al. 2010; Miller et al. 2001, 2003; Govindan et al. 2009). Thus, sperm signal a slow current of oocytes (one every ~25 min) that flows from the ovary, through the spermatheca, and into the uterus. Sperm must fight against this flow to maintain spermathecal positioning or risk expulsion through the vulva along with passing eggs (Kubagawa et al. 2006). Electron micrographs of the spermatheca often show numerous sperm with pseudopods facing the incoming oocyte, eagerly waiting to be the first to contact it (Kosinski et al. 2005). Hermaphrodite sperm are highly efficient at fertilizing oocytes. Very few sperm, if any, are lost from the reproductive tract (Singson 2001). Ovulated oocytes fail to form polar bodies and assemble meiosis II spindles when sperm are absent from the spermatheca (McNally and McNally 2005).

Male sperm face a tougher challenge than hermaphrodite sperm. In contrast to hermaphrodite gonads, male gonads produce sperm continuously throughout adulthood (Klass et al. 1976; L'Hernault 1997). Germline stem cells give rise to primary spermatocytes that divide during meiosis I into two secondary spermatocytes (L'Hernault 2009; Ward et al. 1981; Ellis and Stanfield 2014). Completion of meiosis II produces four spermatids connected to a shared anucleate residual body. Sessile spermatids, which lack actin, tubulin, and ribosomes, are stored in the seminal vesicle until ejaculation. Mitochondria and fibrous body membranous organelles, lysosome-like structures important for trafficking during spermatogenesis, surround densely packed chromatin. During mating, male spermatids are ejaculated from the vas deferens through the hermaphrodite vulva into the uterus (Fig. 9.4) (Ward and Carrel 1979; Singson 2001). Mixing of spermatids with seminal fluid triggers spermiogenesis (Shakes and Ward 1989; Ellis and Stanfield 2014). The SWM-1 protease inhibitor and TRY-5 serine protease coordinate this process in males (Stanfield and Villeneuve 2006; Smith and Stanfield 2011). Seminal fluid from many wild isolates contains material for copulatory plug deposition at the vulva (Hodgkin and Doniach 1997; Palopoli et al. 2008). Potential roles for this plug include preventing mating by other males and sperm loss from the uterus during egg laying. Motile sperm crawl around fertilized eggs, across their eggshells and the uterine walls, to the spermatheca (Kubagawa et al. 2006;

Edmonds et al. 2010; Ward and Carrel 1979). The distance from the vulva to the spermatheca can be up to 250 microns in well-fed adult hermaphrodites. Larger male sperm outcompete hermaphrodite sperm for spermathecal position, almost exclusively resulting in outcross progeny (Ward and Carrel 1979; LaMunyon and Ward 1998).

Nematode sperm provide a striking example of evolutionary novelty. Instead of an acrosome and flagellum, they have a single pseudopod that translocates in a treadmilling fashion, dragging the cell body behind (Fig. 9.4b) (Roberts and Stewart 1995, 2000; Ellis and Stanfield 2014). The pseudopod is thought to contact the oocyte plasma membrane first, fusing with it (Singson et al. 1998; Marcello et al. 2013). Despite morphological differences, *C. elegans* sperm express evolutionarily conserved proteins critical for fertilization, such as Izumo and the transient receptor potential Ca^{2+} channel SPE-41 (Xu and Sternberg 2003; Singaravelu et al. 2015; Nishimura et al. 2015). On the other hand, sperm lack an actin cytoskeleton and motor proteins (Smith 2006; Roberts and Stewart 2000). The pseudopod primarily contains MSP filaments that form during spermiogenesis. pH controls filament assembly at the pseudopod's leading edge and disassembly at its trailing edge, propelling the sperm forward (Italiano et al. 1999; Roberts and Stewart 2012). The regulatory mechanisms that control pH in the pseudopod are not well understood. Excellent reviews on nematode sperm motility provide additional insight and details (Roberts and Stewart 2012; Smith 2006).

MSPs are central to nematode sperm form and function. *msp* genes derive from an ancestral gene called *vap* (Vamp-synaptobrevin-associated protein), which contains an N-terminal MSP domain (Miller et al. 2001; Tsuda et al. 2008). VAPs are broadly expressed, type II membrane proteins with roles in lipid metabolism and cell signaling (Tsuda et al. 2008; Lev et al. 2008; Stefan et al. 2011; Han et al. 2010). A *vap* gene duplication event is likely to have precipitated the evolution of nematode sperm-specific functions. The most parsimonious explanation is that this initial function was a signaling function, such as triggering oocyte maturation. Sperm cannot synthesize proteins de novo, and filament assembly might enhance MSP storage over longer time periods. Its role in motility could be linked to storage. The take-home message is that nematode sperm have peculiar features that derive in part from their dependence on MSPs.

9.4.1 Male Sperm Motility Behavior In Vivo

The *C. elegans* epidermis is transparent, making it possible to view sperm motility in live hermaphrodite uteri. However, tracking single sperm movements in the uterus is too difficult without high-resolution, fluorescence time-lapse imaging. Whether or not sperm move with directional motility toward the spermatheca was not clear until technological improvements were made. Fluorescent male sperm can be generated through integrated transgenic reporters or fluorescent vital dyes, such as Mitotracker (Fig. 9.4c) (Hill and L'Hernault 2001; Kubagawa et al. 2006; Ting et al. 2014). Dyes have the advantage that they tend to be brighter than transgenes, which must evade germline-silencing mechanisms. Dyes can also be applied to different genotypes without the need for crosses. When wild-type males with fluorescent sperm are mated to hermaphrodites, more than 90 % of sperm accumulate at the two spermathecae within an hour after mating (Fig. 9.4c) (Edmonds et al. 2010, 2011; Hoang et al. 2013). Time-lapse videos from freshly mated, anesthetized hermaphrodites enable tracking individual sperm after insemination (Kubagawa et al. 2006; Edmonds et al. 2010; McKnight et al. 2014).

Males inject varying numbers of spermatids during individual mates. These spermatids are deposited in a large mass within the uterus near the vulva, where they undergo spermiogenesis. Nearly all sperm are motile at the time of imaging, which occurs several minutes after insemination. Sperm have the choice of moving toward the anterior or posterior spermatheca (Fig. 9.4). Time-lapse imaging shows that aggregated sperm near the vulva move slower and with higher reversal frequency than sperm closer to the spermatheca. Once sperm exit the vulval region, their motility behavior changes as they exhibit strong directional velocity toward the closest spermatheca. Sperm crawl around fertilized eggs and across uterine epithelial cells, rarely reversing their direction. Analyzing time-lapse videos indicates that sperm move in vivo with an average velocity of about 8–10 microns per minute (Kubagawa et al. 2006; Edmonds et al. 2010; McKnight et al. 2014; Hoang et al. 2013). Sperm directional velocity is about half the total velocity due to rare reversals and obstructing eggs. Sperm accumulate in a bottleneck at the spermathecal–uterine junction, where individual sperm pass through a small valve into the spermatheca. Motility in these masses changes again, this time to short motility bursts with frequent reversals, presumably reflecting a valve-searching behavior. Sperm rarely leave the mass and move back toward the vulva. These sperm motility behaviors are consistent with an active guidance mechanism toward the spermatheca or adjacent ovary.

Sperm also compete with each other for spermathecal positioning. Male sperm are larger than hermaphrodite sperm and displace them from the spermatheca (Ward and Carrel 1979). Comparisons among *Caenorhabditis* species and experimental studies suggest that competition increases sperm size (LaMunyon and Ward 1998, 2002). Size isn't the whole story, however. Males with *comp-1* mutations produce normal-sized sperm that are competitively inferior to wild-type sperm (Hansen et al. 2015). Additionally, studies of interspecies matings show that sperm can sterilize hermaphrodites and females, revealing sperm-mediated sexual conflict (Ting et al. 2014). Increased sperm competition is predicted to favor more aggressive sperm that win access to oocytes. A tradeoff is that aggressive sperm can invade the proximal gonad, disrupting oogenesis and other processes. It is not enough for *C. elegans* sperm to locate the spermatheca. They must also fight with other sperm for positioning to fertilize an incoming oocyte.

9.4.2 A Lipid Guidance Mechanism

The ability to track sperm movement led to methods to identify genes critical for targeting (Fig. 9.4c) (Kubagawa et al. 2006; Edmonds et al. 2010; Hoang et al. 2013; McKnight et al. 2014). Hermaphrodite gene products were disrupted via genetic mutation or RNA-mediated interference, while males were left alone. This strategy permitted the identification of genes that function outside of sperm (i.e., cell nonautonomously) to control their migratory behavior. These genes are likely to influence production of signaling molecules. Sperm density within the spermatheca is greatest at the distal valve, an observation consistent with oocytes secreting a sperm attractant (Kubagawa et al. 2006). In this model, sperm migrate toward oocytes in the proximal gonad, not the spermatheca per se. Their path to the oocyte is blocked by the distal valve or other mechanisms promoting retention in the spermatheca. When ovulation occurs, the distal valve opens and the oocyte enters the spermatheca, contacting the waiting sperm. This model was tested by eliminating the oogonia and examining effects on sperm motility (Kubagawa et al. 2006). Sperm from wild-type males accumulate in the uterus and rarely reach the spermatheca in animals that lack oocytes, such as *glp-4(bn2)*. Time-lapse videos indicate that sperm move with reduced velocity, no directional velocity, and increased reversal frequency.

Mutations in two genes proved critical to discovering the mechanism by which oocytes govern sperm motility (Kubagawa et al. 2006). The first gene encodes the RME-2 low-density lipoprotein receptor. RME-2 is expressed specifically in developing oocytes, where it promotes yolk lipoprotein complex uptake (Grant and Hirsh 1999). Yolk forms in the intestine and flows to the gonad through the pseudocoelom. The second gene encodes the FAT-2 Δ12 fatty acyl desaturase, which is required for formation of 18-carbon (C18) and C20 polyunsaturated fatty acids (PUFAs) (Watts and Browse 2002). Yolk contains significant quantities of C20 PUFAs, including dihomo-gamma-linolenic acid (DGLA), arachidonic acid (AA), and eicosapentaenoic acid (EPA) (Kubagawa et al. 2006). Mutations in *rme-2* or *fat-2* strongly affect sperm motility performance (Kubagawa et al. 2006; Hoang et al. 2013). Wild-type sperm motility behavior in these mutant hermaphrodites resembles that of mutant hermaphrodites lacking oocytes. Microinjection experiments demonstrated that oocyte C20 PUFAs are necessary and sufficient to promote sperm guidance in *fat-2* mutants (Kubagawa et al. 2006). These and other experiments led to the idea that C20 PUFAs are precursors of a sperm attractant produced by oocytes.

In mammals, C20 PUFAs are converted into signaling molecules called prostaglandins, leukotrienes, lipoxins, and other eicosanoids (Funk 2001; Simmons et al. 2004). A series of microinjection experiments and genetic results pointed to prostaglandins as the guidance cues (Edmonds et al. 2010, 2011). Importantly, microinjecting F-series prostaglandins induces a robust increase in sperm velocity in PUFA-deficient hermaphrodites (Edmonds et al. 2010). F-series prostaglandins contain a saturated cyclopentane ring with two hydroxyl groups (Funk 2001). This ring structure is critical for activity (Edmonds et al. 2010). Injected lipids rapidly diffuse throughout the uterus and in many cases extrude back through the vulva, causing dilution. Commercial PGF2α injected at 25 nM is sufficient to stimulate sperm motility, suggesting F-series prostaglandins function at physiological concentrations (Edmonds et al. 2010). A caveat of microinjection experiments is that directional motility is not affected, likely due to rapid diffusion and a short delay between microinjection and imaging. On the other hand, the data imply that prostaglandins are not simply permissive signals. Otherwise, sperm should acquire wild-type motility behavior.

Liquid chromatography coupled to electrospray ionization tandem mass spectrometry (LC-MS/MS) has identified F-series prostaglandin isomers in *C. elegans* extracts (Edmonds et al. 2010; Hoang et al. 2013; McKnight et al. 2014). Oocytes or their precursors synthesize more than ten structurally related F-series prostaglandins, which function collectively and largely redundantly to promote sperm guidance (Hoang et al. 2013). DGLA, AA, omega-3 AA, and EPA are converted into discrete sets of isomers. Although the PGF3 series derived from EPA is the most abundant, the PGF2 series derived from AA is better characterized because individual PGF2α stereoisomers are commercially available. Reverse-phase LC-MS/MS, normal-phase chiral LC-MS/MS, and nanoLC QTOF analyses indicate that *C. elegans* produces very little PGF2α. Instead, major products derived from AA appear to be *8-epi*-PGF2α, *ent*-PGF2α, and an unidentified PGF2α stereoisomer (Hoang et al. 2013; McKnight et al. 2014). There is currently no LC-MS/MS evidence thus far for D- and E-series prostaglandins, thromboxanes, leukotrienes, or lipoxins. *C. elegans* does synthesize an array of epoxy- and hydroxy-C20 PUFA metabolites (Zhou et al. 2015; Ma et al. 2013; Deline et al. 2015). Mutant strains deficient in synthesis of all F-series prostaglandin isomers have nonautonomous defects in sperm motility, including reduced velocity, little to no directional velocity, and increased reversal frequency. Self-derived and male sperm are likely similarly affected. Self-derived sperm are more difficult to monitor, but prostaglandin-deficient strains have a strongly reduced brood size (Kubagawa et al. 2006; Edmonds et al. 2010; Hoang et al. 2013). Self-derived sperm are found throughout the uterus in still images, suggesting these sperm retarget the spermatheca poorly after being swept out of it.

A fascinating aspect of *C. elegans* prostaglandin synthesis is that it occurs without prostaglandin G/H synthase enzymes (a.k.a. Cox). In mammals, Cox enzymes convert AA into the bicyclic endoperoxide PGG2, which is then reduced to PGH2 (Simmons et al. 2004; Chandrasekharan and Simmons 2004). PGD, PGE, and PGF synthases convert PGH2 into respective bioactive forms. Nonsteroidal anti-inflammatory drugs (NSAIDs) target Cox to inhibit prostaglandin synthesis (van der Donk et al. 2002; Vane 1971). The *C. elegans* genome does not encode Cox homologs and other key enzymes in Cox-metabolic pathways. Accordingly, prostaglandin synthesis is refractory to NSAIDs (Edmonds et al. 2010). The biochemical pathway that converts C20 PUFA precursors into specific PGF isomers is not understood at present. Genetic and mass spectrometry approaches are underway to help shed light on the pathway (Prasain et al. 2015). In mice, over 50 % of

PGF2α isomers are formed independent of Cox enzymes (McKnight et al. 2014). Mouse and worm PGF2α isomers co-elute under conditions that separate most PGF2α stereoisomers. Cox-independent PGF2α isomers are also found in zebrafish and human tissues. These and other data raise the possibility that a second, evolutionarily conserved pathway for prostaglandin synthesis exists. As the Cox enzyme and prostaglandins are found in basal animals like cnidarians (Valmsen et al. 2001; Chandrasekharan and Simmons 2004), the most parsimonious explanation is that *C. elegans* lost the Cox pathway during evolution.

In addition to questions related to prostaglandin synthesis, these studies bring up other important questions. Among them are how prostaglandins are spatially restricted and how sperm transduce prostaglandin signals. Although little is known about prostaglandin signal transduction, several genes have been implicated in restricting prostaglandin metabolism or secretion to unfertilized oocytes (Edmonds et al. 2011; Hoang et al. 2013). Sperm ignore fertilized eggs in the uterus, migrating over them in the direction of maturing, unfertilized oocytes in the ovary. Two class 4 cytochrome P450 paralogs called CYP-31A2 and CYP-31A3 may play a role in this process. In *cyp-31A2(tm2711)* hermaphrodites, wild-type sperm move with reduced directional velocity, but total velocity is unaffected (Hoang et al. 2013). LC-MS/MS analysis of *cyp-31A2(tm2711)* extracts shows increased prostaglandin levels, suggesting that CYP-31A2 restricts prostaglandin metabolism (Hoang et al. 2013). CYP-31A2 promotes permeability barrier formation in fertilized eggs (Benenati et al. 2009). This barrier prevents the entry and exit of small molecules (<3000 Da) and is thought to consist of C20 PUFAs and other long-chain fatty acids conjugated to sugars (Stein and Golden 2015; Olson et al. 2012). CYP-31A2 might act after fertilization to divert PUFAs away from prostaglandin synthesis to permeability barrier formation. The permeability barrier could also prevent prostaglandins from being secreted into the extracellular environment. Another cell structure implicated in regulating prostaglandins is oocyte and sheath cell gap junctions (Edmonds et al. 2011), which form in developing oocytes and are lost during ovulation (Starich et al. 2014; Hall et al. 1999). The mechanism by which CYP-31A2 and gap junctions modulate prostaglandin metabolism is not well understood. Nevertheless, it is likely that prostaglandins are spatially regulated. In the next section, we describe evidence for temporal regulation.

9.4.3 Temporal Regulation of Sperm Guidance

Adult hermaphrodites temporally modulate prostaglandin metabolism and, accordingly, sperm guidance in response to nutritional cues and pheromones (Edmonds et al. 2010; McKnight et al. 2014). These regulatory mechanisms are thought to coordinate fertility rates with environmental conditions. When food levels are high and population density is low, hermaphrodites optimize sperm usage through increased oocyte prostaglandin synthesis (McKnight et al. 2014). Male sperm in hermaphrodites starved for as little as 16 h start to exhibit poor motility performance even when males are well fed (Kubagawa et al. 2006; Edmonds et al. 2010). Nutritional signals act via the conserved insulin-signaling pathway (Edmonds et al. 2010). The *C. elegans* genome encodes for more than forty insulin-like peptides but only one receptor (Murphy and Hu 2013; Mukhopadhyay et al. 2006). Upon ligand binding, the DAF-2 insulin receptor activates the phosphoinositide-3 kinase AGE-1, stimulating a phosphorylation cascade mediated by phosphoinositide-3-phosphate (PIP3). PIP3 activates the PDK-1 kinase, which then activates AKT-1, AKT-2, and SGK-2 kinases. These kinases phosphorylate the forkhead box O transcription factor DAF-16, inhibiting its entry into the nucleus and, therefore, its transcriptional activity (Ogg et al. 1997; Lin et al. 2001). *daf-16* loss suppresses many of the defects in *daf-2* mutants, indicating that much of the insulin pathway in worms is mediated through DAF-16 transcriptional activity.

Mutations in *daf-2*, *age-1*, and *akt-1* cause strong sperm guidance defects (Edmonds et al. 2010). Wild-type sperm in *daf-2* mutant uteri move with reduced velocity, little directional velocity, and high reversal frequency, similar to sperm in prostaglandin-deficient uteri. These motility defects are suppressed by *daf-16* loss. Using a temperature-sensitive mutation in *daf-2*, it was demonstrated that insulin signaling is required continuously in adults for sperm guidance. DAF-16 appears to function in the intestine and germline to inhibit oocyte prostaglandin synthesis. In the intestine, DAF-16 represses vitellogenin expression (i.e., ApoB-100 homologs) (Edmonds et al. 2010; DePina et al. 2011), downregulating yolk synthesis important for transporting PUFAs to the oocytes. Overexpressing the DAF-16B isoform in the germline is sufficient to induce sperm guidance defects (Edmonds et al. 2010). Yolk endocytosis is severely compromised in these transgenic hermaphrodites. Thus, one mechanism by which insulin promotes prostaglandin synthesis is through increased flux of PUFA precursors into the oocytes. LC-MS/MS analysis of *daf-2* mutants shows reduced prostaglandin levels. However, *daf-2; daf-16* double mutants have complex patterns of lipid alterations, consistent with tissue-specific regulation (Edmonds et al. 2010).

Exposure to pheromones called ascarosides also impacts prostaglandin synthesis and sperm guidance (McKnight et al. 2014). Hermaphrodites monitor levels of asc-C6-MK, asc$-\Delta$C9, and other ascarosides to interpret population density (Ludewig and Schroeder 2013). As the ascaroside to food ratio increases, amphid sensory neurons called ASI downregulate expression of the DAF-7 TGF-β ligand (Ren et al. 1996), which transduces signals through DAF-1 type I and DAF-4 type II receptors (Gumienny and Savage-Dunn 2013). Mutations in *daf-7*, the *daf-1* and *daf-4* receptors, or downstream *daf-8* and *daf-14* R-Smads impair sperm targeting to the spermatheca (McKnight et al. 2014). Loss of the antagonistic co-Smad *daf-3* suppresses the sperm motility defects. Sperm motility behavior and LC-MS/MS in *daf-1* mutants indicate prostaglandin deficiency. Therefore, DAF-7 TGF-β connects ascaroside exposure to ovarian prostaglandin metabolism. High ascaroside levels are sufficient to trigger sperm guidance defects in adult hermaphrodites (McKnight et al. 2014).

ASI neurons extend ciliated dendrites through the tip of the nose. Their cell bodies sit near the nerve ring, adjacent to the second pharyngeal bulb. Although DAF-7 expression is limited to ASI and nearby ASJ neuron pairs (Meisel et al. 2014; Ren et al. 1996), the DAF-1 and DAF-4 receptors are broadly expressed (Gumienny and Savage-Dunn 2013; Gunther et al. 2000). Mosaic and cell type-specific transgenic studies indicate that TGF-β signals are transduced in the germline and RIM/RIC interneurons (McKnight et al. 2014), which coordinate feeding and fat metabolism (Greer et al. 2008). RIM/RIC interneurons lie close to ASI sensory neurons in the head and both neuron types extend projections to the nerve ring. ASI and RIM/RIC neurons are hypothesized to transduce signals to the gonad in a neuroendocrine fashion (McKnight et al. 2014). A closed circulatory cavity called the pseudocoelom mixes secretions from most tissues, including head neurons and the germline (Gottschalk and Schafer 2006). DAF-7 secreted from ASI could flow directly to the gonad, binding DAF-1 and DAF-4 receptors expressed in the germline. Genetic data suggest that DAF-7 also transduces signals to the gonad indirectly, using secondary insulin and steroid ligands (McKnight et al. 2014). RIM/RIC interneurons might propagate this signaling pathway. In any event, both pathways converge in the oogonia to promote prostaglandin synthesis.

These results bring up a suite of new questions regarding the relationship of environment to fertility. Among those questions is why *C. elegans* reduces sperm performance when resources or the perception of resources is limited. One possibility is that hermaphrodites aim to invest available nutrients into smaller numbers of progeny. High ascaroside levels and low food inhibit egg laying, causing internal embryo hatching within the uterus. Larvae then eat their mom from the inside out, using her as a food source. Larvae that enter the dauer stage can survive without food for several months and exhibit dispersal behaviors (Fielenbach and Antebi 2008; Hu 2007). However, dauer entry requires nutrients; embryos that hatch without food do not form dauers (Cassada and Russell 1975). It has been proposed that internal hatching provides a mechanism to increase dauer entry under starvation conditions (Chen and Caswell-Chen 2004). An alternative possibility is that *C. elegans* may want to limit outcrossing when resources are scarce. In any event, hermaphrodites possess multiple mechanisms that couple environmental conditions to sperm performance. These endocrine mechanisms are mediated by evolutionarily conserved signaling circuits that modulate ovarian prostaglandin synthesis.

9.5 Perspectives

Despite differences in reproductive anatomy among *C. elegans*, *D. melanogaster*, and mammalian species, their sperm face similar challenges to fertilize an oocyte. Our knowledge of the molecular mechanisms is rudimentary, yet these mechanisms are critically important to reproduction and evolution. Studies in mammals are starting to illuminate the difficulty of the sperm's journey through the female reproductive tract. Out of the millions of sperm inseminated, only a small number make it near the oocyte's vicinity. Chemotaxis, rheotaxis, and thermotaxis have all been proposed to aid the sperm. The extent to which these mechanisms mediate sperm guidance in vivo remains to be determined. It is clear that the female reproductive tract plays an instrumental role in controlling sperm adherence and motility behaviors. Accordingly, the mammalian sperm is capable of directional movement in response to multiple sensory modalities.

Drosophila studies highlight the importance of seminal fluid proteins and the female nervous system in regulating sperm storage and release. Similarly, seminal fluid proteins in male *C. elegans* promote spermiogenesis, which is essential for motility. Sperm then interpret prostaglandin positional cues secreted by oocytes. Pheromone exposures and nutritional status influence prostaglandin metabolism. Both the nervous system and the reproductive tract play key roles in controlling the sperm and oocyte encounter. *Drosophila* and *C. elegans* models are helping discover fundamental principles, molecular mechanisms, and evolutionary forces that govern fertilization. Delineating the extent to which ancestral mechanisms change among different animal lineages is critical for understanding our own reproductive biology and developing strategies to manipulate animal fertility rates.

In summary, sperm from diverse animal species process sensory information. It should not be surprising that sperm rely on this information to complete their mission. This chapter describes examples whereby sperm movement within the female reproductive tract is controlled by male proteins, female lipids, and physical forces (i.e., fluid flow). In addition to reproductive biology, these studies are providing insight into how migrating cells, in general, interpret their positions. Basic discoveries, such as novel mechanisms of prostaglandin synthesis and progesterone signaling, are illuminating other fields as well. The future impinges on technological advances, creative minds, and curiosity. As imaging technologies improve, so will our ability to monitor sperm movement in a wide range of female animals. This may be particularly important if animals optimize sperm navigation performance to specific environments. Humans are changing the Earth's environment in unprecedented ways. To limit accumulating damage to animal species, we must understand how these changes are impacting fertility. Human infertility is also a growing problem, and poor sperm navigation might be a major contributing factor.

References

Aitken RJ, Nixon B (2013) Sperm capacitation: a distant landscape glimpsed but unexplored. Mol Hum Reprod 19:785–793

Alonso-Pimentel H, Tolbert LP, Heed WB (1994) Ultrastructural examination of the insemination reaction in Drosophila. Cell Tissue Res 275:467–479

Arthur B, Hauschteck-Jungen E, Nöthiger R, Ward PI (1998) A female nervous system is necessary for normal sperm storage in *Drosophila melanogaster*: a masculinized nervous system is as good as none. Proc R Soc Lond B Biol 265:1749–1753

Austin CR (1952) The capacitation of the mammalian sperm. Nature 170:326

Avenarius MR, Hildebrand MS, Zhang Y, Meyer NC, Smith LL, Kahrizi K, Najmabadi H, Smith RJ (2009) Human male infertility caused by mutations in the CATSPER1 channel protein. Am J Hum Genet 84:505–510

Avila FW, Wolfner MF (2009) Acp36DE is required for uterine conformational changes in mated Drosophila females. Proc Natl Acad Sci USA 106:15796–15800

Avila FW, Sirot LK, Laflamme BA, Rubinstein CD, Wolfner MF (2011) Insect seminal fluid proteins: identification and function. Annu Rev Entomol 56:21–40

Avila FW, Bloch Qazi MC, Rubinstein CD, Wolfner MF (2012) A requirement for the neuromodulators octopamine and tyramine in *Drosophila melanogaster* female sperm storage. Proc Natl Acad Sci USA 109:4562–4567

Bahat A, Tur-Kaspa I, Gakamsky A, Giojalas LC, Breitbart H, Eisenbach M (2003) Thermotaxis of mammalian sperm cells: a potential navigation mechanism in the female genital tract. Nat Med 9:149–150

Bahat A, Eisenbach M, Tur-Kaspa I (2005) Periovulatory increase in temperature difference within the rabbit oviduct. Hum Reprod 20:2118–2121

Benenati G, Penkov S, Muller-Reichert T, Entchev EV, Kurzchalia TV (2009) Two cytochrome P450s in *Caenorhabditis elegans* are essential for the organization of eggshell, correct execution of meiosis and the polarization of embryo. Mech Dev 126:382–393

Bian F, Mao G, Guo M, Mao G, Wang J, Li J, Han Y, Chen X, Zhang M, Xia G (2012) Gradients of natriuretic peptide precursor A (NPPA) in oviduct and of natriuretic peptide receptor 1 (NPR1) in spermatozoon are involved in mouse sperm chemotaxis and fertilization. J Cell Physiol 227:2230–2239

Boden MJ, Varcoe TJ, Kennaway DJ (2013) Circadian regulation of reproduction: from gamete to offspring. Prog Biophys Mol Biol 113:387–397

Brenker C, Goodwin N, Weyand I, Kashikar ND, Naruse M, Krahling M, Muller A, Kaupp UB, Strunker T (2012) The CatSper channel: a polymodal chemosensor in human sperm. EMBO J 31:1654–1665

Byrd DT, Kimble J (2009) Scratching the niche that controls *Caenorhabditis elegans* germline stem cells. Semin Cell Dev Biol 20:1107–1113

Carlson AE, Westenbroek RE, Quill T, Ren D, Clapham DE, Hille B, Garbers DL, Babcock DF (2003) CatSper1 required for evoked Ca2+ entry and control of flagellar function in sperm. Proc Natl Acad Sci USA 100:14864–14868

Cassada RC, Russell RL (1975) The dauerlarva, a post-embryonic developmental variant of the nematode *Caenorhabditis elegans*. Dev Biol 46:326–342

Chandrasekharan NV, Simmons DL (2004) The cyclooxygenases. Genome Biol 5:241

Chang MC (1951) Fertilizing capacity of spermatozoa deposited into the fallopian tubes. Nature 168:697–698

Chang H, Suarez SS (2012) Unexpected flagellar movement patterns and epithelial binding behavior of mouse sperm in the oviduct. Biol Reprod 86(140):1–8

Chasnov JR (2013) The evolutionary role of males in *C. elegans*. Worm 2:e21146

Chen J, Caswell-Chen EP (2004) Facultative vivipary is a life-history trait in *Caenorhabditis elegans*. J Nematol 36:107–113

Chung JJ, Shim SH, Everley RA, Gygi SP, Zhuang X, Clapham DE (2014) Structurally distinct ca (2+) signaling domains of sperm flagella orchestrate tyrosine phosphorylation and motility. Cell 157:808–822

Czaran TL, Hoekstra RF (2004) Evolution of sexual asymmetry. BMC Evol Biol 4:34

Dekel N (2005) Cellular, biochemical and molecular mechanisms regulating oocyte maturation. Mol Cell Endocrinol 234:19–25

Deline M, Keller J, Rothe M, Schunck WH, Menzel R, Watts JL (2015) Epoxides derived from dietary dihomo-gamma-linolenic acid induce germ cell death in *C. elegans*. Sci Rep 5:15417

Denissenko P, Kantsler V, Smith DJ, Kirkman-Brown J (2012) Human spermatozoa migration in microchannels reveals boundary-following navigation. Proc Natl Acad Sci USA 109:8007–8010

Depina AS, Iser WB, Park SS, Maudsley S, Wilson MA, Wolkow CA (2011) Regulation of *Caenorhabditis elegans* vitellogenesis by DAF-2/IIS through separable transcriptional and posttranscriptional mechanisms. BMC Physiol 11:11

Edmonds JW, Prasain JK, Dorand D, Yang Y, Hoang HD, Vibbert J, Kubagawa HM, Miller MA (2010) Insulin/FOXO signaling regulates ovarian prostaglandins critical for reproduction. Dev Cell 19:858–871

Edmonds JW, Mckinney SL, Prasain JK, Miller MA (2011) The gap junctional protein INX-14 functions in oocyte precursors to promote *C. elegans* sperm guidance. Dev Biol 359:47–58

Eisenbach M, Giojalas LC (2006) Sperm guidance in mammals - an unpaved road to the egg. Nat Rev Mol Cell Biol 7:276–285

Eisenbach M, Tur-Kaspa I (1999) Do human eggs attract spermatozoa? Bioessays 21:203–210

Ellis R, Schedl T (2007) Sex determination in the germ line. WormBook 1–13

Ellis RE, Stanfield GM (2014) The regulation of spermatogenesis and sperm function in nematodes. Semin Cell Dev Biol 29:17–30

Ernesto JI, Weigel Munoz M, Battistone MA, Vasen G, Martinez-Lopez P, Orta G, Figueiras-Fierro D, De La Vega-Beltran JL, Moreno IA, Guidobaldi HA, Giojalas L, Darszon A, Cohen DJ, Cuasnicu PS (2015) CRISP1 as a novel CatSper regulator that modulates sperm motility and orientation during fertilization. J Cell Biol 210:1213–1224

Evans JP, Florman HM (2002) The state of the union: the cell biology of fertilization. Nat Cell Biol 4(Suppl):s57–s63

Fielenbach N, Antebi A (2008) *C. elegans* dauer formation and the molecular basis of plasticity. Genes Dev 22:2149–2165

Frezal L, Felix MA (2015) *C. elegans* outside the Petri dish. Elife 4

Funk CD (2001) Prostaglandins and leukotrienes: advances in eicosanoid biology. Science 294:1871–1875

Gao Z, Ruden DM, Lu X (2003) PKD2 cation channel is required for directional sperm movement and male fertility. Curr Biol 13:2175–2178

Gilbert DG, Richmond RC (1981) Studies of esterase 6 in *Drosophila melanogaster*. VI. Ejaculate competitive abilities of males having null or active alleles. Genetics 97:85–94

Gottschalk A, Schafer WR (2006) Visualization of integral and peripheral cell surface proteins in live *Caenorhabditis elegans*. J Neurosci Methods 154:68–79

Govindan JA, Nadarajan S, Kim S, Starich TA, Greenstein D (2009) Somatic cAMP signaling regulates MSP-dependent oocyte growth and meiotic maturation in *C. elegans*. Development 136:2211–2221

Grant B, Hirsh D (1999) Receptor-mediated endocytosis in the *Caenorhabditis elegans* oocyte. Mol Biol Cell 10:4311–4326

Greer ER, Perez CL, Van Gilst MR, Lee BH, Ashrafi K (2008) Neural and molecular dissection of a *C. elegans* sensory circuit that regulates fat and feeding. Cell Metab 8:118–131

Griswold MD (2016) Spermatogenesis: the commitment to meiosis. Physiol Rev 96:1–17

Guidobaldi HA, Teves ME, Unates DR, Anastasia A, Giojalas LC (2008) Progesterone from the cumulus cells is the sperm chemoattractant secreted by the rabbit oocyte cumulus complex. PLoS One 3:e3040

Gumienny TL, Savage-Dunn C (2013) TGF-beta signaling in *C. elegans*. WormBook 1–34

Gunther CV, Georgi LL, Riddle DL (2000) A *Caenorhabditis elegans* type I TGF beta receptor can function in the absence of type II kinase to promote larval development. Development 127:3337–3347

Hall DH, Winfrey VP, Blaeuer G, Hoffman LH, Furuta T, Rose KL, Hobert O, Greenstein D (1999) Ultrastructural features of the adult hermaphrodite gonad of *Caenorhabditis elegans*: relations between the germ line and soma. Dev Biol 212:101–123

Han SM, Cottee PA, Miller MA (2010) Sperm and oocyte communication mechanisms controlling *C. elegans* fertility. Dev Dyn 239:1265–1281

Hansen JM, Chavez DR, Stanfield GM (2015) COMP-1 promotes competitive advantage of nematode sperm. Elife 4

Heifetz Y, Rivlin PK (2010) Beyond the mouse model: using Drosophila as a model for sperm interaction with the female reproductive tract. Theriogenology 73:723–739

Hill KL, L'hernault SW (2001) Analyses of reproductive interactions that occur after heterospecific matings within the genus Caenorhabditis. Dev Biol 232:105–114

Hillers KJ, Jantsch V, Martinez-Perez E, Yanowitz JL (2015) Meiosis. WormBook 1–54

Hirsh D, Oppenheim D, Klass M (1976) Development of the reproductive system of *Caenorhabditis elegans*. Dev Biol 49:200–219

Hoang HD, Prasain JK, Dorand D, Miller MA (2013) A heterogeneous mixture of F-series prostaglandins promotes sperm guidance in the *Caenorhabditis elegans* reproductive tract. PLoS Genet 9:e1003271

Hodgkin J, Doniach T (1997) Natural variation and copulatory plug formation in *Caenorhabditis elegans*. Genetics 146:149–164

Hodgkin J, Horvitz HR, Brenner S (1979) Nondisjunction mutants of the nematode *Caenorhabditis elegans*. Genetics 91:67–94

Hoekstra RF (1987) The evolution of sexes. Experientia 55:59–91

Holt WV, Fazeli A (2016) Sperm storage in the female reproductive tract. Annu Rev Anim Biosci 4:291–310

Hu PJ (2007) Dauer. WormBook (the online review of C. elegans biology) 1–19

Hubbard EJ, Greenstein D (2000) The *Caenorhabditis elegans* gonad: a test tube for cell and developmental biology. Dev Dyn 218:2–22

Hunt PA, Lemaire-Adkins R (1998) Genetic control of mammalian female meiosis. Curr Top Dev Biol 37:359–381

Hunter RH, Nichol R (1986) A preovulatory temperature gradient between the isthmus and ampulla of pig oviducts during the phase of sperm storage. J Reprod Fertil 77:599–606

Iida K, Cavener DR (2004) Glucose dehydrogenase is required for normal sperm storage and utilization in female *Drosophila melanogaster*. J Exp Biol 207:675–681

Italiano JE Jr, Stewart M, Roberts TM (1999) Localized depolymerization of the major sperm protein cytoskeleton correlates with the forward movement of the cell body in the amoeboid movement of nematode sperm. J Cell Biol 146:1087–1096

Jansen RP (1978) Fallopian tube isthmic mucus and ovum transport. Science 201:349–351

Kantsler V, Dunkel J, Blayney M, Goldstein RE (2014) Rheotaxis facilitates upstream navigation of mammalian sperm cells. Elife 3:e02403

Kaupp UB, Kashikar ND, Weyand I (2008) Mechanisms of sperm chemotaxis. Annu Rev Physiol 70:93–117

Kimble J, Crittenden SL (2005) Germline proliferation and its control. WormBook 1–14

Klass M, Wolf N, Hirsh D (1976) Development of the male reproductive system and sexual transformation in the nematode *Caenorhabditis elegans*. Dev Biol 52:1–18

Kosinski M, Mcdonald K, Schwartz J, Yamamoto I, Greenstein D (2005) *C. elegans* sperm bud vesicles to deliver a meiotic maturation signal to distant oocytes. Development 132:3357–3369

Kottgen M, Hofherr A, Li W, Chu K, Cook S, Montell C, Watnick T (2011) Drosophila sperm swim backwards in the female reproductive tract and are activated via TRPP2 ion channels. PLoS One 6:e20031

Kubagawa HM, Watts JL, Corrigan C, Edmonds JW, Sztul E, Browse J, Miller MA (2006) Oocyte signals derived from polyunsaturated fatty acids control sperm recruitment in vivo. Nat Cell Biol 8:1143–1148

L'hernault SW (1997) Spermatogenesis. In: Riddle DL, Blumenthal T, Meyer BJ, Priess JR (eds) *C. elegans* II. Cold Spring Harbor Laboratory Press, Cold Spring Harbor, NY

Lamunyon CW, Ward S (1998) Larger sperm outcompete smaller sperm in the nematode *Caenorhabditis elegans*. Proc Biol Sci 265:1997–2002

Lamunyon CW, Ward S (2002) Evolution of larger sperm in response to experimentally increased sperm competition in *Caenorhabditis elegans*. Proc Biol Sci 269:1125–1128

Lefevre G Jr, Jonsson UB (1962) Sperm transfer, storage, displacement, and utilization in *Drosophila melanogaster*. Genetics 47:1719–1736

Lev S, Ben Halevy D, Peretti D, Dahan N (2008) The VAP protein family: from cellular functions to motor neuron disease. Trends Cell Biol 18:282–290

L'hernault SW (2006) Spermatogenesis. WormBook 1–14

L'hernault SW (2009) The genetics and cell biology of spermatogenesis in the nematode *C. elegans*. Mol Cell Endocrinol 306:59–65

Lin K, Hsin H, Libina N, Kenyon C (2001) Regulation of the *Caenorhabditis elegans* longevity protein DAF-16 by insulin/IGF-1 and germline signaling. Nat Genet 28:139–145

Lishko PV, Botchkina IL, Kirichok Y (2011) Progesterone activates the principal Ca2+ channel of human sperm. Nature 471:387–391

Loppin B, Dubruille R, Horard B (2015) The intimate genetics of Drosophila fertilization. Open Biol 5

Ludewig AH, Schroeder FC (2013) Ascaroside signaling in *C. elegans*. WormBook 1–22

Ma DK, Rothe M, Zheng S, Bhatla N, Pender CL, Menzel R, Horvitz HR (2013) Cytochrome P450 drives a HIF-regulated behavioral response to reoxygenation by *C. elegans*. Science 341:554–558

Manier MK, Belote JM, Berben KS, Novikov D, Stuart WT, Pitnick S (2010) Resolving mechanisms of competitive fertilization success in *Drosophila melanogaster*. Science 328:354–357

Marcello MR, Singaravelu G, Singson A (2013) Fertilization. Adv Exp Med Biol 757:321–350

Masui Y (1985) Meiotic arrest in animal oocytes. In: Monroy A, Metz CB (eds) Biology of fertilization. Academic Press, Orlando, FL

Matova N, Cooley L (2001) Comparative aspects of animal oogenesis. Dev Biol 231:291–320

Mcgraw LA, Gibson G, Clark AG, Wolfner MF (2004) Genes regulated by mating, sperm, or seminal proteins in mated female *Drosophila melanogaster*. Curr Biol 14:1509–1514

Mcknight K, Hoang HD, Prasain JK, Brown N, Vibbert J, Hollister KA, Moore R, Ragains JR, Reese J, Miller MA (2014) Neurosensory perception of environmental cues modulates sperm motility critical for fertilization. Science 344:754–757

Mcnally KL, Mcnally FJ (2005) Fertilization initiates the transition from anaphase I to metaphase II during female meiosis in *C. elegans*. Dev Biol 282:218–230

Meisel JD, Panda O, Mahanti P, Schroeder FC, Kim DH (2014) Chemosensation of bacterial secondary metabolites modulates neuroendocrine signaling and behavior of *C. elegans*. Cell 159:267–280

Meyer BJ (2005) X-Chromosome dosage compensation. WormBook 1–14

Miki K, Clapham DE (2013) Rheotaxis guides mammalian sperm. Curr Biol 23:443–452

Miller MA, Technau U, Smith KM, Steele RE (2000) Oocyte development in Hydra involves selection from competent precursor cells. Dev Biol 224:326–338

Miller MA, Nguyen VQ, Lee MH, Kosinski M, Schedl T, Caprioli RM, Greenstein D (2001) A sperm cytoskeletal protein that signals oocyte meiotic maturation and ovulation. Science 291:2144–2147

Miller MA, Ruest PJ, Kosinski M, Hanks SK, Greenstein D (2003) An Eph receptor sperm-sensing control mechanism for oocyte meiotic maturation in *Caenorhabditis elegans*. Genes Dev 17:187–200

Mukhopadhyay A, Oh SW, Tissenbaum HA (2006) Worming pathways to and from DAF-16/FOXO. Exp Gerontol 41:928–934

Murphy CT, Hu PJ (2013) Insulin/insulin-like growth factor signaling in *C. elegans*. WormBook 1–43

Nance J, Minniti AN, Sadler C, Ward S (1999) spe-12 encodes a sperm cell surface protein that promotes spermiogenesis in *Caenorhabditis elegans*. Genetics 152:209–220

Navarrete FA, Garcia-Vazquez FA, Alvau A, Escoffier J, Krapf D, Sanchez-Cardenas C, Salicioni AM, Darszon A, Visconti PE (2015) Biphasic role of calcium in mouse sperm capacitation signaling pathways. J Cell Physiol 230:1758–1769

Neubaum DM, Wolfner MF (1999) Wise, winsome, or weird? Mechanisms of sperm storage in female animals. Curr Top Dev Biol 41:67–97

Nishimura H, Tajima T, Comstra HS, Gleason EJ, L'hernault SW (2015) The immunoglobulin-like gene spe-45 acts during fertilization in *Caenorhabditis elegans* like the mouse Izumo1 gene. Curr Biol 25:3225–3231

Ogg S, Paradis S, Gottlieb S, Patterson GI, Lee L, Tissenbaum HA, Ruvkun G (1997) The fork head transcription factor DAF-16 transduces insulin-like metabolic and longevity signals in *C. elegans*. Nature 389:994–999

Ohsako T, Yamamoto MT (2011) Sperm of the wasted mutant are wasted when females utilize the stored sperm in *Drosophila melanogaster*. Genes Genet Syst 86:97–108

Olson JH, Xiang X, Ziegert T, Kittelson A, Rawls A, Bieber AL, Chandler DE (2001) Allurin, a 21-kDa sperm chemoattractant from Xenopus egg jelly, is related to mammalian sperm-binding proteins. Proc Natl Acad Sci USA 98:11205–11210

Olson SK, Greenan G, Desai A, Muller-Reichert T, Oegema K (2012) Hierarchical assembly of the eggshell and permeability barrier in *C. elegans*. J Cell Biol 198:731–748

Oren-Benaroya R, Orvieto R, Gakamsky A, Pinchasov M, Eisenbach M (2008) The sperm chemoattractant secreted from human cumulus cells is progesterone. Hum Reprod 23:2339–2345

Palopoli MF, Rockman MV, Tinmaung A, Ramsay C, Curwen S, Aduna A, Laurita J, Kruglyak L (2008) Molecular basis of the copulatory plug polymorphism in *Caenorhabditis elegans*. Nature 454:1019–1022

Pattarini JM, Starmer WT, Bjork A, Pitnick S (2006) Mechanisms underlying the sperm quality advantage in *Drosophila melanogaster*. Evolution 60:2064–2080

Perez-Cerezales S, Boryshpolets S, Afanzar O, Brandis A, Nevo R, Kiss V, Eisenbach M (2015a) Involvement of opsins in mammalian sperm thermotaxis. Sci Rep 5:16146

Perez-Cerezales S, Boryshpolets S, Eisenbach M (2015b) Behavioral mechanisms of mammalian sperm guidance. Asian J Androl 17:628–632

Perez-Cerezales S, Lopez-Cardona AP, Gutierrez-Adan A (2016) Progesterone effects on mouse sperm kinetics in conditions of viscosity. Reproduction 151(5):501–507

Prasain JK, Wilson L, Hoang HD, Moore R, Miller MA (2015) Comparative lipidomics of *Caenorhabditis elegans* metabolic disease models by SWATH non-targeted tandem mass spectrometry. Metabolites 5:677–696

Publicover SJ, Giojalas LC, Teves ME, De Oliveira GS, Garcia AA, Barratt CL, Harper CV (2008) Ca2+ signalling in the control of motility and guidance in mammalian sperm. Front Biosci 13:5623–5637

Qi H, Moran MM, Navarro B, Chong JA, Krapivinsky G, Krapivinsky L, Kirichok Y, Ramsey IS, Quill TA, Clapham DE (2007) All four CatSper ion channel proteins are required for male fertility and sperm cell hyperactivated motility. Proc Natl Acad Sci USA 104:1219–1223

Ren P, Lim CS, Johnsen R, Albert PS, Pilgrim D, Riddle DL (1996) Control of *C. elegans* larval development by neuronal expression of a TGF-beta homolog. Science 274:1389–1391

Roberts TM, Stewart M (1995) Nematode sperm locomotion. Curr Opin Cell Biol 7:13–17

Roberts TM, Stewart M (2000) Acting like actin. The dynamics of the nematode major sperm protein (msp) cytoskeleton indicate a push-pull mechanism for amoeboid cell motility. J Cell Biol 149:7–12

Roberts TM, Stewart M (2012) Role of major sperm protein (MSP) in the protrusion and retraction of Ascaris sperm. Int Rev Cell Mol Biol 297:265–293

Runft LL, Jaffe LA, Mehlmann LM (2002) Egg activation at fertilization: where it all begins. Dev Biol 245:237–254

Schnakenberg SL, Matias WR, Siegal ML (2011) Sperm-storage defects and live birth in Drosophila females lacking spermathecal secretory cells. PLoS Biol 9:e1001192

Schnakenberg SL, Siegal ML, Bloch Qazi MC (2012) Oh, the places they'll go: female sperm storage and sperm precedence in *Drosophila melanogaster*. Spermatogenesis 2:224–235

Seifert R, Flick M, Bonigk W, Alvarez L, Trotschel C, Poetsch A, Muller A, Goodwin N, Pelzer P, Kashikar ND, Kremmer E, Jikeli J, Timmermann B, Kuhl H, Fridman D, Windler F, Kaupp UB, Strunker T (2015) The CatSper channel controls chemosensation in sea urchin sperm. EMBO J 34:379–392

Shakes DC, Ward S (1989) Initiation of spermiogenesis in *C. elegans*: a pharmacological and genetic analysis. Dev Biol 134:189–200

Shen WL, Kwon Y, Adegbola AA, Luo J, Chess A, Montell C (2011) Function of rhodopsin in temperature discrimination in Drosophila. Science 331:1333–1336

Simmons DL, Botting RM, Hla T (2004) Cyclooxygenase isozymes: the biology of prostaglandin synthesis and inhibition. Pharmacol Rev 56:387–437

Singaravelu G, Rahimi S, Krauchunas A, Rizvi A, Dharia S, Shakes D, Smith H, Golden A, Singson A (2015) Forward genetics identifies a requirement for the Izumo-like immunoglobulin superfamily spe-45 gene in *Caenorhabditis elegans* fertilization. Curr Biol 25:3220–3224

Singson A (2001) Every sperm is sacred: fertilization in *Caenorhabditis elegans*. Dev Biol 230:101–109

Singson A, Mercer KB, L'hernault SW (1998) The *C. elegans* spe-9 gene encodes a sperm transmembrane protein that contains EGF-like repeats and is required for fertilization. Cell 93:71–79

Smith H (2006) Sperm motility and MSP. WormBook 1–8

Smith JR, Stanfield GM (2011) TRY-5 is a sperm-activating protease in *Caenorhabditis elegans* seminal fluid. PLoS Genet 7:e1002375

Stanfield GM, Villeneuve AM (2006) Regulation of sperm activation by SWM-1 is required for reproductive success of *C. elegans* males. Curr Biol 16:252–263

Starich TA, Hall DH, Greenstein D (2014) Two classes of gap junction channels mediate soma-germline interactions essential for germline proliferation and gametogenesis in *Caenorhabditis elegans*. Genetics 198:1127–1153

Stefan CJ, Manford AG, Baird D, Yamada-Hanff J, Mao Y, Emr SD (2011) Osh proteins regulate phosphoinositide metabolism at ER-plasma membrane contact sites. Cell 144:389–401

Stein KK, Golden A (2015) The *C. elegans* eggshell. WormBook 1–35

Strunker T, Goodwin N, Brenker C, Kashikar ND, Weyand I, Seifert R, Kaupp UB (2011) The CatSper channel mediates progesterone-induced Ca2+ influx in human sperm. Nature 471:382–386

Suarez SS (2008) Regulation of sperm storage and movement in the mammalian oviduct. Int J Dev Biol 52:455–462

Suarez SS (2016) Mammalian sperm interactions with the female reproductive tract. Cell Tissue Res 363:185–194

Suarez SS, Pacey AA (2006) Sperm transport in the female reproductive tract. Hum Reprod Update 12:23–37

Sun F, Bahat A, Gakamsky A, Girsh E, Katz N, Giojalas LC, Tur-Kaspa I, Eisenbach M (2005) Human sperm chemotaxis: both the oocyte and its surrounding cumulus cells secrete sperm chemoattractants. Hum Reprod 20:761–767

Teves ME, Barbano F, Guidobaldi HA, Sanchez R, Miska W, Giojalas LC (2006) Progesterone at the picomolar range is a chemoattractant for mammalian spermatozoa. Fertil Steril 86:745–749

Ting JJ, Woodruff GC, Leung G, Shin NR, Cutter AD, Haag ES (2014) Intense sperm-mediated sexual conflict promotes reproductive isolation in Caenorhabditis nematodes. PLoS Biol 12: e1001915

Tsuda H, Han SM, Yang Y, Tong C, Lin YQ, Mohan K, Haueter C, Zoghbi A, Harati Y, Kwan J, Miller MA, Bellen HJ (2008) The amyotrophic lateral sclerosis 8 protein VAPB is cleaved, secreted, and acts as a ligand for Eph receptors. Cell 133:963–977

Tung CK, Hu L, Fiore AG, Ardon F, Hickman DG, Gilbert RO, Suarez SS, Wu M (2015) Microgrooves and fluid flows provide preferential passageways for sperm over pathogen Tritrichomonas foetus. Proc Natl Acad Sci USA 112:5431–5436

Valmsen K, Jarving I, Boeglin WE, Varvas K, Koljak R, Pehk T, Brash AR, Samel N (2001) The origin of 15R-prostaglandins in the Caribbean coral *Plexaura homomalla*: molecular cloning and expression of a novel cyclooxygenase. Proc Natl Acad Sci USA 98:7700–7705

Van Der Donk WA, Tsai AL, Kulmacz RJ (2002) The cyclooxygenase reaction mechanism. Biochemistry 41:15451–15458

Vane JR (1971) Inhibition of prostaglandin synthesis as a mechanism of action for aspirin-like drugs. Nat New Biol 231:232–235

Ward S, Carrel JS (1979) Fertilization and sperm competition in the nematode *Caenorhabditis elegans*. Dev Biol 73:304–321

Ward S, Argon Y, Nelson GA (1981) Sperm morphogenesis in wild-type and fertilization-defective mutants of *Caenorhabditis elegans*. J Cell Biol 91:26–44

Ward S, Hogan E, Nelson GA (1983) The initiation of spermiogenesis in the nematode *Caenorhabditis elegans*. Dev Biol 98:70–79

Ward GE, Brokaw CJ, Garbers DL, Vacquier VD (1985) Chemotaxis of *Arbacia punctulata* spermatozoa to resact, a peptide from the egg jelly layer. J Cell Biol 101:2324–2329

Watnick TJ, Jin Y, Matunis E, Kernan MJ, Montell C (2003) A flagellar polycystin-2 homolog required for male fertility in Drosophila. Curr Biol 13:2179–2184

Watts JL, Browse J (2002) Genetic dissection of polyunsaturated fatty acid synthesis in *Caenorhabditis elegans*. Proc Natl Acad Sci USA 99:5854–5859

Williams M, Hill CJ, Scudamore I, Dunphy B, Cooke ID, Barratt CL (1993) Sperm numbers and distribution within the human fallopian tube around ovulation. Hum Reprod 8:2019–2026

Xu XZ, Sternberg PW (2003) A C. elegans sperm TRP protein required for sperm-egg interactions during fertilization. Cell 114:285–297

Yamaguchi R, Muro Y, Isotani A, Tokuhiro K, Takumi K, Adham I, Ikawa M, Okabe M (2009) Disruption of ADAM3 impairs the migration of sperm into oviduct in mouse. Biol Reprod 81:142–146

Yanagimachi R (2009) Germ cell research: a personal perspective. Biol Reprod 80:204–218

Yang Y, Lu X (2011) Drosophila sperm motility in the reproductive tract. Biol Reprod 84:1005–1015

Zarkower D (2006) Somatic sex determination. WormBook 1–12

Zhou Y, Falck JR, Rothe M, Schunck WH, Menzel R (2015) Role of CYP eicosanoids in the regulation of pharyngeal pumping and food uptake in *Caenorhabditis elegans*. J Lipid Res 56:2110–2123

Chapter 10
Cell Fate Maintenance and Reprogramming During the Oocyte-to-Embryo Transition

Christina Fassnacht and Rafal Ciosk

Abstract This chapter reviews our current understanding of the mechanisms that regulate reprogramming during the oocyte-to-embryo transition (OET). There are two major events reshaping the transcriptome during OET. One is the clearance of maternal transcripts in the early embryo, extensively reviewed by others. The other event, which is the focus of this chapter, is the embryonic (or zygotic) genome activation (EGA). The mechanisms controlling EGA can be broadly divided into transcriptional and posttranscriptional. The former includes the regulation of the basal transcription machinery, the regulation by specific transcription factors and chromatin modifications. The latter is performed mostly via specific RNA-binding proteins (RBPs). Different animal models have been used to decipher the regulation of EGA. These models are often biased for the specific type of regulation, which is why we discuss the models ranging from invertebrates to mammals. Whether these biases stem from incomplete understanding of EGA in these models, or reflect evolutionarily distinct solutions to EGA regulation, is a key unresolved problem in developmental biology. As the mechanisms controlling developmental reprogramming can, and in some cases have been shown to, function in differentiated cells subjected to induced reprogramming, our understanding of EGA regulation may have implications for the efficiency of induced reprogramming and, thus, for regenerative medicine.

10.1 Introduction

Cellular differentiation leads to the formation of highly specialized cells, such as neurons and muscles, which fulfill complex functions. Cellular reprogramming is the opposite process, where a specialized cell, with a restricted developmental potential, is turned into a pluripotent cell that can give rise to different types of cells. During development, a complete reprogramming into pluripotency is rare,

C. Fassnacht • R. Ciosk (✉)
Friedrich Miescher Institute for Biomedical Research, Basel, Switzerland
e-mail: rafal.ciosk@fmi.ch

S. Arur (ed.), *Signaling-Mediated Control of Cell Division*, Results and Problems in Cell Differentiation 59, DOI 10.1007/978-3-319-44820-6_10

being restricted to the OET. During this transition, a fertilized oocyte develops into a pluripotent embryo, which has the potential to give rise to an entire new individual. The main cellular processes occurring during the OET (oocyte maturation, ovulation, and fertilization) are discussed in the chapter "Oocyte activation and fertilization: crucial sperm and oocyte factors." Here, we focus on changes in the transcriptome, underlying the transition from maternal to embryonic control of development. In addition, we concentrate on mechanisms operating in the oocytes because unfertilized oocytes display a broad developmental potential, which is evident by their ability to give rise to developmentally plastic parthenogenetic cell lines (Brevini et al. 2012), or even offspring when paternal imprinting is mimicked (Kono et al. 2004). The initial stages of the OET, from oocyte maturation to early embryogenesis, occur in the absence of *de novo* Pol II-mediated transcription and are controlled, posttranscriptionally, until the embryo begins producing its own transcripts during EGA (LaMarca et al. 1975; Moore et al. 1974). Our current understanding of the mechanisms controlling developmental reprogramming during the OET remains very limited. Experiments with pluripotent embryonic stem (ES) cells have shown that pluripotency can be regulated at each step of gene expression. However, which of the mechanisms described in ES cells are employed during development remains to be, to a large extent, determined. During development, a successful OET depends on the ability to maintain the germ cell fate in a developing oocyte, which includes the maintenance of germ line-specific gene expression and the inhibition of soma/embryo-specific expression. Failures in this process can result in a precocious acquisition of pluripotency and somatic-like, teratomatous differentiation. In an early embryo, the regulation of gene expression is reversed, resulting in the activation of somatic/embryonic genes and the inhibition of germ-line genes. In this chapter, we discuss critical factors and emerging mechanisms controlling this dramatic reprogramming of gene expression.

10.2 Developmental Potential of Germ Cells

Germ cells are typically the only cells giving rise to the offspring. This striking property of germ cells led to the formulation of the germplasm theory of heredity, also known as the Weismann theory of heredity, which states that only germ cells transmit genetic information to the next generation (Weismann 1893). Indeed, the development and maintenance of germ cells are crucial for the survival of species with sexual reproduction, and germ cell-specific genome surveillance mechanisms have evolved to ensure genome integrity. Germ cells originate from primordial germ cells (PGCs) that are usually specified during embryonic development (for a recent review on the specification of germ cell fate in different organisms, see Strome and Updike 2015). PGCs and also adult germ cells have the potential to give rise to other types of cells, as evident by the different types of pluripotent cell lines derived from germ cells at different stages of development (Kerr et al. 2006). For example, embryonic germ (EG) cells can be derived from mouse PGCs between the embryonic days E8.5 and E13.5 (Cantone and Fisher 2013). The EG cells are in many aspects like blastocyst-derived embryonic stem (ES) cells in that they can generate chimeras and display germ line transmission (Smith 2001). The pluripotency of germ cells is also manifested by the properties of a specific germ cell tumor (GCT) called teratoma. In human teratomas, germ cells abnormally differentiate into somatic cell types, such as neurons, muscles, hair, or teeth (Ulbright 2005). They can arise in ovary and testis, with testicular GCTs accounting for the most common solid malignancy found in males between the ages of 15 and 40 years (Hussain et al. 2008). Origins of GCT development during human and mouse gametogenesis have been recently reviewed in Dolci et al. (2015). They include epigenetic remodeling, as well as defects in proliferation, apoptosis, and the mitotic to meiotic switch. Undifferentiated cells from teratomas can also be grown ex vivo as embryonic carcinoma (EC) cells and can be differentiated into cells of all three germ layers in culture (Smith 2001). The formation of teratomas shows that to ensure normal development, there need to be repressive mechanisms operating in adult germ cells to control pluripotency. It is mostly the work from the nematode *C. elegans* that contributed to the identification of pluripotency repressors, which will be discussed later. A final evidence for the underlying pluripotency of germ cells is the ability of the oocyte cytoplasm to reprogram somatic nuclei. This was experimentally demonstrated by using the cytoplasm of an enucleated *Xenopus* oocyte to reprogram an intestine nucleus (Gurdon and Uehlinger 1966). Since then, reprogramming has been mostly studied using cell culture. This revealed some important regulators, most notably the pluripotency transcription factors Oct4, Sox2, Klf4, and c-Myc (Takahashi and Yamanaka 2006). However, the reprogramming remains a very inefficient process, indicating that repressive mechanisms stabilizing the differentiated state may exist (Pasque et al. 2011). Arguably, understanding the mechanisms controlling the natural reprogramming during OET may be useful for improving the in vitro reprogramming techniques.

10.3 Developmental Reprogramming to Pluripotency During the OET

Two major processes occur during the OET to reprogram the transcriptome. One is the maternal clearance, the degradation of many maternal mRNAs and proteins deposited into the oocyte. The mechanisms behind the maternal clearance have been recently reviewed (Barckmann and Simonelig 2013; Walser and Lipshitz 2011). The other process is the embryonic genome activation (EGA), the onset of transcription in the early embryo. Here, we concentrate on the advances that have been made in understanding the timing, dynamics, and regulation of EGA using different model organisms. Embryonic transcription initiates at a species-specific time, which ranges in vertebrates from 1 to 2 cell cycles in mouse to cell cycles 6–9 in *Xenopus* and zebrafish embryos (Lee et al. 2014). Though, from an absolute time perspective, EGA occurs rather later in the mouse, as the first cell division occurs a long time after fertilization. Among invertebrates, embryonic transcription starts at cell cycle 2 in the nematode *C. elegans* and at cell cycle 6 in the fruit fly *D. melanogaster* (Lee et al. 2014). A common sequence in EGA across model organisms is the activation of transcription in two waves. In the first wave, only a few embryonic mRNAs are transcribed, whereas in the second, later wave, the bulk of transcription occurs. The dynamics of these two waves vary between organisms. In mouse and in *C. elegans*, the second wave of transcription rapidly proceeds the first, whereas in the fruit fly there is a long pause between the waves, with the first at cell cycle 6 and the major wave only at cell cycle 13, where the embryo consists already of roughly 6000 cells (Lee et al. 2014).

Different models have been proposed to explain the regulation of EGA, including the nucleocytoplasmic (N/C) ratio model, the maternal clock model, and the transcript abortion model. The N/C ratio model postulates that for EGA to begin, an EGA repressor in the cytoplasm must be titrated away, as the nuclear to cytoplasmic ratio increases during cell divisions (Newport and Kirschner 1982). One candidate repressor to be regulated by titration is xDnmt1 in *Xenopus* early embryos, whose molecular function will be discussed later. The maternal clock model, in contrast, is cell-cycle independent and hypothesizes that the accumulation or increase in activity of an EGA activator to a certain threshold is needed to start EGA, rather than the titration of a repressor (Lee et al. 2014). One example for the regulation by a maternal clock is TAF-4, a *C. elegans* TFIID subunit, whose activity is controlled by a fertilization-dependent mechanism. The transcript abortion model proposes that the properties of the cell cycle control early gene expression. In many organisms, the first embryonic divisions occur very fast, cycling only between S and M phases and lacking the gap phases of the cell cycle. The transcript abortion model postulates that during such fast cell cycles, zygotic transcripts are aborted and can only be fully transcribed when the cell cycles lengthen (Tadros and Lipshitz 2009). The abortion of a nascent long transcript was, indeed, observed during early divisions in *Drosophila* (Shermoen and O'Farrell 1991). Further experimental evidence comes from studies, where a block of the cell cycle leads to premature EGA in *Xenopus* (Kimelman et al. 1987) and *Drosophila* embryos (Edgar and Schubiger 1986). It is tempting to speculate that EGA regulation by the cell cycle is more important in organisms with fast embryonic divisions, like *Drosophila* with early mitotic cycles of 8 min only, whereas could be less relevant in organisms with slow embryonic divisions, like mice, where the first two cell cycles take 2 days. However, the short length of early transcribed genes is conserved among species (Heyn et al. 2014), including mice, leaving the questions open if gene length is also restricting gene expression in early mouse embryos. In summary, the three general EGA models are not mutually exclusive, and it is possible that a combination of them allows the tight regulation of EGA. The recent identification of critical EGA repressors and activators has shed some light into the molecular mechanisms controlling EGA. When considering the major changes that have to occur from a transcriptionally silent oocyte, with tightly packed DNA, to an early embryo undergoing rapid DNA replication and transcription, it is perhaps not surprising that mechanisms, which globally regulate DNA packaging and transcription, have been identified to be crucial for the reprogramming during OET. In addition, although little is known about posttranscriptional regulation of reprogramming, recent examples highlight the contribution of RNA regulation to a successful OET. For the sake of clarity, examples of the transcriptional and posttranscriptional regulation of reprogramming will be presented in two separate chapters.

10.3.1 Transcriptional Regulation of Reprogramming

10.3.1.1 Histone Variants

Changes in the chromatin regulate the accessibility of transcription factors to DNA and, thus, transcription. Heterochromatin contains tightly packed DNA with a limited access by the transcription machinery. In contrast, euchromatin is loosely packed, allowing a relatively easy access to DNA. Chromatin is established by the packing of DNA into nucleosomes and is influenced by nucleosome composition and epigenetic modifications on histone tails of the nucleosomes. Nucleosomes consist of four histone core proteins (H2A, H2B, H3, H4), which can be exchanged by variant histones with unique regulatory properties. Histone exchange, from repressive gamete-specific histones to somatic histones, is likely contributing to the establishment of a permissive chromatin state in embryos. It was shown in mouse that the repressive macroH2A variant in the female pronucleus is progressively lost after fertilization and during EGA onset (Chang et al. 2005). Similarly, studies from *Drosophila* showed that an early embryonic and germ line-specific H1 variant, named dBigH1, is exchanged for a somatic version during embryonic development, coinciding with the transcriptional activation (Perez-Montero et al. 2013). Consequently, the loss of dBigH1 leads to a precocious activation of Pol II-dependent transcription in the early embryo (Perez-Montero et al. 2013).

10.3.1.2 Chromatin Remodelers

Another type of chromatin regulation that influences gene expression is the nucleosome positioning, which is mediated by chromatin remodeling complexes. The finding that the remodelers contribute to EGA came from the mouse, as the loss of maternal Brg1 protein, a catalytic subunit of the SWI/SNF chromatin remodeler complex, leads to a reduced transcription of 30 % of embryonic genes and embryonic arrest (Bultman et al. 2006). However, the question remains whether Brg1 contributes to a general opening of chromatin or whether Brg1 might be specifically targeted to the promoters of embryonic genes, making them accessible for the transcription machinery.

10.3.1.3 Epigenetic Modifications: Histone Marks

As indicated earlier, chromatin can be also influenced by specific modifications of histone tails in the nucleosomes, either promoting or repressing transcription. An enrichment of the activating histone H3K4me3 mark was found on embryonic gene promoters, preferentially on those with housekeeping roles that are expressed in *Xenopus* and zebrafish embryos after EGA onset (Akkers et al. 2009; Vastenhouw et al. 2010; Lindeman et al. 2011). In contrast, the repressive histone H3K27me3 mark was present on the promoters of later expressed genes with developmental roles. This suggests a role for these histone modifications in distinguishing earlier vs. later expressed embryonic genes. Another interesting finding is the appearance of these histone marks only at the time of active transcription in the early embryos (Li et al. 2014; Vastenhouw et al. 2010; Akkers et al. 2009), indicating the presence of a relatively unmodified, naïve chromatin state of the embryo before EGA in several species. Therefore, histone marks likely do not contribute to the transcriptional regulation of the first embryonic genes, but rather direct gene expression later during embryonic development.

10.3.1.4 Epigenetic Modifications: DNA Methylation

In mammals, also DNA methylation is thought to contribute to EGA regulation. DNA methylation is the addition of a methyl group to cytosines in CpG dinucleotides, and methylation of promoter sequences is thought to have a repressive effect on transcription. Consistent with a repressive function, studies in mouse have reported a global DNA demethylation shortly after fertilization and during EGA, which might contribute to an open pluripotent chromatin state (Paranjpe and Veenstra 2015). However, in other vertebrate species, the opposite trend was reported: zebrafish and *Xenopus* embryonic genomes appear to be globally hypermethylated during EGA (Paranjpe and Veenstra 2015). In addition, no obvious correlation between DNA methylation and transcriptional repression of genes seems to exist in *Xenopus* embryos (Bogdanovic et al. 2011). The DNA methyltransferase xDnmt1 was identified in *Xenopus* as an EGA repressor (Stancheva and Meehan 2000), though its silencing function appears to be independent of its catalytic activity, and it is suggested that Dnmt1 acts as a direct transcriptional repressor before the onset of EGA (Dunican et al. 2008). In sum, the contribution of DNA methylation dynamics to EGA regulation may vary from species to species.

10.3.1.5 Alternative Promoters

An important contribution towards understanding the link between chromatin modifications and transcription comes from studies using cap analysis of gene expression (CAGE) in early zebrafish development to map transcription start sites (TSSs) (Nepal et al. 2013; Haberle et al. 2014). CAGE data showed differences in the TSS usage between maternal and embryonic transcripts and identified an A/T-rich (W-box) motif upstream the maternal TTSs and G/C-rich regions downstream of zygotic TSSs, with the two motifs often co-existing in promoters of constitutively expressed genes. This suggests that specific sequences, presumably via recruitment of specific transcription factors, direct the dynamic use of maternal and embryonic TSSs during the OET. Interestingly, H3K4me3-marked nucleosomes are positioned specifically at G/C-rich embryonic sequences, which provide a link between chromatin remodeling and sequence information. It remains to be determined which transcription factors bind these motifs. The use of alternative promoters during the OET has also been reported from a study in mouse, where retrotransposons that are highly expressed in mature oocytes and early embryos can serve as alternative promoters and first exons for embryonic genes (Peaston et al. 2004). This results in the production of early embryo-specific chimeric transcripts. The production of such chimeric transcripts has been reported for a number of host genes, and it remains to be shown how widespread this type of regulation is.

10.3.1.6 Transcription Factors

The availability of both general (pioneering) and more specific transcription factors has been found to contribute to the regulation of embryonic gene expression. In addition, work from different model organisms has shown that the regulation of Pol II initiation and elongation, as well as of components of the pre-initiation complex (PIC), contributes to EGA control (Zurita et al. 2008). The activity of Pol II can be regulated via phosphorylation of serine residues of tandem repeats on the C-terminal domain (CTD). Particularly, the phosphorylation of Serines 2 and 5 of the CTD is crucial during the initiation and elongation of gene transcription. Pol II phosphorylation in mouse correlates well with transcription states during the OET. Transcriptionally silent oocytes show decreasing levels of CTD phosphorylation, whereas after fertilization, phosphorylation increases again before EGA (Bellier et al. 1997). Another level of Pol II and PIC component regulation is their cellular localization. Pol II and PIC components have been shown to localize to the oocyte cytoplasm in many organisms, and only after their nuclear translocation in early embryos, the general transcription machinery becomes active. One example is the sequestration of TAF-4, the *C. elegans* TFIID subunit, by two repressors called OMA-1 and OMA-2 in the cytoplasm of the one- and two-cell embryo (Guven-Ozkan et al. 2008). Responsible for the inhibitory function of OMA-1/-2 is the fertilization-dependent phosphorylation by MBK-2. This phosphorylation ultimately leads to OMA-1/-2 degradation at the four-cell stage and the release of TAF-4 into the nucleus, where assembly of TFIID can occur and transcription begin.

The establishment of an open chromatin state and the activation of the basal transcription machinery during OET provide permissive mechanisms for EGA to occur. However, it is expected that other repressive and activating transcription factors provide instructive mechanisms to gene expression and mediate fine-tuning of transcription during embryo development. The identification of transcription factors with an intrinsic chromatin remodeling function, like Zelda in *D. melanogaster*, suggests that both regulatory mechanisms can even occur at the same time. Zelda is a maternally deposited transcription factor that binds to a heptamer motif, called TAGteam motif, in early embryonic gene promoters and activates their transcription (Liang et al. 2008; De Renzis et al. 2007; ten Bosch et al. 2006). Loss of Zelda leads to a failure in the activation of over a hundred embryonic genes, and defects in cellularization and pattern formation in the early embryo (Liang et al. 2008). In addition to the activation of these early genes, Zelda associates with the promoters of a thousand genes, whose transcription occurs later in embryonic development, suggesting a broader role for Zelda in licensing genes for transcription (Harrison et al. 2011; Nien et al. 2011). Indeed, Zelda was recently found to mediate chromatin accessibility and can therefore be seen as a pioneer transcription factor, facilitating transcription factor binding genome-wide (Schulz et al. 2015). Also in zebrafish, transcription factors that likely possess pioneering activity have been discovered to contribute to EGA. They include the pluripotency transcription factors Nanog, Pou5f3 (formerly Pou5f1, an Oct4 homolog), and Sox19b (a Sox2 homolog), which have been found before to mediate reprogramming to iPS cells (Takahashi and Yamanaka 2006). Their combined loss leads to a failure in the activation of more than 75 % of early embryonic genes and developmental arrest of zebrafish embryos before gastrulation (Lee et al. 2013). The observed correlation between the chromatin association of Pou5f3 and Sox2, and the enrichment in H3K4me3 and Pol II binding, suggests that these transcription factors, like in ES and iPS cells, also function during early development as pioneering factors to facilitate *de novo* gene expression (Leichsenring et al. 2013). In addition, it must be noted that many EGA activators, like Zelda, and the pluripotency transcription factors, Nanog, Pou5f3, and Sox19b, also contribute to the maternal mRNA clearance by activating the expression of factors mediating mRNA degradation; for example, the expression of microRNA miR-430 is activated by Nanog (Lee et al. 2013). Thus, these transcription factors provide a connection between the mechanisms driving EGA and maternal product degradation.

Concluding, transcriptional regulation of reprogramming plays a critical role during the OET (the involved factors are summarized in Table 10.1 and illustrated in Fig. 10.1). The regulation of chromatin and basal transcription machinery provides mechanisms to globally control transcriptional competence in germ cells and early embryos. However, open remaining questions are: How is specificity of gene regulation achieved? What marks embryonic genes as active and maternal genes as inactive? What induces the switch from maternal to embryonic gene expression? What determines the level and spatial regulation of embryonic gene expression? The dynamic use of maternal and embryonic transcription start sites provides one exciting possibility how differential transcription can be achieved. However, the molecular mechanisms underlying the differential expression, for example transcription factors recognizing the motifs, remain to be determined. Also the

Table 10.1 Transcriptional regulation of reprogramming during the OET

Type of regulation	Name	Organism	Function during OET	Molecular function	Reference
Histone variants	macroH2A	*M. musculus*	EGA repressor	Repressive H2A variant in germ line and early embryo	Chang et al. (2005)
	dBigH1	*D. melanogaster*	EGA repressor	Repressive H1 variant in germ line and early embryo	Perez-Montero et al. (2013)
Chromatin remodeler	Brg1	*M. musculus*	EGA activator	Catalytic subunit of the SWI/SNF chromatin remodeler complex	Bultman et al. (2006)
Epigenetic modifier	xDnmt1	*X. laevis*	EGA repressor	DNA methyltransferase activity is not required for EGA repression, might act directly as transcriptional repressor	Stancheva and Meehan (2000), Dunican et al. (2008)
Alternative promoters	unknown	*D. rerio*	both	Maternal A/T-rich and embryonic G/C-rich TSSs lead to differential activation of genes during OET	Nepal et al. (2013), Haberle et al. (2014)
	Retrotransposons, especially LTRs	*M. musculus*	EGA activator	Retrotransposons act as alternative promoters in oocytes and embryos	Peaston et al. (2004)
Basal transcription machinery	TAF-4	*C. elegans*	EGA activator	TFIID subunit that is repressed before EGA	Guven-Ozkan et al. (2008)
TF with pioneer activity	Zelda	*D. melanogaster*	EGA activator	Activates transcription of embryonic genes, likely contains pioneer activity	Liang et al. (2008), De Renzis et al. (2007), ten Bosch et al. (2006), Harrison et al. (2011), Nien et al. (2011)
	Nanog, Pou5f3, Sox19b	*D. rerio*	EGA activator	Activate transcription of embryonic genes, likely contain pioneer activity	Lee et al. (2013), Leichsenring et al. (2013)

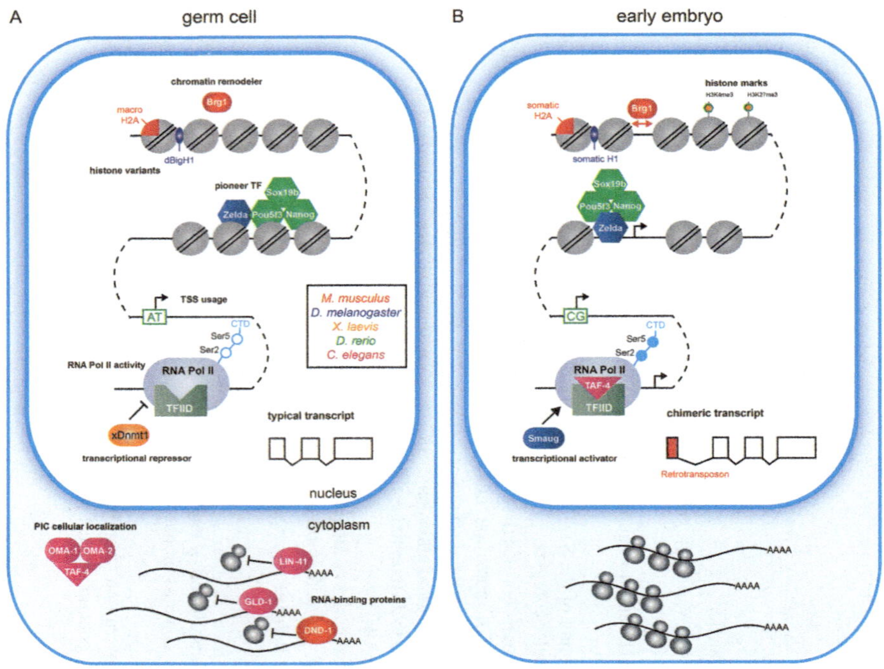

Fig. 10.1 Integration of reprogramming-controlling mechanisms from different species in stereotypical cells. (**a**) Germ cell: chromatin is in a closed conformation, mRNA transcription is inactive, mRNAs are translationally repressed. (**b**) Early embryos: mRNA transcription of embryonic genes from open chromatin can occur, activating (H3K4me3) and repressive (H3K27me3) histone marks are deposited to regulate gene expression, mRNAs are translated into proteins. Colors of the involved factors match to the color of the respective model organism (indicated in the *square box*). References are listed in Tables 10.1 and 10.2. Abbreviations: *TF*, transcription factor; *TSS*, transcription start site; *RNA Pol II*, RNA Polymerase II; *PIC*, pre-initiation complex; and *CTD*, C-terminal domain

identification of the TAG team motif in early *Drosophila* genes and the identification of Zelda, as the associating TF, provided some insights into embryonic gene regulation. However, such motifs in embryonic gene promoters have not been identified in other model organisms. This raises the question how widespread and conserved the utilization of embryo-specific TF-binding motifs is across organisms.

10.3.2 Posttranscriptional Regulation of Reprogramming

10.3.2.1 RNA-Binding Proteins

As described earlier, germ cells have the developmental capacity to develop into cells of all three germ layers, which, in disease, is manifested in teratomas.

Posttranscriptional regulation appears to play a predominant role in preventing such a precocious onset of pluripotency in germ cells. In mouse, as in worms, posttranscriptional mechanisms involving the activity of RNA-binding proteins (RBPs) were found to maintain the germ cell fate and inhibit precocious reprogramming. In mouse, DND-1 (dead end 1) prevents teratoma formation in the embryonic germ cells (Youngren et al. 2005). Loss of DND-1 leads to a failure in downregulating pluripotency genes, like Nanog or Oct4, and defective cell cycle arrest (Cook et al. 2011). Defects in cell cycle arrest likely result from a failure to protect mRNAs encoding cell cycle regulators, like $p27^{Kip1}$ and $p21^{Cip1}$, from microRNA-mediated degradation (Kedde et al. 2007). Comparable to the function of DND-1 in mouse, two *C. elegans* RBPs named GLD-1 (defective in germ line development 1) and LIN-41 (abnormal cell lineage 41) have been found to inhibit teratoma formation in the adult germ cells. Interestingly, GLD-1 and LIN-41 prevent germ cell reprogramming at two consecutive phases of oogenic differentiation. While GLD-1/Quaking functions earlier, maintaining the germ-line fate during the pachytene stage of meiosis I (Ciosk et al. 2006), LIN-41/TRIM-71 performs this function later, during the diplotene and diakinesis stages (Tocchini et al. 2014). Why different RBPs are employed at different developmental stages is not clear. One possibility is that LIN-41 takes the control from GLD-1 so that some of the GLD-1 mRNA targets, with OET-promoting functions, may be translated (Scheckel et al. 2012). Irrespective of the molecular details, common to *gld-1* and *lin-41* loss-of-function mutants is a premature entry of the meiotic cells into mitosis, followed by precocious EGA and teratomatous differentiation (Biedermann et al. 2009; Tocchini et al. 2014; Ciosk et al. 2006). In these mutants, the expression of early embryonic genes takes place before the expression of differentiation-specific genes (Tocchini et al. 2014). Moreover, just like in normal embryogenesis, the teratomatous differentiation, at least into muscles, depends on the maternally supplied transcription factor, PAL-1/Caudal, a master regulator for muscle fate in embryos (Baugh et al. 2005). All these observations suggest that the *gld-1* and *lin-41* teratomas reflect the function of GLD-1 and LIN-41 in delaying the onset of embryonic events until after fertilization. How precisely they do it remains unclear. One key GLD-1 target is the *cye-1* mRNA encoding cyclin E (Biedermann et al. 2009). By repressing this mRNA, and consequently preventing cyclin E translation, GLD-1 regulates the activity of a key cyclin E partner, the cell cycle-driving kinase CDK-2. In *gld-1* mutants, CDK-2 is critical for the precocious mitosis, which is somehow linked with EGA (Biedermann et al. 2009). Similarly, in *lin-41* germ lines, the translational derepression of CDC-25.3, the activator of another cell cycle kinase called CDK-1, has been suggested to contribute to the premature entry into mitosis (Spike et al. 2014). Although in this case the connection between mitosis and EGA remains to be tested, GLD-1 and LIN-41 both repress premature mitosis, which appears to be linked to the transcriptional remodeling manifested by EGA and teratoma formation. Thus, a critical question for the future is whether there is a causal relation between the cell cycle and reprogramming and, if yes, what is the underlying mechanism.

Although the most-studied posttranscriptional regulators act as the repressors of reprogramming, RBPs could also act as activators. In *Drosophila*, another RBP, Smaug, has been reported to act as an EGA activator. The loss of Smaug leads to pleiotropic developmental defects in embryos, including the failure to start high-level embryonic transcription, possibly due to defects in the activation of the basal transcription machinery (Benoit et al. 2009). However, whether Smaug directly regulates the transcription machinery, or whether the failure to activate transcription is indirectly due to additional functions of Smaug in maternal mRNA degradation (Tadros et al. 2007), remains to be seen.

10.3.2.2 P Granules

C. elegans P granules are germ line-specific ribonucleoprotein granules with, typically, perinuclear localization, similar to germ granules or nuage in other organisms. It has been observed that in *gld-1* and *lin-41* mutant germ lines, but also in germ lines undergoing a direct germ-to-soma-conversion (Tursun et al. 2011), the P granules disappear from differentiating cells. However, whether the loss of P granules is the cause or consequence of the somatic differentiation of germ cells remains controversial. A recent study, aimed at answering this question, used simultaneous knockdown of RBPs that nucleate P granule formation (Updike et al. 2014). This resulted in the expression of some somatic genes and differentiation into neurons after the ectopic expression of a neural fate master regulator called CHE-1. Thus, it seems that compromising P granules enhances the ability of germ cells to be reprogrammed into somatic cells, but, alone, is not sufficient to cause an OET-like transition, as observed in *gld-1* or *lin-41* mutant germ lines. Also, while in some cases the association of a given RBP with P granules may be important for the RBP activity (Voronina et al. 2012), in other cases, including GLD-1 or LIN-41, the significance of the association with P granules (if any) remains unknown. Finally, compromising P granules could affect the biogenesis of endo-siRNAs (endogenous small interfering RNAs), including those facilitating the expression of germ-line genes (Gu et al. 2009; Cecere et al. 2014; Claycomb et al. 2009). Thus, the connection between P granules and the maintenance of the germ cell fate may be indirect.

In summary, posttranscriptional regulation appears to contribute to both EGA/pluripotency repression in germ cells and EGA activation in early embryos (involved factors are summarized in Table 10.2 and illustrated in Fig. 10.1). RBPs have been found to serve as cytoplasmic roadblocks to reprogramming and thereby prevent a precocious pluripotent state in germ cells. Their identification in mice and worms suggests that this is likely a conserved regulatory paradigm. This is further supported by the common use of posttranscriptional regulation in oocytes, as oocyte

Table 10.2 Posttranscriptional regulation of reprogramming during the OET

Type of regulation	Name	Organism	Function during OET	Molecular function	Reference
RBPs	DND-1	*M. musculus*	EGA repressor	Translational activator of the cell-cycle genes $p27^{Kip1}$ and $p21^{Cip1}$, translational repressor of pluripotency genes like Nanog	Youngren et al. (2005), Cook et al. (2011)
	GLD-1	*C. elegans*	EGA repressor	Translational repressor of the cell-cycle gene CYE-1/Cyclin E	Ciosk et al. (2006), Biedermann et al. (2009)
	LIN-41	*C. elegans*	EGA repressor	Translational repressor	Tocchini et al. (2014)
	Smaug	*D. melanogaster*	EGA activator	Possibly a direct activator of the basal transcription machinery, maternal mRNA degradation	Benoit et al. (2009), Tadros et al. (2007)
Germ granules	PGL-1, PGL-3, GLH-1, GLH-4	*C. elegans*	Somatic gene repressor	Restrict the reprogramming of germ cells into somatic cells	Updike et al. (2014)

development is associated with the entry into a transcriptionally silent state in most species. Common targets regulated by these RBPs are cell cycle genes, which indicate that they could be conserved targets for reprogramming regulation. The role of RNP granules in germ cell identity maintenance remains less well understood, as their impairment likely disturbs the function of many cytoplasmic regulators.

10.4 Conclusion

The reprogramming occurring during the OET is a complex process, and we are only at the beginning of understanding the underlying mechanisms. Work from various model organisms has shown that key processes appear to be conserved during the OET, like the reactivation of the cell cycle and transcriptional competence in early embryos, or the occurrence of EGA in two waves. The mechanisms controlling these processes work at the transcriptional and posttranscriptional levels and may vary between organisms. Thus, one future challenge is to address the conservation of the identified players and mechanisms. The discovery of the pioneering transcription factor Zelda, and its promoter-binding motif in embryonic genes, enhanced our knowledge about the transcriptional regulation of embryonic genes in *Drosophila*. However, DNA-binding motifs specific to embryonic genes, or the transcription factors binding to them, have not been identified in other model organisms, which leaves the question about the conservation of this mechanism open. Moreover, while some of the identified players establish a global transcriptional competence in early embryos, the mechanisms mediating specificity of gene expression remain poorly understood. Furthermore, posttranscriptional regulation during the OET remains largely unexplored, although it is expected to play an important role during the transcriptionally silent period preceeding EGA. So far, a few RNA-binding proteins have been identified to contribute to EGA regulation. However, the precise molecular mechanisms and conservation remain to be explored. In addition to RBPs, testing the importance of other types of RNA regulation for reprogramming will be of future interest. Once the individual regulatory mechanisms are understood, the next challenge will be to understand the interplay between those mechanisms. While some regulatory hierarchies have been understood, for example, the regulatory axis in the *C. elegans* oocyte connecting the posttranslational modification of OMA-1 and OMA-2 proteins to the cytoplasmic retention of the transcription factor TAF-4, the connections between other types of regulatory mechanisms remain to be determined. Dissecting the mechanisms underlying reprogramming during the OET will not only enhance the understanding of one of the most important developmental events but can also help in understanding and treating diseases, in which uncontrolled reprogramming can lead to teratomatous differentiation or undifferentiated tumors. Moreover, identification of novel players controlling reprogramming is of potential interest for the production of induced pluripotent stem (iPS) cells. The production of iPS cells still remains inefficient, with 0.01 %–6 % efficiency by transcription factor overexpression (Pasque et al. 2011), and a huge effort goes into developing strategies to achieve a higher reprogramming efficiency. Basic research on reprogramming can contribute to this field by uncovering new molecular players and pathways, potentially with conserved human functions.

References

Akkers RC, van Heeringen SJ, Jacobi UG, Janssen-Megens EM, Francoijs KJ, Stunnenberg HG, Veenstra GJ (2009) A hierarchy of H3K4me3 and H3K27me3 acquisition in spatial gene regulation in Xenopus embryos. Dev Cell 17(3):425–434. doi:10.1016/j.devcel.2009.08.005

Barckmann B, Simonelig M (2013) Control of maternal mRNA stability in germ cells and early embryos. Biochim Biophys Acta 1829(6–7):714–724. doi:10.1016/j.bbagrm.2012.12.011

Baugh LR, Hill AA, Claggett JM, Hill-Harfe K, Wen JC, Slonim DK, Brown EL, Hunter CP (2005) The homeodomain protein PAL-1 specifies a lineage-specific regulatory network in the *C. elegans* embryo. Development 132(8):1843–1854. doi:10.1242/dev.01782

Bellier S, Chastant S, Adenot P, Vincent M, Renard JP, Bensaude O (1997) Nuclear translocation and carboxyl-terminal domain phosphorylation of RNA polymerase II delineate the two phases of zygotic gene activation in mammalian embryos. EMBO J 16(20):6250–6262. doi:10.1093/emboj/16.20.6250

Benoit B, He CH, Zhang F, Votruba SM, Tadros W, Westwood JT, Smibert CA, Lipshitz HD, Theurkauf WE (2009) An essential role for the RNA-binding protein Smaug during the Drosophila maternal-to-zygotic transition. Development 136(6):923–932. doi:10.1242/dev.031815

Biedermann B, Wright J, Senften M, Kalchhauser I, Sarathy G, Lee MH, Ciosk R (2009) Translational repression of cyclin E prevents precocious mitosis and embryonic gene activation during C. elegans meiosis. Dev Cell 17(3):355–364. doi:10.1016/j.devcel.2009.08.003

Bogdanovic O, Long SW, van Heeringen SJ, Brinkman AB, Gomez-Skarmeta JL, Stunnenberg HG, Jones PL, Veenstra GJ (2011) Temporal uncoupling of the DNA methylome and transcriptional repression during embryogenesis. Genome Res 21(8):1313–1327. doi:10.1101/gr.114843.110

Brevini TA, Pennarossa G, Vanelli A, Maffei S, Gandolfi F (2012) Parthenogenesis in non-rodent species: developmental competence and differentiation plasticity. Theriogenology 77(4):766–772. doi:10.1016/j.theriogenology.2011.11.010

Bultman SJ, Gebuhr TC, Pan H, Svoboda P, Schultz RM, Magnuson T (2006) Maternal BRG1 regulates zygotic genome activation in the mouse. Genes Dev 20(13):1744–1754. doi:10.1101/gad.1435106

Cantone I, Fisher AG (2013) Epigenetic programming and reprogramming during development. Nat Struct Mol Biol 20(3):282–289. doi:10.1038/nsmb.2489

Cecere G, Hoersch S, O'Keeffe S, Sachidanandam R, Grishok A (2014) Global effects of the CSR-1 RNA interference pathway on the transcriptional landscape. Nat Struct Mol Biol 21(4):358–365. doi:10.1038/nsmb.2801

Chang CC, Ma Y, Jacobs S, Tian XC, Yang X, Rasmussen TP (2005) A maternal store of macroH2A is removed from pronuclei prior to onset of somatic macroH2A expression in preimplantation embryos. Dev Biol 278(2):367–380. doi:10.1016/j.ydbio.2004.11.032

Ciosk R, DePalma M, Priess JR (2006) Translational regulators maintain totipotency in the Caenorhabditis elegans germline. Science 311(5762):851–853. doi:10.1126/science.1122491

Claycomb JM, Batista PJ, Pang KM, Gu W, Vasale JJ, van Wolfswinkel JC, Chaves DA, Shirayama M, Mitani S, Ketting RF, Conte D Jr, Mello CC (2009) The Argonaute CSR-1 and its 22G-RNA cofactors are required for holocentric chromosome segregation. Cell 139(1):123–134. doi:10.1016/j.cell.2009.09.014

Cook MS, Munger SC, Nadeau JH, Capel B (2011) Regulation of male germ cell cycle arrest and differentiation by DND1 is modulated by genetic background. Development 138(1):23–32. doi:10.1242/dev.057000

De Renzis S, Elemento O, Tavazoie S, Wieschaus EF (2007) Unmasking activation of the zygotic genome using chromosomal deletions in the Drosophila embryo. PLoS Biol 5(5):e117. doi:10.1371/journal.pbio.0050117

Dolci S, Campolo F, De Felici M (2015) Gonadal development and germ cell tumors in mouse and humans. Semin Cell Dev Biol 45:114–123. doi:10.1016/j.semcdb.2015.10.002

Dunican DS, Ruzov A, Hackett JA, Meehan RR (2008) xDnmt1 regulates transcriptional silencing in pre-MBT Xenopus embryos independently of its catalytic function. Development 135(7):1295–1302. doi:10.1242/dev.016402

Edgar BA, Schubiger G (1986) Parameters controlling transcriptional activation during early Drosophila development. Cell 44(6):871–877

Gu W, Shirayama M, Conte D Jr, Vasale J, Batista PJ, Claycomb JM, Moresco JJ, Youngman EM, Keys J, Stoltz MJ, Chen CC, Chaves DA, Duan S, Kasschau KD, Fahlgren N, Yates JR 3rd, Mitani S, Carrington JC, Mello CC (2009) Distinct argonaute-mediated 22G-RNA pathways direct genome surveillance in the C. elegans germline. Mol Cell 36(2):231–244. doi:10.1016/j.molcel.2009.09.020

Gurdon JB, Uehlinger V (1966) "Fertile" intestine nuclei. Nature 210(5042):1240–1241

Guven-Ozkan T, Nishi Y, Robertson SM, Lin R (2008) Global transcriptional repression in C. elegans germline precursors by regulated sequestration of TAF-4. Cell 135(1):149–160. doi:10.1016/j.cell.2008.07.040

Haberle V, Li N, Hadzhiev Y, Plessy C, Previti C, Nepal C, Gehrig J, Dong X, Akalin A, Suzuki AM, van IJcken WF, Armant O, Ferg M, Strahle U, Carninci P, Muller F, Lenhard B (2014) Two independent transcription initiation codes overlap on vertebrate core promoters. Nature 507(7492):381–385. doi:10.1038/nature12974

Harrison MM, Li XY, Kaplan T, Botchan MR, Eisen MB (2011) Zelda binding in the early Drosophila melanogaster embryo marks regions subsequently activated at the maternal-to-zygotic transition. PLoS Genet 7(10):e1002266. doi:10.1371/journal.pgen.1002266

Heyn P, Kircher M, Dahl A, Kelso J, Tomancak P, Kalinka AT, Neugebauer KM (2014) The earliest transcribed zygotic genes are short, newly evolved, and different across species. Cell Rep 6(2):285–292. doi:10.1016/j.celrep.2013.12.030

Hussain SA, Ma YT, Palmer DH, Hutton P, Cullen MH (2008) Biology of testicular germ cell tumors. Expert Rev Anticancer Ther 8(10):1659–1673. doi:10.1586/14737140.8.10.1659

Kedde M, Strasser MJ, Boldajipour B, Oude Vrielink JA, Slanchev K, le Sage C, Nagel R, Voorhoeve PM, van Duijse J, Orom UA, Lund AH, Perrakis A, Raz E, Agami R (2007) RNA-binding protein Dnd1 inhibits microRNA access to target mRNA. Cell 131 (7):1273–1286. doi:10.1016/j.cell.2007.11.034

Kerr CL, Shamblott MJ, Gearhart JD (2006) Pluripotent stem cells from germ cells. Methods Enzymol 419:400–426. doi:10.1016/S0076-6879(06)19016-3

Kimelman D, Kirschner M, Scherson T (1987) The events of the midblastula transition in Xenopus are regulated by changes in the cell cycle. Cell 48(3):399–407

Kono T, Obata Y, Wu Q, Niwa K, Ono Y, Yamamoto Y, Park ES, Seo JS, Ogawa H (2004) Birth of parthenogenetic mice that can develop to adulthood. Nature 428(6985):860–864. doi:10.1038/nature02402

LaMarca MJ, Fidler MC, Smith LD, Keem K (1975) Hormonal effects on RNA synthesis by stage 6 oocytes of Xenopus laevis. Dev Biol 47(2):384–393

Lee MT, Bonneau AR, Giraldez AJ (2014) Zygotic genome activation during the maternal-to-zygotic transition. Annu Rev Cell Dev Biol 30:581–613. doi:10.1146/annurev-cellbio-100913-013027

Lee MT, Bonneau AR, Takacs CM, Bazzini AA, DiVito KR, Fleming ES, Giraldez AJ (2013) Nanog, Pou5f1 and SoxB1 activate zygotic gene expression during the maternal-to-zygotic transition. Nature 503(7476):360–364. doi:10.1038/nature12632

Leichsenring M, Maes J, Mossner R, Driever W, Onichtchouk D (2013) Pou5f1 transcription factor controls zygotic gene activation in vertebrates. Science 341(6149):1005–1009. doi:10.1126/science.1242527

Li XY, Harrison MM, Villalta JE, Kaplan T, Eisen MB (2014) Establishment of regions of genomic activity during the Drosophila maternal to zygotic transition. eLife 3. doi:10.7554/eLife.03737

Liang HL, Nien CY, Liu HY, Metzstein MM, Kirov N, Rushlow C (2008) The zinc-finger protein Zelda is a key activator of the early zygotic genome in Drosophila. Nature 456(7220):400–403. doi:10.1038/nature07388

Lindeman LC, Andersen IS, Reiner AH, Li N, Aanes H, Ostrup O, Winata C, Mathavan S, Muller F, Alestrom P, Collas P (2011) Prepatterning of developmental gene expression by modified histones before zygotic genome activation. Dev Cell 21(6):993–1004. doi:10.1016/j.devcel.2011.10.008

Moore GP, Lintern-Moore S, Peters H, Faber M (1974) RNA synthesis in the mouse oocyte. J Cell Biol 60(2):416–422

Nepal C, Hadzhiev Y, Previti C, Haberle V, Li N, Takahashi H, Suzuki AM, Sheng Y, Abdelhamid RF, Anand S, Gehrig J, Akalin A, Kockx CE, van der Sloot AA, van Ijcken WF, Armant O, Rastegar S, Watson C, Strahle U, Stupka E, Carninci P, Lenhard B, Muller F (2013) Dynamic regulation of the transcription initiation landscape at single nucleotide resolution during vertebrate embryogenesis. Genome Res 23(11):1938–1950. doi:10.1101/gr.153692.112

Newport J, Kirschner M (1982) A major developmental transition in early Xenopus embryos: I. characterization and timing of cellular changes at the midblastula stage. Cell 30(3):675–686

Nien CY, Liang HL, Butcher S, Sun Y, Fu S, Gocha T, Kirov N, Manak JR, Rushlow C (2011) Temporal coordination of gene networks by Zelda in the early Drosophila embryo. PLoS Genet 7(10):e1002339. doi:10.1371/journal.pgen.1002339

Paranjpe SS, Veenstra GJ (2015) Establishing pluripotency in early development. Biochim Biophys Acta 1849(6):626–636. doi:10.1016/j.bbagrm.2015.03.006

Pasque V, Jullien J, Miyamoto K, Halley-Stott RP, Gurdon JB (2011) Epigenetic factors influencing resistance to nuclear reprogramming. Trends Genet 27(12):516–525. doi:10.1016/j.tig.2011.08.002

Peaston AE, Evsikov AV, Graber JH, de Vries WN, Holbrook AE, Solter D, Knowles BB (2004) Retrotransposons regulate host genes in mouse oocytes and preimplantation embryos. Dev Cell 7(4):597–606. doi:10.1016/j.devcel.2004.09.004

Perez-Montero S, Carbonell A, Moran T, Vaquero A, Azorin F (2013) The embryonic linker histone H1 variant of Drosophila, dBigH1, regulates zygotic genome activation. Dev Cell 26 (6):578–590. doi:10.1016/j.devcel.2013.08.011

Robert VJ, Mercier MG, Bedet C, Janczarski S, Merlet J, Garvis S, Ciosk R, Palladino F (2014) The SET-2/SET1 histone H3K4 methyltransferase maintains pluripotency in the Caenorhabditis elegans germline. Cell Rep 9(2):443–450. doi:10.1016/j.celrep.2014.09.018

Scheckel C, Gaidatzis D, Wright JE, Ciosk R (2012) Genome-wide analysis of GLD-1-mediated mRNA regulation suggests a role in mRNA storage. PLoS Genet 8(5):e1002742. doi:10.1371/journal.pgen.1002742

Schulz KN, Bondra ER, Moshe A, Villalta JE, Lieb JD, Kaplan T, McKay DJ, Harrison MM (2015) Zelda is differentially required for chromatin accessibility, transcription factor binding, and gene expression in the early Drosophila embryo. Genome Res 25(11):1715–1726. doi:10.1101/gr.192682.115

Shermoen AW, O'Farrell PH (1991) Progression of the cell cycle through mitosis leads to abortion of nascent transcripts. Cell 67(2):303–310

Smith AG (2001) Embryo-derived stem cells: of mice and men. Annu Rev Cell Dev Biol 17:435–462. doi:10.1146/annurev.cellbio.17.1.435

Spike CA, Coetzee D, Eichten C, Wang X, Hansen D, Greenstein D (2014) The TRIM-NHL protein LIN-41 and the OMA RNA-binding proteins antagonistically control the prophase-to-metaphase transition and growth of Caenorhabditis elegans oocytes. Genetics 198 (4):1535–1558. doi:10.1534/genetics.114.168831

Stancheva I, Meehan RR (2000) Transient depletion of xDnmt1 leads to premature gene activation in Xenopus embryos. Genes Dev 14(3):313–327

Strome S, Updike D (2015) Specifying and protecting germ cell fate. Nat Rev Mol Cell Biol 16 (7):406–416. doi:10.1038/nrm4009

Tadros W, Goldman AL, Babak T, Menzies F, Vardy L, Orr-Weaver T, Hughes TR, Westwood JT, Smibert CA, Lipshitz HD (2007) SMAUG is a major regulator of maternal mRNA destabilization in Drosophila and its translation is activated by the PAN GU kinase. Dev Cell 12 (1):143–155. doi:10.1016/j.devcel.2006.10.005

Tadros W, Lipshitz HD (2009) The maternal-to-zygotic transition: a play in two acts. Development 136(18):3033–3042. doi:10.1242/dev.033183

Takahashi K, Yamanaka S (2006) Induction of pluripotent stem cells from mouse embryonic and adult fibroblast cultures by defined factors. Cell 126(4):663–676. doi:10.1016/j.cell.2006.07.024

ten Bosch JR, Benavides JA, Cline TW (2006) The TAGteam DNA motif controls the timing of Drosophila pre-blastoderm transcription. Development 133(10):1967–1977. doi:10.1242/dev.02373

Tocchini C, Keusch JJ, Miller SB, Finger S, Gut H, Stadler MB, Ciosk R (2014) The TRIM-NHL protein LIN-41 controls the onset of developmental plasticity in Caenorhabditis elegans. PLoS Genet 10(8):e1004533. doi:10.1371/journal.pgen.1004533

Tursun B, Patel T, Kratsios P, Hobert O (2011) Direct conversion of C. elegans germ cells into specific neuron types. Science 331(6015):304–308. doi:10.1126/science.1199082

Ulbright TM (2005) Germ cell tumors of the gonads: a selective review emphasizing problems in differential diagnosis, newly appreciated, and controversial issues. Mod Pathol 18(Suppl 2): S61–S79. doi:10.1038/modpathol.3800310

Updike DL, Knutson AK, Egelhofer TA, Campbell AC, Strome S (2014) Germ-granule components prevent somatic development in the C. elegans germline. Curr Biol 24(9):970–975. doi:10.1016/j.cub.2014.03.015

Vastenhouw NL, Zhang Y, Woods IG, Imam F, Regev A, Liu XS, Rinn J, Schier AF (2010) Chromatin signature of embryonic pluripotency is established during genome activation. Nature 464(7290):922–926. doi:10.1038/nature08866

Voronina E, Paix A, Seydoux G (2012) The P granule component PGL-1 promotes the localization and silencing activity of the PUF protein FBF-2 in germline stem cells. Development 139 (20):3732–3740. doi:10.1242/dev.083980

Walser CB, Lipshitz HD (2011) Transcript clearance during the maternal-to-zygotic transition. Curr Opin Genet Dev 21(4):431–443. doi:10.1016/j.gde.2011.03.003

Weismann A (1893) The germ-plasm: A theory of heredity. Charles Scribner's Sons, New York

Youngren KK, Coveney D, Peng X, Bhattacharya C, Schmidt LS, Nickerson ML, Lamb BT, Deng JM, Behringer RR, Capel B, Rubin EM, Nadeau JH, Matin A (2005) The Ter mutation in the dead end gene causes germ cell loss and testicular germ cell tumours. Nature 435 (7040):360–364. doi:10.1038/nature03595

Zurita M, Reynaud E, Aguilar-Fuentes J (2008) From the beginning: The basal transcription machinery and onset of transcription in the early animal embryo. Cell Mol Life Sci 65 (2):212–227. doi:10.1007/s00018-007-7295-4

Index

A

Adipocyte factors, 82
Aneuploidy
  centromere-spindle attachment defects, 154, 155
  cohesion maintenance, 152–153
  crossover distribution, 152
  crossover formation, 150–151
  crossover position, 153, 154
  meiosis, 149–150
  outcomes, 155, 156
  "weak checkpoint," 153
Artificial oocyte activation (AOA), 274–275

C

Caenorhabditis elegans (*C.elegans*)
  female reproductive tract
    hermaphrodite gonad, 211
    lipid guidance mechanism, 215–217
    male sperm motility behavior in vivo, 213–214
    spermatids, 211, 212
    sperm guidance, 217–219
    SWM-1 protease inhibitor, 212
    TRY-5 protease inhibitor, 212
    vap, 213
  germ cell specification and PGC development
    asymmetric segregation, 181
    sequential translational activation, 182
    transcriptional quiescence, 182
  germ line development
    LET-60 RAS/MPK-1 ERK signaling pathway, 104–106
    oocyte growth and development, 106–107
    oocyte meiotic maturation, 107–109
    oogenic meiotic prophase progression, 102–103
  meiotic progression and oocyte maturation
    maternal mRNAs, 190–191
    multiple RNA-binding proteins act, 189–190
  mitosis–meiosis decision, 183–186
Calcium bomb theory, 266–267
Cap cells (CpCs), 15, 44–47, 70, 77
Cdc6, in mouse and *Xenopus laevis*
  homologs and isoforms of, 248–249
  meiotic division, 252–254
  motifs and domains, 249
  M-phase, 254–255
  S-phase
    antagonistic activities control, 251
    Cdc6 levels and localization, 252
    chromatin-bound formation, 252–253
    degradation mechanism, 250
    first and second meiotic division, 249–250
    GVBD, 250–251
    northern blot analysis, 250
    pre-RCs formation and initiation, 250, 251
    primary oocytes, 250
    stabilizing mechanism, 251
    Western blot analysis, 250
Chemotaxis, 202–203, 206, 208, 220
Chromatin regulation
  in somatic niche cells regulate signaling pathways
    cap cells, 15
    chromatin modifiers, 15–16

S. Arur (ed.), *Signaling-Mediated Control of Cell Division*, Results and Problems in Cell Differentiation 59, DOI 10.1007/978-3-319-44820-6

Chromatin regulation (*cont.*)
chromatin remodeling factors, 19
escort cells, 15
follicular stem cells, 15
histone-modifying enzymes, 17–19
terminal filament, 15
Ub-H2B, 5–6
Cystoblast (CB), 3, 44–47, 70, 186

D

Decapentaplegic (Dpp) signaling molecules, 4, 46, 186
Distal tip cell (DTC), 37–38, 71, 183, 185, 211
DNA methylation, 234
Double-strand breaks (DSBs)
formation
CRA-1 and XND-1, 131
crossover pathway, 127, 129
Dmc1/Rad51, 127, 129
H3K4me3, 130–131
hot spots, 127, 129
non crossover, 127, 129
nucleases Exo1 and Dna2, 127, 128
PRDM9, 129
proteins, 127, 130
recombination nodules, 129
replication, 127, 128
RPA, 127–129
topoisomerase-like protein Spo11 and MRX complex, 127, 128
processing, 131–132
Drosophila
female reproductive tract, 208–210
germ cell specification and PGC development, 178–180
germ line development
meiotic prophase I progression, 110–112
oocyte meiotic maturation, 111, 113
stages, 109–110
mitosis–meiosis decision, 186–187
oogenesis, 2–3, 110
ovary, 68–70
asymmetric divisions, 44–45
organization, 42–44
plasticity, 47–48
proliferation *vs.* differentiation signals, 45–47
self-renewal/differentiation decision, 186–187
testis, 69, 71
DSBs. *See* Double-strand breaks (DSBs)
DTC. *See* Distal tip cell (DTC)

E

Embryonic genome activation (EGA), 231–232
Epigenetics
definition, 1–2
female *Drosophila* germ line, 2
histone modifications, 2, 4–5
mode of function, 3
modifications, 234
in oocyte specification and maintenance
histone-modifying enzymes, 11–12
histone variants, 11–12
nuclear organization, 12–14
nucleosome remodeling complex, 13
phenotypic effects, 2
Escort cells (ECs), 43, 44, 46, 47, 70

F

Female reproductive tract
in *C. elegans*
hermaphrodite gonad, 211
lipid guidance mechanism, 215–217
male sperm motility behavior in vivo, 213–214
spermatids, 211, 212
sperm guidance, 217–219
SWM-1 protease inhibitor, 212
TRY-5 protease inhibitor, 212
vap, 213
in Drosophila, 208–210
mammals
capacitation, 205–206
CatSper, 206–207
chemoattractant molecules, 206
ejaculation, 204–205
fluid viscosity, 207
general features, 204
video microscopy and computer image analysis, 206
Follicle-stimulating hormone (FSH), 79–81, 83
Follicular stem cells (FSCs), 15
Fusome, 45

G

Germarium, 2, 3, 42, 44, 110
Germinal vesicle breakdown (GVBD). *See* Nuclear envelope breakdown (NEBD)
Germ line development
C. elegans
LET-60 RAS/MPK-1 ERK Signaling Pathway, 104–106

oocyte growth and development, 106–107
oocyte meiotic maturation, 107–109
oogenic meiotic prophase progression, 102–103
Drosophila female
meiotic prophase I progression, 110–112
oocyte meiotic maturation, 111, 113
stages, 109–110
mammalian females
LH-induced MAPK activation and meiotic resumption, 116–117
luteinizing hormone, 116
prophase I arrest, 114–115
meiotic progression and maturation, 117–118
Germline stem cell lineages
C. elegans gonad, 70–72
cell type-specific requirements, 83, 84
Drosophila ovary, 68–70
Drosophila testis, 69, 71
mouse testis, 70, 72
nutritional control
additional interorgan communication influences, 80–83
conserved nutrient sensing pathways, 73, 74
diet-dependent hormones, 76–80
TOR signaling, 73–75
physiological regulation, 83, 85–86
Germ line stem cells (GSCs)
adhesion molecules, 35
C. elegans hermaphrodite
anatomy of the gonad, 37–38
plasticity, 42
proliferation *vs.* differentiation signaling, 38–42
cell signaling maintainance, 36–37
Drosophila oogenesis, 2–3
Drosophila ovariole
asymmetric divisions, 44–45
organization, 42–44
plasticity, 47–48
proliferation *vs.* differentiation signals, 45–47
lineage tracing experiments, 36
mouse testis
plasticity, 54
seminiferous tubule, 48–50
spermatogenesis, 50–51
niche, 33
self-renewal and differentiation
Bam, 4
Dpp signaling/Gbb, 4
gene expression, 4
histones and histone-modifying enzymes, 5–7
intrinsic and extrinsic regulatory factors, 4
nuclear organization, 8–10
nucleosome remodeling complexes, 7–8
Stat92e, 4
symmetric/asymmetric division, 33–35
transient amplifying cells, 35
transit-amplifying cells, 35
Germplasm, 178–179
Germplasm theory of heredity, 230
Glass bottom boat (Gbb), 4
Glial cell line-derived neurotrophic factor (GDNF), 52
GSCs. *See* Germ line stem cells (GSCs)

H

H2B monoubiquitination (Ub-H2B), 5–6
Histone marks, 234
Homologous recombination (HR)
crossover formation, 134–137
DSB formation
CRA-1 and XND-1, 131
crossover pathway, 127, 129
Dmc1/Rad51, 127, 129
H3K4me3, 130–131
hot spots, 127, 129
non crossover, 127, 129
nucleases Exo1 and Dna2, 127, 128
PRDM9, 129
proteins, 127, 130
recombination nodules, 129
replication, 127, 128
RPA, 127–129
topoisomerase-like protein Spo11 and MRX complex, 127, 128
HR. *See* Homologous recombination (HR)

I

Inner germarium sheath cells, 44
Inner sheath cells, 44
In vitro maturation (IVM), 273–274
IP_3 receptor (IP_3R), 269–270

L

Leydig cells, 49–50
Liquid chromatography coupled to electrospray ionization tandem mass spectrometry (LC-MS/MS), 216
Luteinizing hormone (LH), 79–80, 114–117

M

Major sperm protein (MSP), 108, 211, 212
MAP kinase (MAPK) pathway, 104, 105, 149
Maternal clearance, 231
Maternal clock model, 232
Meiotic prophase I
  aneuploidy
    centromere-spindle attachment defects, 154, 155
    cohesion maintenance, 152–153
    crossover distribution, 152
    crossover formation, 150–151
    crossover position, 153, 154
    meiosis, 149–150
    outcomes, 155, 156
    "weak checkpoint," 153
  chiasma, 126
  chromosome structure
    sister chromatid cohesion, 137–42
    synaptonemal complex (*see* Synaptonemal complex protein)
  homologous recombination
    crossover formation, 134–137
    DSB formation, 127–131
    DSB processing, 131–132
    strand invasion, 132–134
  signaling pathways
    ATM/Tel1 and ATR/Mec1, 147
    Chk2, 148
    MAPK, 149
    PLKs, 148
  stages, 126
  unique chromosomal processes, 126
Mitosis–meiosis decision
  in *C. elegans,* complex RNA regulatory network guides, 183–186
  in *Drosophila,* self-renewal/differentiation decision, 186–187
  in mouse testis, 187–188
Mouse
  Cdc6
    homologs and isoforms of, 248–249
    meiotic division, 252–254
    motifs and domains, 249
    M-phase, 254–255
    S-phase (*see* S-phase)
  synaptonemal complex protein, 142, 143
  testis
    mitosis-meiosis decision, 187–188
    plasticity, 54
    seminiferous tubule, 48–50
    spermatogenesis, 50–51

N

Neutral competition, 35–36
Nonsteroidal anti-inflammatory drugs (NSAIDs), 216
Nuclear envelope breakdown (NEBD), 113
Nucleocytoplasmic (N/C) ratio model, 232
Nutritional control
  additional interorgan communication influences, 80–83
  conserved nutrient sensing pathways, 73, 74
  diet-dependent hormones
    conserved signaling pathways, 77
    20E, 76, 77
    ecdysone signaling controls, 77
    hypothalamic-pituitary-gonadal axis, 79, 80
    ILPs, 76–77
    insulin/insulin-like signaling, 76
    insulin-like growth factors 1 and 2, 76
    Leydig cell testosterone production, 79
    PPARγ, 77
    RARs and RXR, 78–79
    vitellogenesis control, 78
  TOR signaling, 73–75

O

OET. *See* Oocyte-to-embryo transition (OET)
Oocyte activation and fertilisation
  AOA, 274–275
  Ca^{2+}, 270–272
  calmodulin-dependent protein kinase II, 272
  definition, 260
  infertility, 272–273
  IVM, 273–274
  oocyte factors
    IP_3R, 269–270
    PIP_2, 269
    PKCs, 270
  sperm and oocyte membranes, 264–265
  sperm and zona pellucida
    acrosome-reacted sperm, 261–262
    hardening, 264
    in mammals, 263
    ZP-glycoproteins, 260–261
  sperm factors
    calcium bomb theory, 266–267
    PAWP/WBP2N, 268
    PLCζ, 267–268
    receptor theory, 266–267
    SOAF, 267
    sperm factor theory, 266, 267

Oocyte-to-embryo transition (OET)
EGA, 232–233
germplasm theory of heredity, 230
maternal clearance, 231
maternal clock model, 232
N/C ratio model, 232
PGCs and adult germ cells, 230–231
posttranscriptional regulation
P granules, 240–241
RNA-binding proteins, 238–240
transcript abortion model, 232
transcriptional regulation
alternative promoters, 234–235
chromatin remodelers, 233
DNA methylation, 234
histone marks, 234
histone variants, 233
reprogramming-controlling mechanisms, 236–238

P
Pachytene cellular organization, 105
Peroxisome proliferator-activated receptor γ (PPARγ), 77
PGCs. *See* Primordial germ cells (PGCs)
Phospholipase A2 (PLA2), 261
Phospholipase c zeta (PLCζ), 267–268
Polo-like kinases (PLKs), 111, 148
Positive rheotaxis, 204, 207
Primordial germ cells (PGCs)
C. elegans
LET-60 RAS/MPK-1 ERK Signaling Pathway, 104–106
oocyte growth and development, 106–107
oocyte meiotic maturation, 107–109
oogenic meiotic prophase progression, 102–103
Drosophila female
meiotic prophase I progression, 110–112
oocyte meiotic maturation, 111, 113
stages, 109–110
origin, 230
Protein kinase Cs (PKCs), 270

R
Rachis/core, 102
RAS/ERK signaling pathway, 107
Receptor theory, 266–267
Retinoic acid receptors (RARs), 78

S
Seminiferous tubule, 48–50
Signal transducer and activator of transcription 92e (Stat92e), 4
Somatic niche, 3
Spectrosome, 44
Sperm
factors
calcium bomb theory, 266–267
PAWP/WBP2N, 268
PLCζ, 267–268
receptor theory, 266–267
SOAF, 267
sperm factor theory, 266, 267
and oocyte membranes, 264–265
and zona pellucida
acrosome-reacted sperm, 261–262
hardening, 264
in mammals, 263
ZP-glycoproteins, 260–261
Spermatogonial stem cells (SSCs), 50, 72
Sperm-borne oocyte activation factor (SOAF), 267
Sperm factor theory, 266, 267
Sperm navigation mechanisms
C. elegans hermaphrodite reproductive tract
DTC, 211
hermaphrodite gonad, 211
lipid guidance mechanism, 215–217
male sperm motility behavior in vivo, 213–214
spermatids, 211, 212
sperm guidance, 217–219
SWM-1 protease inhibitor, 212
TRY-5 protease inhibitor, 212
vap, 213
chemotaxis, 202–203
competitive race model, 202
Drosophila female reproductive tract, 208–210
external fertilizers, 202
in vivo sperm guidance mechanisms, 202, 203
mammalian female reproductive tract
capacitation, 205–206
CatSper, 206–207
chemoattractant molecules, 206
ejaculation, 204–205
fluid viscosity, 207
general features, 204
video microscopy and computer image analysis, 206
meiosis, 201–202
perspectives, 219–220

Sperm navigation mechanisms (*cont.*)
positive rheotaxis, 204, 207
sexual reproduction, 201
thermotaxis, 204, 207
S-phase
antagonistic activities control, 251
Cdc6 levels and localization, 252
chromatin-bound formation, 252–253
degradation mechanism, 250
first and second meiotic division, 249–250
GVBD, 250–251
northern blot analysis, 250
pre-RCs formation and initiation, 250, 251
primary oocytes, 250
stabilizing mechanism, 251
Western blot analysis, 250
Stem cell factor (SCF), 53
Synaptonemal complex protein
C. elegans, 142, 143
in *D. melanogaster,* 142, 143
dynamics, pairing and signaling, 142–145
in mouse, 142, 143
posttranslational modification, 146
recombination, 145
synapsis of homologous chromosomes, 142

T

Target of Rapamycin (TOR) signaling, 73–75
Teratoma, 231
Terminal filament (TF) cells, 44, 45
Thermotaxis, 204, 207
Transcript abortion model, 232
Transcriptional control, of gene expression
germ cells, 177–178
germline specification, 176–177
meiosis and gametogenesis, 177
new mRNA transcription, 177
Transcriptional regulation
alternative promoters, 234–235
chromatin remodelers, 233
DNA methylation, 234
histone marks, 234
histone variants, 233
reprogramming-controlling mechanisms, 236–238
Transit-amplifying (TA) cells, 35
Translational control
germ cell specification and PGC development
in *C. elegans,* 180–182
in *Drosophila,* 178–180
in vertebrates, 182–183
meiotic progression and oocyte maturation
in *C. elegans*, 189–191
in *Xenopus*, 191–193
mitosis-meiosis decision
in *C. elegans*, 183–186
in *Drosophila,* 186–187
in mouse testis, 187–188

V

Vamp-synaptobrevin-associated protein (*vap*), 213

W

Weismann theory of heredity. *See* Germplasm theory of heredity

X

Xenopus laevis, Cdc6
homologs and isoforms of, 248–249
meiotic division, 252–254
motifs and domains, 249
M-phase, 254–255
S-phase
antagonistic activities control, 251
Cdc6 levels and localization, 252
chromatin-bound formation, 252–253
degradation mechanism, 250
first and second meiotic division, 249–250
GVBD, 250–251
northern blot analysis, 250
pre-RCs formation and initiation, 250, 251
primary oocytes, 250
stabilizing mechanism, 251
Western blot analysis, 250

